Introduction to Microsystem Technology

The Wiley Microsystem and Nanotechnology Series

Series Editor:
Ronald Pething

Introduction to Microsystem Technology
A Guide for Students
Gerald Gerlach and **Wolfram Dötzel**

Introduction to Microsystem Technology

A Guide for Students

Gerald Gerlach
Dresden University of Technology

Wolfram Dötzel
Chemnitz University of Technology

Translated by **Dörte Müller**
powerwording.com

Telephone (+44) 1243 779777

Email (for orders and customer service enquiries): cs-books@wiley.co.uk
Visit our Home Page on www.wileyeurope.com or www.wiley.com

Other Wiley Editorial Offices

John Wiley & Sons Inc., 111 River Street, Hoboken, NJ 07030, USA

Jossey-Bass, 989 Market Street, San Francisco, CA 94103-1741, USA

Wiley-VCH Verlag GmbH, Boschstr. 12, D-69469 Weinheim, Germany

John Wiley & Sons Australia Ltd, 42 McDougall Street, Milton, Queensland 4064, Australia

John Wiley & Sons (Asia) Pte Ltd, 2 Clementi Loop #02-01, Jin Xing Distripark, Singapore 129809

John Wiley & Sons Canada Ltd, 6045 Freemont Blvd, Mississauga, Ontario, L5R 4J3, Canada

Wiley also publishes its books in a variety of electronic formats. Some content that appears
in print may not be available in electronic books.

Library of Congress Cataloging-in-Publication Data

Gerlach, Gerald.
[Einführung in die Mikrosystemtechnik. English]
Introduction to microsystem technology : a guide for students /
Gerald Gerlach, Wolfram Dötzel ; translated by Dörte Müller.
p. cm.
Includes bibliographical references and index.
ISBN 978-0-470-05861-9 (cloth)
1. Microelectromechanical systems – Design and construction. 2.
System design. 3. Microelectronics. I. Dötzel, Wolfram. II. Title.
TK7875.G46 2008
621 – dc22

2007050172

British Library Cataloguing in Publication Data

A catalogue record for this book is available from the British Library

ISBN 978-0-470-05861-9

Typeset in 10/12 Times by Laserwords Private Limited, Chennai, India

Contents

Preface

Silicon is a fascinating material. It can be produced with high purity and as quasi-perfect single crystals. Adding small amounts of doping atoms can change resistivity by more than six orders of magnitude resulting in highly conductive or quasi-insulating silicon areas. It takes only simple technologies to generate quasi defect-free, highly insulating layers on silicon surfaces. In addition to photolithography – which allows simultaneous pattern transfer for thousands and millions of components – the named characteristics caused a rapid development of microelectronics which continues even today.

In 1953, CHARLES S. SMITH discovered in the Bell Telephone Laboratories the piezoresistive effect in semiconductors. This discovery started a unique development towards miniaturized silicon sensors. Initially, the sensors were predominantly mechanical; later, further functions were added. Such functions referred to actuator and signal-processing components, on the one hand, and to microfluidic, microoptical, magnetic and many other components on the other hand. Miniaturization and integration have thus opened up a new technical field which has many names. In Germany, we talk about microsystem technology; in the US, it is called MicroElectroMechanical Systems (MEMS) and in Japan Micromachining. In 1959, the US physician and Nobel prize winner RICHARD P. FEYNMAN gave – at the California Institute of Technology – a speech that was to become famous: 'There is plenty of room at the bottom'.[1] For the first time, attention was drawn to the possibility of miniaturized machines. In 1982, KURT E. PETERSEN presented his article 'Silicon as a mechanical material'[2] and was the first to describe the entire potential of microsystem technology. Since then, the development has been outstanding with a large number of new technologies and applications being created. At the same time, several textbooks and technical publications about this subject have come out.

The present publication constitutes a textbook for university students and provides an introduction to the basic aspects of microsystem technology. In spite of the necessary restrictions regarding the size of the book, the authors have nevertheless tried to take into account the complexity of the matter. Therefore, the subjects treated range from typical materials of microsystem technology via the most important microtechnical manufacturing techniques and technological methods of system integration all the way to the design of function components and complete systems. This approach takes into account both functional and technological aspects of microtechnologies as well as functional and technological aspects of system technologies that characterize microsystem technology. In order to solve problems at the micro- and nano-level, engineers working in the field

[1]Feynman, R.P. (1992) There is plenty of room at the bottom. *Journal of Microelectromechanical Systems* 1(1992)1, pp. 60–6.

[2]Petersen, K.E. Silicon as a mechanical material. *Proceedings of the IEEE*, vol. 70, no. 5, pp. 420–57, 1982.

of microsystem technology need to have interdisciplinary knowledge from several areas including the capability of thinking on the system level.

The present publication is based on lectures given at the Technical Universities in Dresden and Chemnitz, Germany. It was originally written for students of microsystem technology, but has also proved to be very useful as a compendium or reference for experts working with microsystem technology. Due to its textbook character, the book contains more than 100 examples and exercises. The examples illustrate and consolidate the facts presented in the text. The exercises at the end of each chapter allow the reader actively to test the acquired knowledge. The following Internet addresses can be used to check the results:

http://www.tu-chemnitz.de/etit/microsys/mst-buch2006 or
http://ife.et.tu-dresden.de/mst-buch2006

In the next section, you will find lists of symbols and abbreviations used in the book. The appendices comprise important physical constants and specific physical derivations that provide a better understanding of the text. Each chapter has its own reference section.

This textbook is the English version of the original German textbook which was published two years ago. Following the German publication, we have received many comments and recommendations that have been proof of the readers' interest in this field and which have allowed us to include several improvements in the current version. For that, we would like to express our gratitude to all critical readers.

Microsystem technology is a fast-developing scientific and engineering field which is attracting more and more industrial attention. Several generations of commercialized products are already on the market; new generations are under preparation in research labs. Because of new developments, microsystem technology will change substantially in the future. In order to provide an impression of what this field will look like in the ten or twenty years, we have added a new chapter that describes future trends and challenges regarding microsystem technology. We hope that this outlook will encourage readers to approach the existing technical problems that have to be surmounted for further developments in this area.

We would like to thank the co-authors who have contributed to this publication their comprehensive knowledge in special areas that are, nonetheless, of outstanding importance to the current development of microsystem technology. In particular, we would like to mention the commitment of our co-author Nam-Trung Nguyen who produced the final version of all the manuscript parts. Heike Collasch and Alexey Shaporin have diligently worked on the production of text and figures and we would like to extend our thanks for this. We would also like to thank those companies and institutions that allowed us to use images and who are individually named in the figure captions. Furthermore, we would like to thank Carl Hanser Verlag who published the original German version of the book, and in particular Christine Fritzsch, for her wonderful cooperation, but also for her patience when repeatedly faced with delays due to the authors' workload. We are deeply grateful to the Wiley staff for their never-faltering support during the entire process from deciding to publish an extended English version all the way through to the final book.

We would also like to express our appreciation to Dörte Müller who did a great job in translating the manuscript from German into English

The companion website for this book is (www.wiley/go/gerlach_microsystem).

Gerald Gerlach
Wolfram Dötzel
Dresden and Chemnitz
Germany

List of Symbols

a	dimension; acceleration; lattice constant; coefficient of the trial function; constant; directional component; shortening
b	width; integer; constant; positional parameter; directional component
c	spring stiffness; elastic stiffness coefficient; integer; directional component; specific heat capacity
d	distance; diameter; integer; piezoelectric coefficients
e	unit vector; unit charge; natural number
f	frequency; function; modal load
f_F	FANNING friction factor
g	acceleration of gravity
h	thickness; integer; height; PLANCK constant; directional component
i	counting index
j	imaginary unit; counting index
k	BOLTZMANN constant; damping constant; factor; integer; coefficient; reaction order; directional component; WEIBULL's modulus; counting index
k_S	specific reaction constant
k^2	coupling factor
l	integer; length; directional component
m	valence; dimension of a matrix; coefficient; mass; directional component; counting index
$\dot{m}$	mass flow
n	number; charge carrier density; compliance; directional component; number of pieces; particle density; counting index; number of degrees of freedom; number of model parameters
p	pressure; partial pressure; directional component; vapor pressure
p	hole density
p^{++}	very large hole density (highly doped)
q	weight of form function; distance between mass points; heat flow density
r	distance/length ratio; number; warping radius; local vector; friction impedance; relative resistance change
s	compliance coefficient; complex frequency; layer thickness; gap distance; standard uncertainty; displacement; distance
t	duration; time
u	displacement; measurement uncertainty/standard deviation; undercutting; flow velocity
v	velocity

w	number of springs
x	displacement; coordinate; temperature relation
y	displacement; coordinate
z	coordinate
A	activity; anisotropy; area; channel cross-section; oxidation constant; transducer constant
B	width; bow; magnetic field; oxidation constant
C	capacity; constant; heat capacity; costs
Ca	CAUCHY number; capillary number
D	damping constant; defect density; diffusion coefficient; diameter; dielectric displacement
E	YOUNG's modulus; electric field strength; energy
E_a	activation energy
E_g	band gap energy
EI	stiffness
Eu	EULER number
F	failure probability; force; real load; visibility factor; transformation coefficient
FC	fixed costs
FP	function parameter
FPD	Focal Plane Deviation
G	geometric parameter; proximity distance; shear modulus; transformation coefficient
H	height
I	areal moment of inertia; ion flow; current;
J	flow density; flow measurand; mass moment of inertia; current density
K	K-factor; concentration; stiffness
Kn	KNUDSEN number
L	diffusion length; inductivity; channel length; line width; inertia
L_0	evaporation heat
M	bending moment; mass; material parameter
MP	model parameter
N	number of critical masks levels
Ne	NEWTON number
P	effect; process parameter
Pe	PÉCLET number
Q	charge; resonance quality; heat quantity
R	etch rate; warping radius; resistance
R_a	roughness
R_m	ultimate strength
R_p	projected range
Re	REYNOLDS number
S	entropy; selectivity; sensitivity; particle flow density
T	taper; temperature
TB	thermal budget
TCO	temperature coefficient of the offset voltage
TCS	temperature coefficient of the sensitivity

TTV	total thickness variation
V	voltage; volume
VAR	variance
VC	variable costs
W	energy; half channel width; warp; section modulus
We	WEBER number
X	driving quantity
Y	yield; transducer/gyrator constant
α	thermal coefficient of longitudinal expansion; heat transfer coefficient; geometric relation; angle
α_A	evaporation coefficient
α_S	SEEBECK coefficient
β	angle
γ	area-related adhesion energy; compressibility factor; surface tension; thermal coefficient of volume expansion; transducer coefficient; angle
ε	strain; dielectric constant; emittivity; permittivity
η	dynamic viscosity; efficiency
ϑ	temperature
λ	similarity constant; free path length; scale factor; scaling factor; thermal conductivity; wavelength
μ	mobility; dynamic viscosity
ν	kinematic viscosity; POISSON's ratio
ξ	damping measure
π	similarity parameter; piezoresistive coefficient
ρ	absorption coefficient; density; resistivity
σ	stress; surface tension; real part of complex frequencies; STEFAN-BOLTZMANN constant; variance
τ	charge carrier lifetime; oxidation time constant; torsional stress; time constant
υ	EULER angle in transformed coordinate system
φ	EULER angle in transformed coordinate system; contact angle; angle; angle displacement
ω	angular frequency
Δ	deviation; change, difference
Π	product
Σ	sum
Φ	evaporation rate; natural oscillation modes; form function; heat flow
ψ	EULER angle in transformed coordinate system
Ω	yaw rate; excitation frequency; angular frequency

Indices

a	average
ab	depositing
ad	admissible
amb	ambient

b	bending; width
c	coercive; critical; short circuit; edge
cd	conduction
ch	characteristic
chip	chip
coup	coupling
cv	convection
d	thickness; dynamic; dissipating
dep	deposition
e	self-; excess
eff	effective
el	electric
etch	etching
exc	exciting
fluid	fluidic
g	generating
h	thickness-related; hybrid; hydraulic
i	internal
in	input
l, ℓ	longitudinal; no-load
m	mass; miniaturized; monolithic
max	maximum
mech	mechanical
min	minimum
n	electron; spring
out	output
p	hole
pol	polarization
r	remanence, radiation
rel	relative
rot	rotational
t	torsional
targ	target
th	thermal
tot	total
trans	transducer
u	displacement; undercutting
x, y, z	coordinate direction
A	atom
ACC	accelerometer
B	wetted; cantilever beam
BL	boundary layer
Bond	bonding
C	capacitor
CB	counterbody
D	deformation; breakdown
Diff	diffusion

EQUI	equilibrium
Eu	corner undercutting
F	fiber; film; fluid
FA	at the boundary fluid-air
G	glass transformation
Glas	glass
H	heat
I	pulse
K	kinetic; contact; spheric
L	longitudinal; lateral
N	final; full scale
Ox	oxide
P	period; polarization
PC	planar coil
React	reaction
S	melting; sensor; boiling; radiation; substrate; surface
Si	silicon
SF	at the boundary solid-fluid
SA	at the boundary solid-air
T	tuning; transversal; particle
U	undercutting
V	vertical
W	resistor
0	amplitude; initial value; reference; resonance; ambient
1, 2, 3	coordinate directions
=	static
~	alternating

List of Abbreviations

ABM	Analog Behavioral Model
ABS	Anti-lock Braking System
AFM	Atomic Force Microscope
AIM	Air-gap Insulated Microstructures
AMS	Analogue und Mixed-Signal-Systems
APCVD	Atmospheric Pressure Chemical Vapor Deposition
ASTM	American Society for Testing and Material
B	Bow
BEM	Boundary Element Method
BESOI	Bond-and-Etch-back Silicon on Insulator
BHF	Buffered Hydrofluoric Acid
BiCMOS	Bipolar and CMOS Technology on the same Chip
Bio-MEMS	MEMS for Bio Engineering
BOX	Buried Oxide
BS	Back Side
CMOS	Complementary Metal-Oxide-Semiconductor
CMP	Chemical Mechanical Polishing
CNT	Carbon Nanotubes
CSD	Chemical Solution Deposition
CSP	Chip Size Packaging
CVD	Chemical Vapor Deposition
C&W	Chip and Wire
DI	Deionized
DIL	Dual In Line
DIN	Deutsches Institut für Normung (German Institute for Standardization)
DLP	Digital Light Processing
DMD	Digital Micromirror Device
DNQ	Diazonaphtoquinone
DNS	Deoxyribonucleid Acid (DNA)
DRIE	Deep Reactive Ion Etching
DSMC	Direct Simulation Monte Carlo
EAP	Electroactive Polymers
ECD	Electrochemical Deposition
EDP	Ethylenediamine and Pyrocatechol
EP	Epoxy Resin
ESP	Electronic Stability Program

F	Flat
FBAR	Film Bulk Acoustic Resonator
FC	Flip Chip, Fixed Combs
FED STD	US Federal Standard
FE	Finite Element
FEM	Finite Element Method
FFT	Fast Fourier Transformation
FIB	Focused Ion Beam
FPA	Focal Plane Array
FPD	Focal Plane Deviation
FS	Front Side
FZ	Float Zone
GBL	Gamma-Butyrolacetone
GLV	Grating Light Valve
GPS	Global Positioning System
HARMST	High Aspect Ratio Micro System Technologies
HART	High Aspect Ratio Trenches
HF	Hydrofluoric Acid
IC	Integrated Circuit
IPA	Isopropyl Alcohol
ISO	International Organization for Standardization
KTFR	Kodak's Thin Film Resist
LCP	Liquid Crystal Polymer
LIGA	Lithography, Galvanics (Electroplating), Abformung (engl.: Moulding)
LPCVD	Low Pressure Chemical Vapor Deposition
LTO	Low Temperature Oxide
MBE	Molecular Beam Epitaxy
MC	Movable Combs
MD	Molecular Dynamics
MEMS	Micro Electro Mechanical System
MF	Main Flat
MOCVD	Metal Organic Chemical Vapor Deposition
MOD	Metal Organic Destruction
MOEMS	Micro Opto Electro Mechanical System
MOMBE	Metal Organic Molecular Beam Epitaxy
N	Notch, North pole
NPCVD	Normal Pressure Chemical Vapor Deposition
OLED	Organic Light Emitting Diode
OP	Operating Point
PA	Polyamide
PC	Polycarbonate
PDMS	Polydimethylsiloxane
PE	Polyethylene
PECVD	Plasma Enhanced Chemical Vapor Deposition
PGMEA	Propylglycol Methylether Acetate
PI	Polyimide
PLD	Pulsed Laser Deposition

PMMA	Polymethyl Methacrylate
PP	Polypropylene
ppm	Parts per Million (10^{-6})
PSG	Phosphorsilicate Glass
PTFE	Polytetrafluorethylene
PVC	Polyvinyl Chloride
PVD	Physical Vapor Deposition
PVDF	Polyvinylidene Fluoride
PZT	Lead Zirconate Titanate
RCA	Radio Corporation of America
RF	Radio Frequency
RIE	Reactive Ion Etching
RT	Room Temperature
RTA	Rapid Thermal Annealing
S	South Pole
SAW	Surface Acoustic Wave
SC	Standard Clean
SCREAM	Single Crystal Reactive Etching and Metallization
SDB	Silicon Direct Bonding
SEMI	Semiconductor and Materials International
SF	Secondary Flat
SFB	Silicon Fusion Bonding
SIMOX	Separation by Implanted Oxygen
SIMPLE	Silicon Micromachining by Plasma Etching
SMA	Shape Memory Alloy
SOD	Silicon on Dielectric
SOG	Silicon on Glass
SOI	Silicon on Insulator
SPT	Scalable Printing Technology
TAB	Tape Automated Bonding
TC	Thermocompression
TEOS	Tetraethoxysilane
TMAH	Tetramethylammonium Hydroxide
TS	Thermosonic
US	Ultrasonic
UV	Ultraviolet
VHDL	Very High Speed Integrated Circuit Hardware Description Language

1

Introduction

A radical change in the entire field of electronics began in 1947 when the transistor was invented; 11 years later in 1958 the first integrated semiconductor circuit was built. Ever since, electronics has turned almost completely into semiconductor electronics. Microelectronic manufacturing methods make it possible simultaneously to produce large numbers of similar components with dimensions that are much too small for precision mechanics. The discovery of the piezoresistive effect in 1953 (Figure 1.1, Table 1.1) created the precondition for also applying semiconductor materials and microelectronic production methods to non-electronic components. The first description of how to use a silicon membrane with integrated piezoresistors as mechanical deformation body dates back to 1962.

Uncountable, new miniaturized function and form elements, components and fabrication procedures have since been introduced, combining electrical and non-electrical functions and using semiconductor production technologies or even especially developed microtechnologies (Figure 1.2).

PHYSICAL REVIEW VOLUME 94, NUMBER 1 APRIL 1, 1954

Piezoresistance Effect in Germanium and Silicon

CHARLES S. SMITH
Bell Telephone Laboratories, Murray Hill, New Jersey
(Received December 30, 1953)

Uniaxial tension causes a change of resistivity in silicon and germanium of both *n* and *p* types. The complete tensor piezoresistance has been determined experimentally for these materials and expressed in terms of the pressure coefficient of resistivity and two simple shear coefficients. One of the shear coefficients for each of the materials is exceptionally large and cannot be explained in terms of previously known mechanisms. A possible microscopic mechanism proposed by C. Herring which could account for one large shear constant is discussed. This so called electron transfer effect arises in the structure of the energy bands of these semiconductors, and piezoresistance may therefore give important direct experimental information about this structure.

INTRODUCTION

THE effect of pure hydrostatic pressure on resistance has been extensively studied, notably by Bridgman, who also made the first piezoresistance measurements[1] known to us on several polycrystalline from two to six are reported for pressure experiments;[9] these values agree fairly well for a number of metals with a simple calculation[9] of the change of mobility produced by the change in the amplitude of thermal vibrations with volume v,

Figure 1.1 Publication by CHARLES S. SMITH regarding the discovery of the piezoresistive effect in the semiconductors germanium and silicon (*Physical Review* 94 (1954), pp. 42–9). Reproduced by permission of the American Physical Society

Introduction to Microsystem Technology: A Guide for Students Gerald Gerlach and Wolfram Dötzel

Table 1.1 Milestones of the development of microsystem technology (selection) [GERLACH05]

Year	Event
1939	pn-junctions in semiconductors (W. SCHOTTKY)
23-12-1947	Invention of the transistor (J. BARDEEN, W.H. BRATTAIN, W. SHOCKLEY; Bell Telephone Laboratories, Nobel Prize 1948)
1953	Discovery of the piezoresistive effect in semiconductors (C.S. SMITH; Case Institute of Technology and Bell Telephone Laboratories, respectively)
1957	First commercial planar transistor (Fairchild Semiconductor)
1958	Production of the first integrated semiconductor circuit (J.S. KILBY; Texas Instruments, Nobel prize 2000)
1959	First planar silicon circuit (R. NOYCE; Fairchild Semiconductor)
1962	Silicon wafer with integrated piezoresistors as deformation bodies (O.N. TUFTE, P.W. CHAPMAN, D. LONG)
1965	Surface micromachining: resonant acceleration sensitive field effect transistor (H.C. NATHANSON, R.A. WICKSTROM; Westinghouse)
1967	Anisotropic deep etching in silicon (H.A. WAGGENER and his team; R.M. FINNE, D.L. KLEIN)
1968	Development of anodic bonding (D.I. POMERANTZ)
1973	Integration of silicon pressure sensors with bipolar signal processing electronics (Integrated Transducers)
1974	First mass production of pressure sensors (National Semiconductor)
1977	First silicon piezoresistive acceleration sensor (L.M. ROYLANCE, J.B. ANGELL; Stanford University)
1979	Microsystem on a silicon wafer: gas chromatograph for air analysis (S.C. TERRY, J.H. JERMAN, J.B. ANGELL)
1983	Pressure sensors with digital sensor signal processor (Honeywell)
1985	Development of the LIGA technology (W. EHRFELD and his team)
1986	Development of silicon direct bonding (M. SHIMBO and his team)
1988	Commercial application of silicon direct bonding: 1000 pressure sensors on one 100 mm Si wafer (NovaSensor)
1988	Freely movable micromechanical structures (R.S. MULLER as well as W.S.N. TRIMMER and their teams)
1991	Market volume of micropressure sensors exceeds 1 billion US$
1992	Near-surface micromachining: SCREAM process (N.C. MACDONALD; Cornell University, Ithaca)
1993	Projection display: 768 × 576 mirror array (Texas Instruments)
1994	First commercial acceleration sensor in surface micromachining (Analog Devices)

The term ‘microsystem technology’ has been used for a wide range of miniaturized technical solutions as well as for the corresponding manufacturing technologies and it has no universally acknowledged definition or differentiation.

Similar to microelectronics, the nonelectric domain uses the terms micromachining/micromechanics, microfluidics or microoptics. Until the mid-1980s, the main focus of research and development was on miniaturized sensors, occasionally also on microactuators. Only after that were examples of complex miniaturized systems, such as micromechanical systems (MEMS) or microsystems, in general, introduced (e.g. gas chromatograph, ink-jet nozzles, force-balanced sensors, analysis systems).

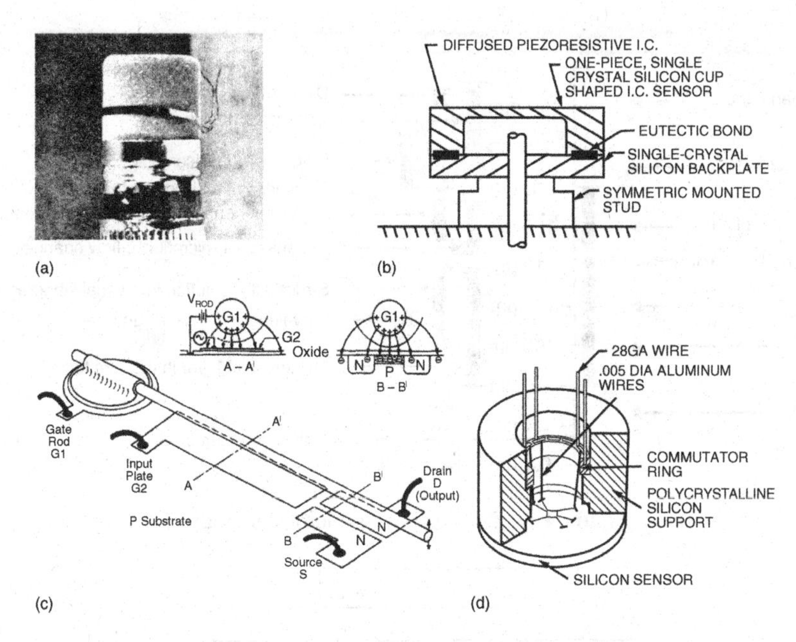

Figure 1.2 Early microsystems: (a) piezoresistive silicon wire resistance strain gauge, wrapped around a pencil (Micro Systems, approximately 1961). (b) integrated piezoresistive pressure sensor. (c) first micromechanically produced surface acceleration sensor (through accelerating a displaced cantilever as a gate of a field effect transistor). (d) pressure sensor with a diffused resistance structure in a single crystalline silicon plate.
(a) Reproduced from Mallon, J. R. Fiftieth birthday of piezoresistive sensing: progenitor to MEMS, http://www.rgrace.com/Conferences/detroit04xtra/mems/memvent.doc
(b) Peake, E. R. *et al.* (1969) Solid-state digital pressure transducer. *IEEE Trans. Electron Devices* 16, pp. 870–6. Reproduced by permission of © IEEE
(c) Nathanson, H. C., Wichstrom, R. A. (1965) A resonant gate silicon surface transistor with high-Q band-pass properties. *Applied Physics Letters* 7(4), pp. 84–86. Reproduced by permission of American Institute of Physics
(d) Kotnik, J. T., Hamilton, J. H. (1970) Pressure transmitter employing a diffused silicon sensor. *IEEE Transactions on Industrial Electronics* IECI-17(4), pp. 285–91. Reproduced by permission of © IEEE

1.1 WHAT IS A MICROSYSTEM?[1]

This book will use the term 'microsystem technology' to mean the following:

> Microsystem technology comprises the design, production and application of miniaturized technical systems with elements and components of a typical structural size in the range of micro and nanometers.

A microsystem can be characterized by the semantics of its word components 'micro' and 'system':

- Components or elements of microsystems have a typical size in the submillimeter range and these sizes are determined by the components' or elements' functions. In general,

[1]Portion of text is reproduced by permission of The American Physical Society.

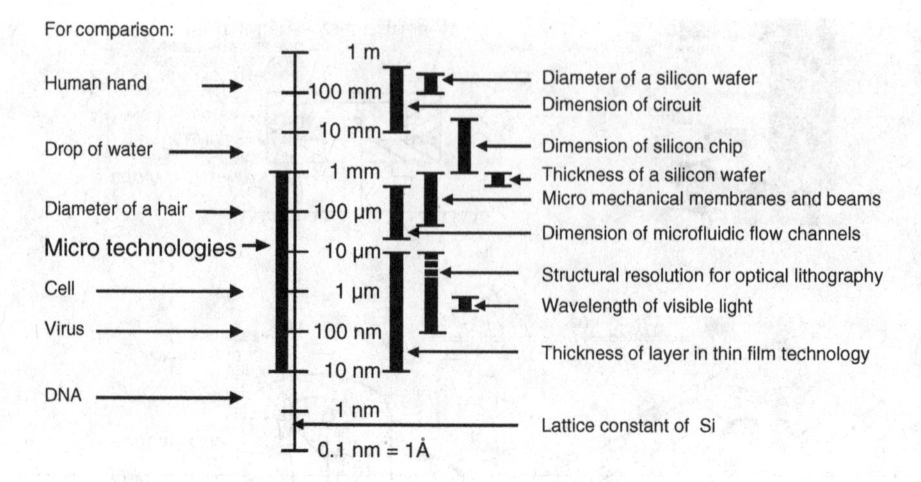

Figure 1.3 Dimensions in microtechnology

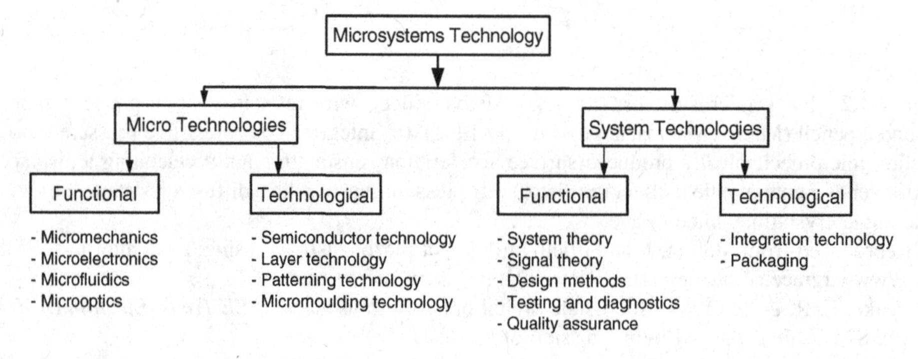

Figure 1.4 Microsystem technology

the size lies in the range between micrometers and nanometers (Figure 1.3). Such small structural sizes can be achieved by directly using or adapting manufacturing methods of semiconductor technology as well as through specifically developed manufacturing processes that are close to microelectronics (Figure 1.4).

- Recently, nanotechnology has started enjoying massive public attention. The prefix 'nano' is used there in two respects. On the one hand, nanotechnology can be applied to downscaling typical sizes, such as the thickness of function layers, from the micrometer down to the nanometer range. Today, typical gate thickness in microelectronic CMOS transistors is only a few dozen nanometers. Here, the term nanotechnology (nanoelectronics, nanoelectronic components) is used for an extremely diminished microtechnology where the known description and design procedures can be applied. On the other hand, the term nanotechnology is used for procedures and components which are only found at a certain miniaturized level. Examples are quantum effects (e.g. quantum dots) as well as tunnel effect devices or single-electron components. This textbook will not address such components.
- Microsystems consist of several components which, in turn, consist of function and form elements (Figure 1.5). The components have specific functions, e.g. sensor, actuator,

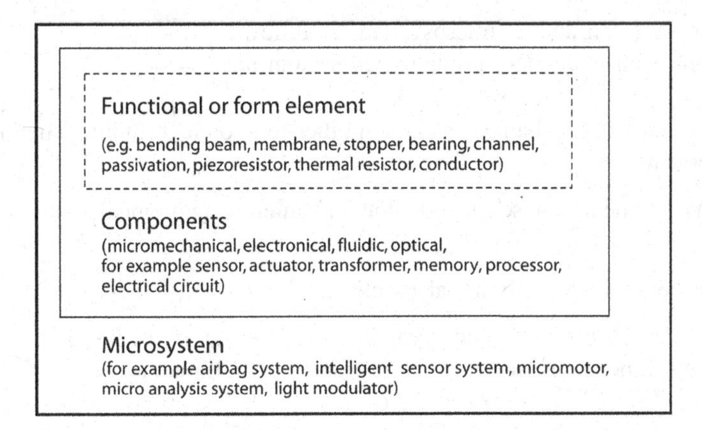

Figure 1.5 Terminological hierarchy in microsystem technology

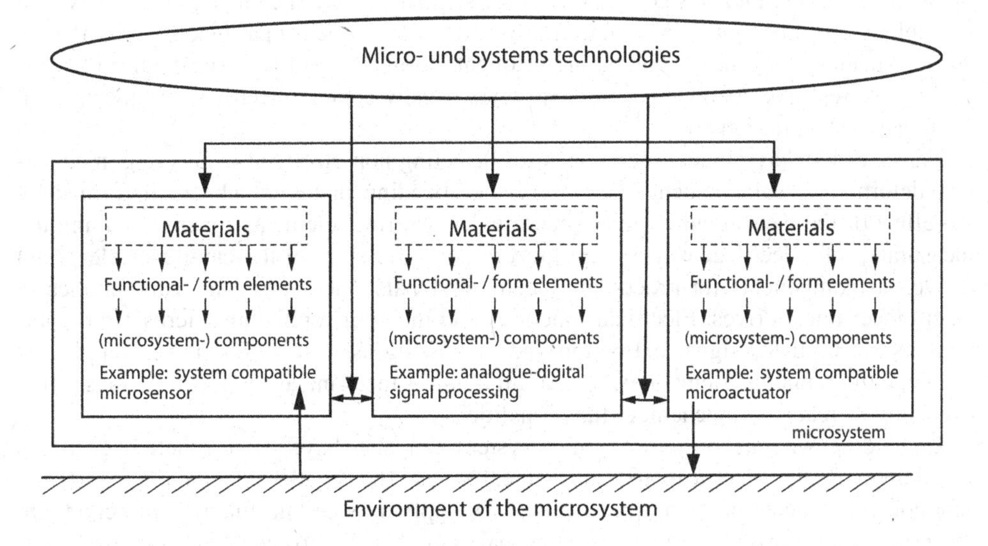

Figure 1.6 Structure and integration of microsystems

transformation, memory or signal processing functions and they can be constructively autonomous entities (e.g. an integrated circuit). Microsystems include both nonelectric and (micro-)electronic as well as electrical components. The system character is due to the fact that the system can only fulfil the total function if the components interact as a complex miniaturized unity.

Figure 1.6 shows the typical design of microsystems. Sensors and actuators as well as signal processing components that are suitable for system integration are – via appropriate interfaces – integrated with each other but also with the microsystem's environment, e.g. with a technical process that has to be controlled. The individual components consist each of a number of function and form elements that can be produced using corresponding materials and applying micro- and system technology. Microsystem technology is also used for the functional integration of the system components.

In summary, we can define 'microsystem' as follows:
A microsystem is an integrated, miniaturized system that

- comprises electrical, mechanical and even other (e.g. optical, fluidic, chemical, biological) components;
- is produced by means of semiconductor and microtechnological manufacturing processes;
- contains sensor, actuator and signal functions;
- comprises function elements and components in the range of micro- and nanometers and has itself dimensions in the range of micro- or millimeters.

This definition does not strictly distinguish between micro- and nanosystems. As microelectronics already uses ultrathin layers of only a few nanometers it has crossed the line to nanotechnology. Piezoresistive resistors are standard function elements in microsystem technology and they act as conduction areas for a two-dimensional electron gas if they are less than 10 nm thick.[2] The resulting quantum effects lead to a substantial increase in the piezoresistive coefficients. Microsystems usually contain electrical and mechanical components as a minimum.

Thus, sensors have function elements for detecting non-electrical values (e.g. mechanical deformation values such as cantilevers or bending plates which are deformed by the effect of the measurand force or pressure), transformer elements for transforming the measurand into electrical values (e.g. piezoresistive resistors in the cantilever elements) as well as components for processing electrical signals. Vice versa, the same applies to electromechanical drives. Electrical functions und their corresponding microsystem components are used for signal extraction and processing as well as for power supply. At the same time, microsystems have – as a minimum – mechanical support functions, often even further reaching mechanical functionalities.

Coinciding with its purpose, a microsystem can also have other function elements in addition to the electrical and mechanical ones. The smallness of a microsystem's function components is often a prerequisite for applying a certain function principle. On the other hand, however, miniaturization makes a coupling to technical systems in our 'macroworld' more difficult. Therefore, complete microsystems often have dimensions in the range of millimeters which clearly facilitates their integration into other systems. Here the transition from the micro- to macroworld already takes place in the packaging of microsystems. However, even here the term microsystem is commonly used.

1.2 MICROELECTRONICS AND MICROSYSTEM TECHNOLOGY

The development of microsystem technology is the immediate result of microelectronics which shows two major drawbacks:

[2]Ivanov, T. (2004) *Piezoresistive Cantilever with an Integrated Bimorph Actuator*. PhD Thesis. University of Kassel.

1. Microelectronics is limited to electronic devices and the integration of electronic functions. Usually, it is not possible to process non-electrical values. Building complex systems that are able to use sensors to read signals from the system environment and to affect the environment via actuators requires a combination of microelectronic components and classical components produced by precision mechanics. This reduces the miniaturizing potential and the level of integration that can be reached. Thus, reliability decreases.

2. Basically, the manufacturing process of semiconductor technology can only be used to produce two-dimensional but not three-dimensional structures. However, a number of functions – especially nonelectrical ones – require three-dimensional function components and their three-dimensional integration.

There are several reasons for the close connection of the development of microsystem technology with microelectronics:

- Within microtechnologies such as micromechanics, microfluidics, microoptics etc., microelectronics has an outstanding position. Given the current state of the art, microsystems without microelectronic components for processing analogous or digital signals appear not to be meaningful.
- Only semiconductor and thin film technology provides manufacturing processes that are able to produce structures in the range of micro- and nanometers. And there are additional advantages of microelectronic manufacturing processes that can be used: the parallel processing of identical elements or components within one and the same manufacturing process as well as the use of completely new physical-chemical procedures which differ substantially from classical manufacturing technologies.
- Often, microsystem technology uses materials that are used in microelectronics. Both microsystem technology and microelectronics are dominated by silicon which has excellent characteristics in comparison with compound semiconductors, for instance. On the one hand, this is due to the fact that electronic components are very important in microsystems and therefore silicon particularly is qualified for integration technologies. On the other hand, silicon can be produced with the highest chemical purity and crystal perfection. And a large number of technological procedures and sensoric as well as actuating effects rely in particular on these crystal features.

Table 1.2 compares typical characteristics of microelectronics and microsystem technology. The given characteristics show that microsystem technology will even in the future mainly use microelectronic technologies. Large production numbers and a high standardization of components in microelectronics are due to the programming options of microprocessors and microcomputers as well as memory circuits. Therefore, silicon-based technology was able to attract development prospects that by far exceed those of microsystem technology. Due to the diversity and heterogeneity of microsystem technology, it will not be possible to find similarly standardized applications with similarly high production numbers. The only option here is to use highly developed fabrication methods of semiconductor technology. Original technological developments, such as the LIGA technology, are rather the exception. Currently, the following developments can be discerned regarding the use of microelectronic manufacturing methods:

Table 1.2 Comparison of typical characteristics of microelectronics and microsystem technology

Criterion	Microelectronics	Microsystem technology
Components	standardized (e.g. memories, processors)	heterogeneous
Production numbers	$10^5 \ldots 10^8$	$10^2 \ldots 10^6$
Applications	electronical	mechanical, electronical, fluidic, optical, ...
Structural dimension	two-dimensional	three-dimensional
Design	automated	heterogeneous with limited design support

- transfer of two-dimensional structuring processes on to three-dimensional applications (e.g. surface and near-surface micromachining, see Chapter 4);
- development of modified system integration technologies (e.g. packaging, see Chapter 5);
- further development und adaptation of microelectronic design methods to complex heterogeneous systems, which are characteristic of microsystem technology (see Chapter 8).

1.3 AREAS OF APPLICATION AND TRENDS OF DEVELOPMENT

Initially, the development of microsystem technology was related to the study (1953–58) and commercialization (1958–72) of piezoresistive sensors [GERLACH05]. Since then, the range has become dramatically wider [KOVACS98]. In the beginning, the focus was on the advantages of miniaturization for new automotive applications (measuring manifold pressure of combustion engines in order to reduce emissions) and invasive biomedical sensors, for instance. Today, new areas of application are of major interest, allowing for large production numbers, low cost per unit and high reliability.

Currently, the following are important examples of the application of microsystems:

- Automation technology: Modern cars contain a large number of new systems for improved driving safety and comfort. Microsystem technologies can be used to produce large volumes with low system cost and high reliability. Examples are acceleration sensors for ABS and airbag applications, yaw rate sensors for driving stability and airflow sensors for controlling air conditioning. More than half of all microsystem applications is used by the automotive industry.
- Medical technology: Microsystems with dimensions in the range of micro and millimeters can be widely used for invasive applications. Important examples are catheters for measuring heart pressure, probes for minimal invasive diagnostics and therapy as well as dosing systems.

- Environmental technology, gene technology and biotechnology: Microanalysis and micro-dosing systems can be used for the chemical and biotechnological analysis of gases and fluids. Microreactors can be used for chemical processes involving very small volumes and other uncommon conditions.
- Microfluid systems: Ink jet nozzles can be produced at low cost using miniaturized integration of electrical, mechanical and fluidic functions.
- Nanotechnology: The production, manipulation and characterization of nanostructures require tools for ultra-precise movement and positioning. Systems that are based on scanning tunneling and atomic force effects often use miniaturized cantilevers with tips in the range of nanometers. Microsystem technology can effectively produce such tools.

Considering the further development of microsystem technology, the following trends can be discerned:

- Microsystem solutions require mainly applications with large production numbers.
- The manufacturing of microsystems increasingly uses commercial semiconductor processes. The development of special technologies is only possible when large production numbers justify the costs or when there are no alternatives to microsystems and therefore a high per-unit price can be realized. This is the case of minimal-invasive medical applications, for instance.
- Reliability and lifetime of microsystems as well as long-term stability and accuracy become ever more important, particularly regarding industrial applications of chemical and biological sensors and analysis systems.
- The economic rather than the technological framework decides which integration technologies are used for producing microsystems. Whereas in the past the monolithic integration was a main goal, today almost exclusively hybrid integration is used for smaller production numbers. Nevertheless, there are substantial efforts to further develop monolithic integration methods, in particular those that try to integrate microtechnologies into commercial semiconductor manufacturing processes (e.g. CMOS processes). The three-dimensional design and integration of alternative methods of microsystem technology then takes place mainly as a back-end process following the conventional microelectronic manufacturing process.
- Microsystems are based mainly on microelectronic materials and methods. Therefore, a main issue of the development of microsystems is the design process. Currently, there is a substantial need for developing appropriate design tools for modeling and simulating complex and heterogeneous systems (see Chapter 8).

1.4 EXAMPLE: YAW RATE SENSOR

In the following, we want to use an example to present a complex microsystem. We will show how a microsystem is constructed with various components and function elements.

We also want to show that the characteristics of the material, manufacturing technology and design are closely interconnected.

1.4.1 Structure and Function

The rotating speed Ω (yaw rate) is an important parameter of bodies moving in space. Yaw rate sensors are therefore very important for the driving stability of cars. Figure 1.7 and Figure 1.9 show a yaw rate sensor that is based on the CORIOLIS principle.

The sensor consists of a silicon chip. Methods of bulk micromachining (see Section 4.6) are used to etch away two masses suspended on springs. In the magnetic constant field B_z, an alternating current I_y induces through conducting lines sinusoidal Lorentz forces F_x on the cantilevers. These forces cause masses m to oscillate harmoniously with speed v_x in direction x (compare the effect chain in Figure 1.8). When the two masses oscillate in opposite direction, the sensor chip remains, in total, motionless and can be translatory considered to be fixed. Both cantilevers and masses consist of single crystalline silicon (compare Section 3.2) and show therefore a long-time stable, ideal elastic behavior. If a yaw rate Ω_z operates vertically to the direction of the oscillation, CORIOLIS force F_x results in direction y.

In this example, a capacitive accelerometer is used to evaluate this force. It has the design of an interdigital finger structure with fingers that are fixed to the masses and other fingers between those fingers. The movable fingers are attached to a middle link that is suspended flexibly over the cantilevers. CORIOLIS force F_y can thus displace mass m_{ACC} of the accelerometer by a distance s_y, which in turn changes capacity ΔC between movable and fixed fingers in relation to their distance from each other. The capacity change of the accelerometer can be transformed into an electronic signal, e.g. voltage change $\Delta V(\Omega)$ or frequency change $\Delta f(\Omega)$.

Methods of surface micromachining (see Section 4.11) are used for producing the accelerometer on a silicon mass m. Thin films with a thickness in the range of a few micrometers are separated and selectively structured. The characteristics of these layers determines the behavior of the accelerometer.

Methods of packaging (see Chapter 5) are used to integrate the sensor chip into a casing and thus to guarantee a safe and reliable operation of the sensor. The casing consists of a glass substrate with a coefficient of linear thermal extension adapted to the silicon as well as of an accordingly shaped silicon cover.

1.4.2 Function Components and Elements

The yaw rate sensor consists of two components (Figure 1.8):

- Electrically oscillated mass m (electromechanical actuator). Mass m and the four cantilevers n_x are the function elements and form an oscillatory mechanical resonant system.
- Yaw-rate related interdigital capacitor (accelerometer). This mechanical-electric sensor structure consists of oscillating mass m with a firmly attached finger structure, an

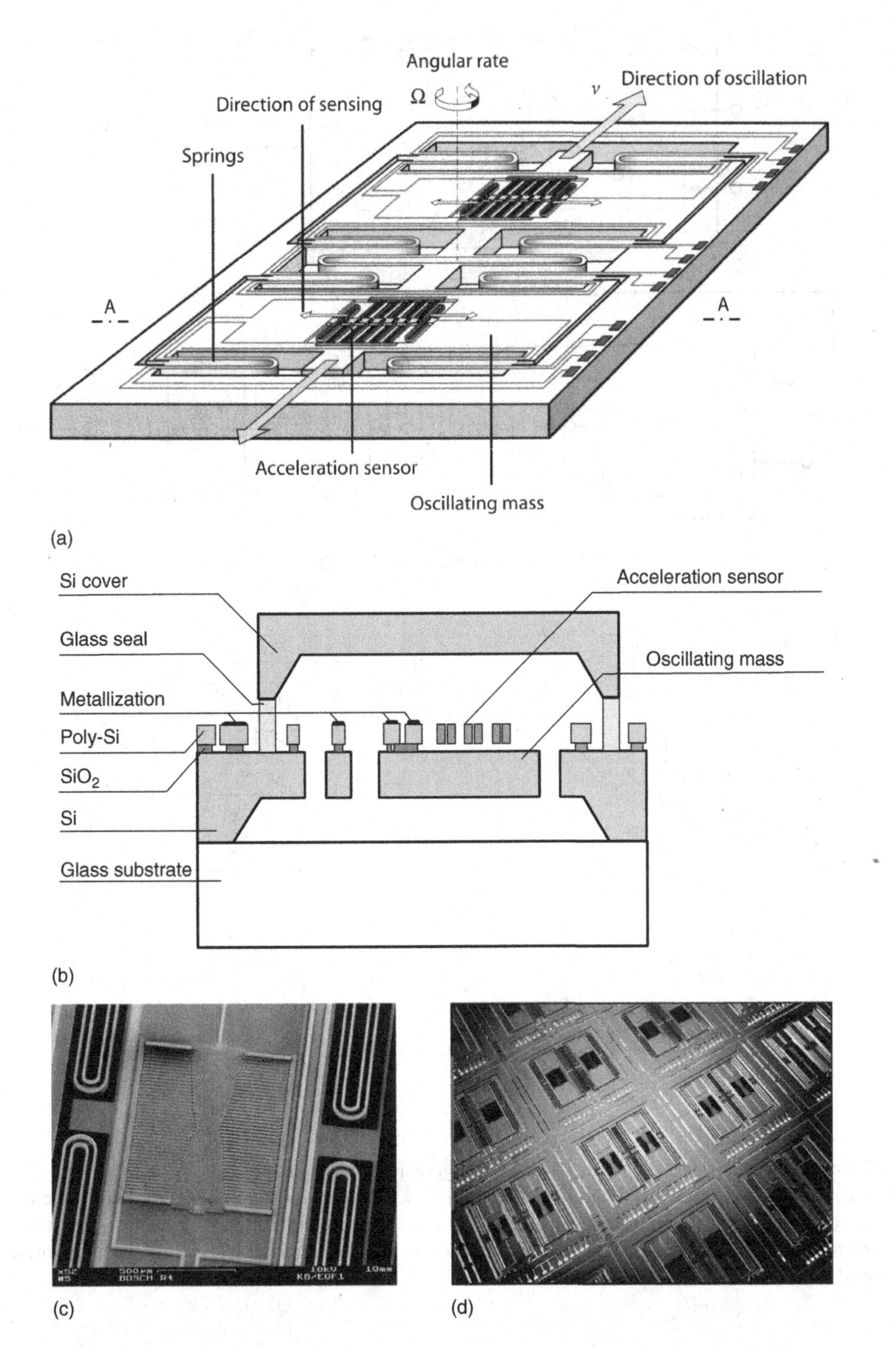

Figure 1.7 Yaw rate sensor: (a) 3D image; (b) cross section A-A; (c) microscope image; (d) several sensors on-wafer. Reproduced by permission of Robert Bosch GmbH, Germany

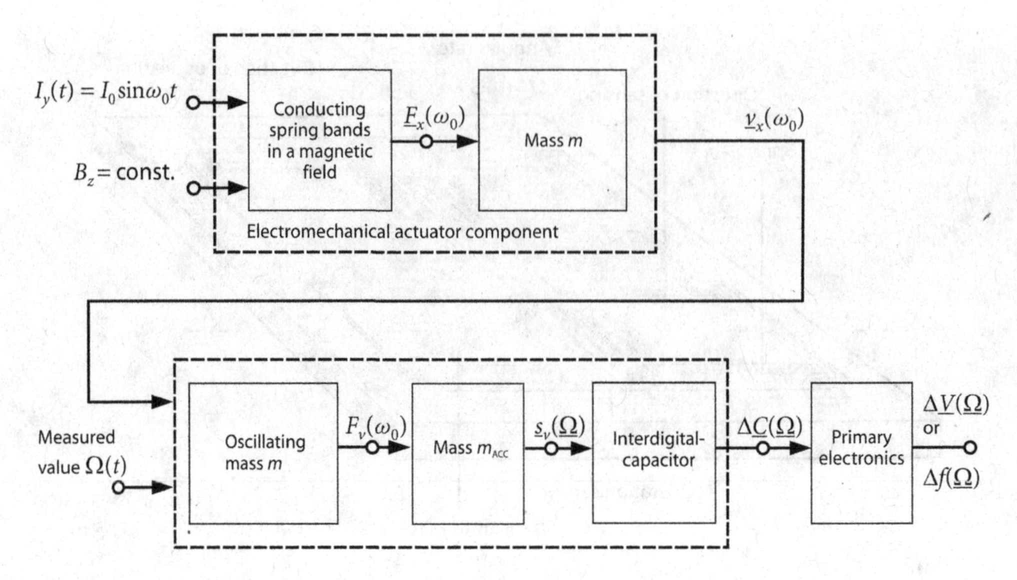

Figure 1.8 Effect chain of the yaw rate sensor in Figure 1.7

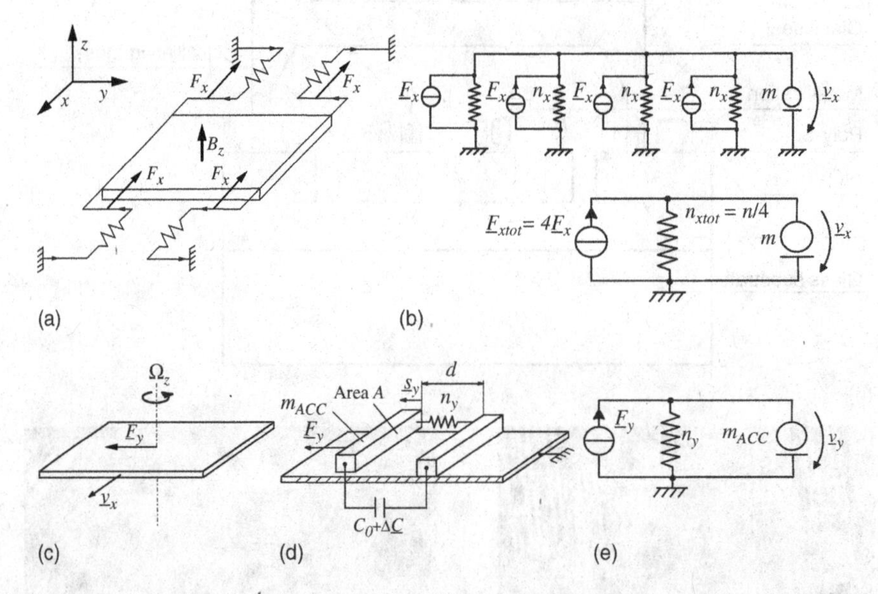

Figure 1.9 Details of the yaw rate sensor in Figure 1.7: (a) oscillating mass m with current I on the cantilevers; (b) mechanical block diagram of the oscillating mass affected by LORENTZ force F_x; (c) CORIOLIS force F_y as a result of yaw rate Ω; (d) detailed presentation of the capacitive finger structure of the accelerometer on oscillating mass m; (e) mechanical scheme of the interdigital transformer in (d)

elastically suspended (compliance n_y) movable finger structure (mass m_{ACC}) and the capacitor electrodes formed by those.

The underlying physical principles can be used to describe the effect of function elements, function components and the entire microsystem. In the following, we will show this for the individual function elements.

Figure 1.9a schematically represents the function elements of the electro-mechanical actuator that induces sinusoidal oscillation motions of the accelerometer. Mass m is suspended by four cantilevers n_x. They form a parallel-motion joint that restricts translatory displacement in direction x. There are conducting lines on the cantilevers that carry an alternating current $I_y(t) = I_0 \cdot \sin \omega_0 t$, where ω_0 is the angular frequency of the excitation current.

If the magnetic field operates in direction z, on each of the cantilevers with an orientation in direction y and a length ℓ_y the following LORENTZ force will result according to the law of BIOT-SAVART

$$\vec{F_x} = I\ell_y(\vec{e_y} \times \vec{B}) \text{ or} \tag{1.1}$$

$$F_x(t) = I(t) \cdot \ell_y \cdot B_z \tag{1.2}$$

in positive or negative direction x, respectively.

Force F_x periodically deforms the corresponding cantilever part n_x and contributes to displacing mass m. If the current flows in the correct direction through all four cantilever elements, the partial forces add up as is shown in the mechanical scheme in Figure 1.9b. The resulting total force $F_{x\text{tot}}$ amounts to four times the individual forces:

$$\underline{F}_{x\text{tot}} = 4\underline{F}_x. \tag{1.3}$$

Total compliance $n_{x\text{tot}}$ amounts to only one-fourth of the individual compliances n_x:

$$n_{x\text{tot}} = n_x \| n_x \| n_x \| n_x = n_x/4. \tag{1.4}$$

Thus, the displacement of mass m is

$$\underline{v}_x = \underline{F}_{x\text{tot}} \cdot \left(j\omega_0 n_{x\text{tot}} \| \frac{1}{j\omega_0\, m} \right) = \underline{F}_{x\text{tot}} \cdot \frac{j\omega_0 n_{x\text{tot}} \cdot \dfrac{1}{j\omega_0 m}}{j\omega_0 n_{x\text{tot}} + \dfrac{1}{j\omega_0 m}}$$

$$= \underline{F}_{x\text{tot}} \cdot j\omega_0 n_{x\text{tot}} \frac{1}{1 - \omega_0^2 m n_{x\text{tot}}} \tag{1.5}$$

or applying Equation (1.2)

$$\underline{v}_x = j\omega_0 n_{x\text{tot}} \ell_y B_z \underline{I} \frac{1}{1 - \omega_0^2 m n_{x\text{tot}}}. \tag{1.6}$$

The imaginary unit j describes the phase displacement between current $I(t)$ and oscillation rate $v_x(t)$ of the mass. The last term of Equation (1.6) derives from the fact that

m_x and n_{xtot} form a resonant system. For $\omega_0^2 \cdot mn_{xtot} = 1$, $\underline{v}_x$ would be infinite. In practice, such systems are operated in resonance as they require only little power input to reach large oscillation amplitudes. The displacement, however, cannot become infinitely large due to damping effects, which were neglected before. Assuming that in Figure 1.9b there is a friction $r = \underline{F}_r/\underline{v}_x$ parallel to n_{xtot} and m, the term $j\omega_0 m$ in Equation (1.5) would have to be replaced by $(j\omega_0 m + r)$. Including resonance quality $Q = 1/\,\omega_0 nr$, Equations (1.5) and (1.6) result in

$$\underline{v}_x = j\omega_0 n_{xtot}\ell_y B_z \underline{I}\frac{1}{1-\omega_0^2 mn_{xtot} + j\frac{1}{Q}} \tag{1.7}$$

or for $\omega_0^2 = 1/mn_{xtot}$

$$\underline{v}_x = \omega_0 n_{xtot}\ell_y B_z \underline{I} \cdot Q. \tag{1.8}$$

Quality Q typically has a value of approximately 10 in air and of $10^4 \ldots 10^6$ in vacuum.

Due to yaw rate $\Omega(t)$, CORIOLIS force

$$\underline{\vec{F}}_y = 2m(\underline{\vec{v}}_x \times \underline{\vec{\Omega}}_z). \tag{1.9}$$

operates on the oscillating mass m.

For an arrangement according to Figure 1.9c there results for the scalar components

$$\underline{F}_y = 2m\ \underline{v}_x \cdot \underline{\Omega}_z = \omega_0 n_{xtot}\ell_y B_z Q\underline{I} \cdot \underline{\Omega}_z. \tag{1.10}$$

The interdigital structure of the accelerometer located on mass m can be represented according to Figure 1.9d. The result is an oscillatory system consisting of mass m_{ACC} of the movable fingers and total compliance n_y of the finger suspension. Analogous to the system in Figure 1.9a, there results the mechanical scheme in Figure 1.9e. Displacement $\underline{s}_y$ of the movable finger electrodes is thus

$$\underline{s}_y = \frac{\underline{v}_y}{j\omega} = \underline{F}_y n_y \cdot \frac{1}{1-\omega^2 m_{ACC} n_y}. \tag{1.11}$$

Rotating speed Ω corresponds here to the yaw rate frequency which, in practical applications, can be replaced by complex angular frequencies as the measurand $\Omega(t) = \Omega_z(t)$ can have any time function.

Including Equations (1.10) and (1.11), the yaw-rate related displacement of the left movable finger in Figure 1.9d results in

$$\frac{\Delta \underline{C}}{C_0} \approx -\frac{\underline{s}_y}{d} = \omega_0 Q n_{xtot} n_y \ell_y \frac{1}{d} B_z \underline{I}(\omega_0) \frac{1}{1-\omega^2 m_{ACC} \cdot n_y} \cdot \underline{\Omega}(\omega). \tag{1.12}$$

The following conclusions can be drawn from Equation (1.12):

- Large capacity changes can be achieved by using elastically suspended masses m (high compliance n_x, large ℓ_y).

- A high quality factor Q can be reached in vacuum. This requires integration methods for the system construction that correspond to Figure 1.7b and can be carried out in vacuum in order to avoid subsequent evacuation.
- Distance d between the finger elements of the interdigital capacitor should lie in the range of micrometer. It has to substantially exceed maximum displacement s_y, though.
- B_z is limited due to the magnetic flux density of typical hard magnets and lies in the range below 1 T.
- Feeding current I of the electro-mechanical actuator is limited by resistance losses in the conducting lines and by the small feeding voltages that are frequently used in microsystem technology and microelectronics.

EXERCISES

1.1 Explain the function principles of the microsystems in Figure 1.2.

1.2 Characterize function elements and components of the following microsystems:
 (a) piezoresistive pressure sensor (Section 7.2.5, Figure 7.23);
 (b) micropump (Section 6.2.4, Figure 6.13);
 (c) bolometer array.

1.3 Search the Internet for examples of microsystems in the areas of automotive sensors (pressure sensors, crash sensors, air flow sensors, yaw rate sensors), medical catheter sensors, ink-jet technology and micro-analysis systems.

REFERENCES

[GERLACH05] Gerlach, G., Werthschützky, R. (2005) 50 Jahre Entdeckung des piezoresistiven Effekts – Geschichte und Entwicklung piezoresistiver Sensoren. *Technisches Messen* **72** (2005) 2, pp. 53–76, and 50 years of piezoresistive sensors – History and state of the art in piezoresistive sensor technology. In: Sensor 2005, 12th *International Conference, Nürnberg, 10–12 May 2005. Proceedings*, vol. I. AMA 2005, pp. 11–16.

[KOVACS98] Kovacs, G.T.A. (1998) *Micromachined Transducers Sourcebook*. Boston: WCB/McGraw-Hill.

2
Scaling and Similarity

Regarding the definition of microsystems in Chapter 1, we have underlined that microsystem technology uses form and function elements with dimensions in the range of micro- and nanometers. These dramatically reduced dimensions in comparison to mechanical and precision engineering, have largely affected their functions, if not their overall functionality. In microelectronics, scaling laws (e.g. the famous MOORE's law)[1] had to be taken into account for further integration and miniaturization. Similarly, the reduced dimensions in micro- and nanosystems have an important impact on the selection and application of certain function principles as well as on the structure of such elements and devices. Mechanical engineering, for instance, is dominated by electromagnetic drives (e.g. electrical engines). In microsystem technology, on the other hand, electrostatic drives are used most commonly. In the following, we will therefore look at how miniaturizing and scaling affect the function of function elements. We will see to what extent physical effects can be transferred from the macro- to a microworld, i.e. whether they are similar. In practice, this similarity is often stated as a dimensionless number.

2.1 SCALING

The transfer function of the individual function unit has to be analyzed in order to arrive at the effect that scaling has on the microsystem's elements and devices. For sensors, for instance, we have to look at sensitivity S

$$S = \frac{\text{output quantity}}{\text{input quantity}} = \frac{V_{\text{out}}}{x}, \tag{2.1}$$

where V_{out} is the output voltage of the sensor and x the measurand (e.g. pressure p, acceleration a). Often it is desirable for an actuator that the applied driving force (e.g. electrical energy) generates a maximum output value (e.g. controller output, mechanical energy). Analogously to sensitivity, this ratio is here called activity A

$$A = \frac{\text{driven quantity}}{\text{driving quantity}} \tag{2.2}$$

[1]Moore, Gordon E. (1965). Cramming more components onto integrated circuits, Electronics Magazine, Vol. 38, Number 8, April 19, 1965

Introduction to Microsystem Technology: A Guide for Students Gerald Gerlach and Wolfram Dötzel

In the following, we will use the example of electrostatic drives in order to analyze the effect that scaling has on the functionality of micromechanical devices. These drives consist of a fixed and a movable electrode (Figure 2.1).

The electrostatic field in such a capacitor arrangement has a potential energy of

$$W_{\mathrm{el}} = \frac{1}{2}\frac{Q^2}{C} = \frac{1}{2}C \cdot V^2 = \frac{\varepsilon \cdot \ell \cdot b}{2d}V^2, \tag{2.3}$$

where V is the drive voltage, Q the charge at the electrodes, C the capacity of the arrangement, ε the permittivity of the medium in the capacitor (e.g. air $\varepsilon = \varepsilon_0$). The force for the two arrangements in Figure 2.2 results from the gradient of energy W_{el} through the general relationship

$$\vec{F}(\vec{u}) = -\mathrm{grad}\ W_{\mathrm{el}}(\vec{u}) \tag{2.4}$$

(where $\vec{u}$ is the position vector) in

$$F_1 = \frac{\varepsilon \cdot \ell}{2d}V^2, \tag{2.5}$$

$$F_3 = \frac{\varepsilon \ell b}{2d^2}V^2. \tag{2.6}$$

Looking at the geometric dimensions of a macrosystem (index 0) with ℓ_0, b_0 and d_0 and those of a miniaturized system (index m) with $\ell_{\mathrm{m}} = \lambda_\ell \cdot \ell_0$, $b_{\mathrm{m}} = \lambda_b \cdot b_0$ and $d_{\mathrm{m}} = \lambda_d \cdot d_0$, applying Equations (2.5) and (2.6), it follows that

$$F_{1\mathrm{m}} = \frac{\lambda_\ell}{\lambda_{\mathrm{d}}} \cdot \frac{\varepsilon_0 \ell_0}{2d_0}V^2 = \frac{\lambda_\ell}{\lambda_{\mathrm{d}}}F_{10}, \tag{2.7}$$

$$F_{3\mathrm{m}} = \frac{\lambda_\ell \lambda_{\mathrm{b}}}{\lambda_{\mathrm{d}}^2} \cdot \frac{\varepsilon \ell_0 b_0}{2d_0^2}V^2 = \frac{\lambda_\ell \lambda_{\mathrm{b}}}{\lambda_{\mathrm{d}}^2}F_{30}. \tag{2.8}$$

Miniaturization means that in Equations (2.7) and (2.8), it is valid for the reduction coefficient λ_i that $0 < \lambda_i \leq 1$. The values for λ_i should be selected in a way that sensitivity S or activity A, respectively, in Equations (2.1) and (2.2) will become large values. In the case of the static displacement of electrostatic drives, this corresponds to

$$A \sim \frac{F_{i\mathrm{m}}}{V^2} \sim \frac{F_{i\mathrm{m}}}{F_{i0}}. \tag{2.9}$$

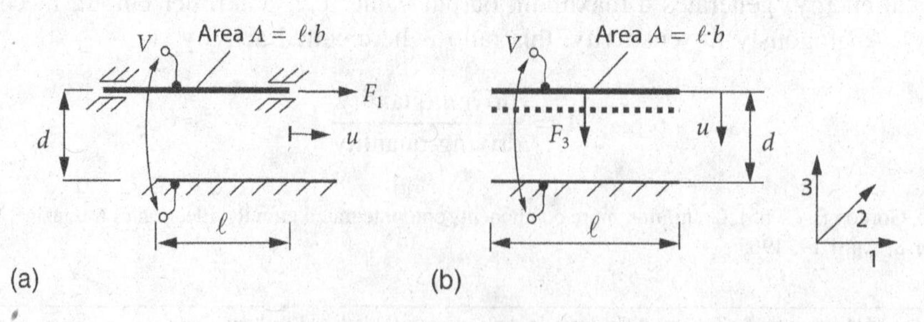

Figure 2.1 Electrostatic drives

For the examples in Figure 2.1, applying Equations (2.7) and (2.8) results in

$$\frac{F_{1\mathrm{m}}}{F_{10}} = \frac{\lambda_\ell}{\lambda_\mathrm{d}}, \tag{2.10}$$

$$\frac{F_{3\mathrm{m}}}{F_{30}} = \frac{\lambda_\ell \lambda_\mathrm{b}}{\lambda_\mathrm{d}^2}. \tag{2.11}$$

For miniaturization, due to $\lambda_i \leq 1$, all dimensions with a reduction coefficient that can be found in the denominator are of interest. This means that the selected distance $d_\mathrm{m} = \lambda_d \cdot d_0$ should be small and area $A = \ell \cdot b$ should be large. In practice, often interdigital (finger) structures are used to reach a very large electrode area, and the distance between movable and fixed electrodes is minimized (see Section 7.2.1).

Example 2.1 The effect of scaling on the sensitivity of capacitive sensors

Figure 2.2 shows the basic structure of capacitive sensors that are used for measuring pressure, acceleration and yaw rate. The corresponding measurand x always results in the displacement of a spring-suspended electrode that is the counterelectrode of capacitor C. In this case, the flexible suspension is a parallel-motion spring consisting of a double bending beam with length ℓ_B, width b_B and thickness h_B. The capacitor area is $A_\mathrm{K} = \ell_\mathrm{K} b_\mathrm{K}$. Due to the effect of the measuring force, the basic distance d_K of the capacitor is reduced by electrode displacement u. Using the relationship for capacity C according to Equation (2.13) and using Equation (2.12) for the double bending beam, relative capacity

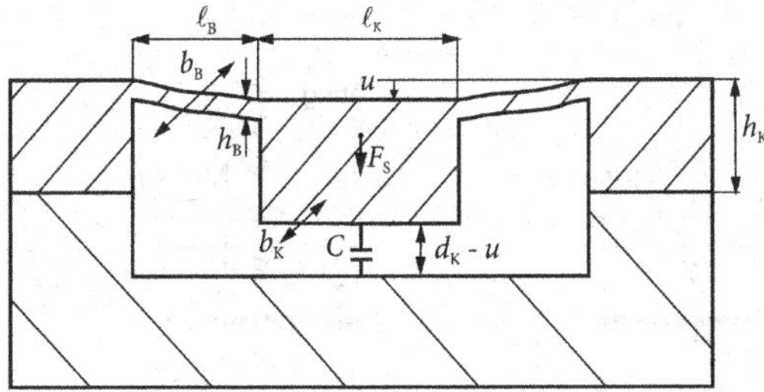

Compliance: $n = \frac{u}{F_\mathrm{S}} = \frac{1}{2Eb_\mathrm{B}}\left(\frac{\ell_\mathrm{B}}{h_\mathrm{B}}\right)^3$ (2.12)

Measuring capacitor: $C = \varepsilon \cdot \frac{b_\mathrm{K}\ell_\mathrm{K}}{d_\mathrm{K} - u} = \varepsilon \frac{b_\mathrm{K}\ell_\mathrm{K}}{d_\mathrm{K}} \frac{1}{1 - \frac{u}{d_\mathrm{K}}} \approx \varepsilon \frac{b_\mathrm{K}\ell_\mathrm{K}}{d_\mathrm{K}}\left(1 + \frac{u}{d_\mathrm{K}}\right)$ (2.13)

Relative capacity change : $\frac{\Delta C}{C_0} \approx \frac{u}{d_\mathrm{K}} = \frac{1}{2E} \frac{1}{d_\mathrm{K} b_\mathrm{B}}\left(\frac{\ell_\mathrm{B}}{h_\mathrm{B}}\right)^3 F_\mathrm{S}$ (2.14)

Figure 2.2 Structure of capacitive sensors. Index K movable capacitor electrode, Index B bending beam, $\ell_\mathrm{B}, b_\mathrm{B}, h_\mathrm{B}$ length, width and thickness of bending beam, $\ell_\mathrm{K} \cdot b_\mathrm{K}$ capacitor area, d_K distance of capacitor electrode, u displacement of movable capacitor electrode, C capacity, F_S sensor force

change $\Delta C/C_0$ results from Equation (2.14). In quasi-static state, different sensor forces F_S (Table 2.1) and the corresponding relative capacity changes $\Delta C/C_0$ (Equations (2.15) to (2.17)) result for the different measurands pressure p, acceleration a and yaw rate Ω. Ratio $\Delta C/C_0$ of the respective measurand p, a or Ω corresponds to sensitivity S in Equation (2.1).

Table 2.1 considers the case that all geometric dimensions of the sensor are reduced by factor $\lambda < 1$. This has a different effect for each of the three different sensors:

- Capacitive pressure sensors do not show any change in sensitivity. In this case, miniaturization basically can reduce component dimensions.
- Miniaturization of acceleration and yaw rate sensors of any dimension reduces sensitivity. The reason is that sensitivity is correlated to mass m and, with a decrease of all dimensions at factor λ, decreases totally by λ^3. Capacitor distance d_K which determines the relative capacity change only compensates this effect by factor λ.

Equations (2.15) to (2.17) show that for increasing or maintaining of sensitivity, only certain dimensions should be reduced. This refers to all values that are part of the denominators of Equations (2.15) to (2.17), in particular to electrode distance d_K and thickness h_B of the bending beam.

Table 2.1 Capacitive sensors according to basic structure in Figure 2.2

Measurand	Pressure p	Acceleration a	Yaw rate Ω
Sensor force	$F_S = A_K p$ $F_S = b_K \ell_K p$	$F_S = m \cdot a$ $F_S = \rho \cdot b_K \ell_K h_K a$	$\vec{F_S} = 2m(\vec{v} \times \vec{\Omega})$ $F_S = 2\rho b_K \ell_K h_K v_0 \Omega$
Sensor arrangement	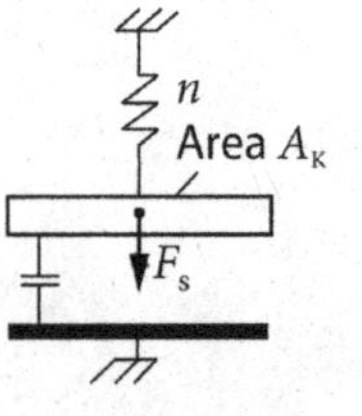 n Area A_K F_s	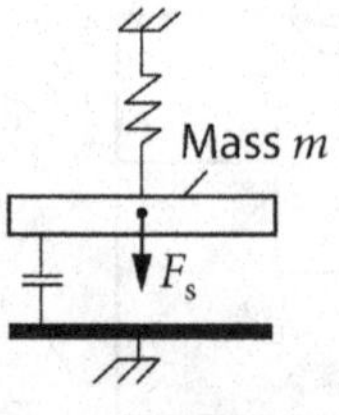 Mass m F_s	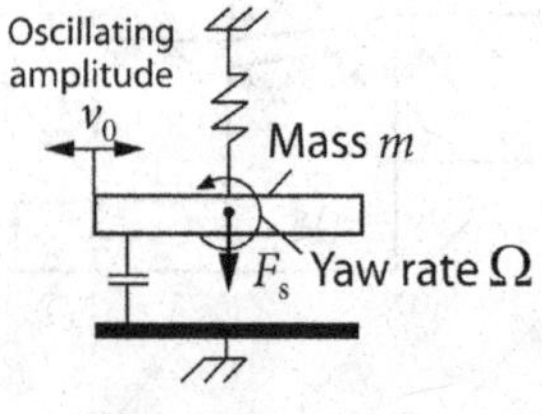 Oscillating amplitude v_0 Mass m F_s Yaw rate Ω
Miniaturization of all dimensions	$b_{Km} = \lambda d_K$ $\lambda_{Km} = \lambda \ell_K$ $d_{Km} = \lambda d_K$	$b_{Bm} = \lambda b_B$ $\ell_{Bm} = \lambda \ell_B$ $h_{Bm} = \lambda h_B$	$v_{0m} = \lambda v_0$
$\dfrac{\Delta C}{C_0}$	$\dfrac{(\ell_B/h_B)^3 b_K \ell_K p}{2E d_K b_B}$ (2.15)	$\dfrac{\rho(\ell_B/h_B)^3 b_K \ell_K h_K a}{2E d_K b_B}$ (2.16)	$\dfrac{\rho(\ell_B/h_B)^3 b_K \ell_K h_K v_0 \Omega}{E d_K b_B}$ (2.17)
$\dfrac{(\Delta C)_m}{\Delta C}$	1	λ	λ^2

Source: Hierold, C. (2004) From micro- to nanosystems: mechanical sensors go nano. *Journal of Micromechanics and Microengineering* 14, pp. 1–11

Example 2.2 Effect of scaling on the force of electrostatic drives

Section 7.2.1 describes in detail the characteristics of electrostatic drives. They can be considered a reversion of the capacitive sensors in the previous example. According to Figure 2.1 and Equation (2.5), the electrostatic force is

$$F_1 = \frac{\varepsilon \cdot \ell}{2d} V^2 = \frac{\varepsilon \cdot \ell \cdot d}{2} \cdot \left(\frac{V}{d}\right)^2. \tag{2.18}$$

Also in this case, the geometric dimensions for the macro- and microsystem will be ℓ_0 and d_0 or $\ell_m = \lambda_\ell \cdot \ell_0$ and $d_m = \lambda_d \cdot d_0$, respectively.

For small capacitor gaps, the PASCHEN effect applies: The critical breakdown field strength $E_D = V/d$ increases with decreasing plate distance. For $d \geq 1$ mm applies $E_D \approx 3 \cdot 10^6$ V/m; for $d \approx 1$ μm applies $E_D \approx 1.5 \cdot 10^8$ V/m. An approximation of this can be represented as

$$E_{Dm} \propto \lambda_d^{-0.5} \cdot E_{D0} \tag{2.19}$$

Force F_1 thereby scales as follows:

$$F_{1m} = \frac{\varepsilon \cdot \lambda_\ell \ell_0 \cdot \lambda_d \cdot d_0}{2} \cdot \frac{E_{D0}^2}{\lambda_d} = \lambda_\ell \cdot F_{10}. \tag{2.20}$$

That means that force F_{1m} decreases proportionally to decrease λ_ℓ of the dimensions. We also want to look at the ratio of electrostatic force F_1 to force F_m for the acceleration of inert mass m

$$F_m = m \cdot g = \rho \cdot a^3 \cdot g \tag{2.21}$$

where g is the acceleration due to gravity, ρ the density and a the dimension of the mass. Dimension a scales with

$$a_m = \lambda_a \cdot a_0, \tag{2.22}$$

This results in

$$F_{mm} = \lambda_a^3 \cdot F_{m0} \tag{2.23}$$

Thus, the ratio of electrostatic and acceleration force in microsystems becomes

$$\frac{F_{1m}}{F_{mm}} = \frac{\lambda_\ell}{\lambda_a^3} \cdot \frac{F_{10}}{F_{m0}}. \tag{2.24}$$

In a situation where $\lambda_\ell = \lambda_a = \lambda$,

$$\frac{F_{1m}}{F_{mm}} = \frac{1}{\lambda^2} \cdot \frac{F_{10}}{F_{m0}}. \tag{2.25}$$

This means that miniaturization ($0 < \lambda \leq 1$) increases the electrostatic force by the diminution squared compared to the acceleration force. Therefore, the electrostatic transformer principle is an interesting form of power generation in the micro-range.

In general, we can draw the following conclusions:

- The surface correlates to the second power and the volume to the third power of the dimension. For miniaturizing function elements, surface effects become therefore increasingly important as compared to volume effects.
- Power generation principles that apply surface forces are therefore more suitable for micromechanic actuators. This is an important reason for the dominant use of electrostatic drives in microsystem technology.
- Miniaturizing micro-mechanical components can result in dimensions of only a few 10 nm. As the lattice constant of silicon of $a = 0.543$ nm and the corresponding distance between two atoms $\sqrt{3} \cdot a/4 = 0.235$ nm (distance between two base atoms at (0, 0, 0) and (a/4, a/4, a/4), see Figure 3.2), it can be assumed that there is a material continuum also for miniaturized function elements. Quantum effects are unlikely to occur (yet), unless dimensions decrease to only a few nanometers.
- In the micro-world down to a few hundred atom layers, the same physical laws rule as in the macro-world. However, the relationship of the impact of various effects can change substantially!

2.2 SIMILARITY AND DIMENSIONLESS NUMBERS

Even though systems may have different dimensions, they may show a similar physical behavior. We will use Figure 2.3 as an example. Two bodies are geometrically similar if all corresponding (homologous) lines have the same ratio:

$$\frac{\ell_{m1}}{\ell_{01}} = \frac{\ell_{m2}}{\ell_{02}} = \frac{\ell_{m3}}{\ell_{03}} = \lambda = \text{const.} \tag{2.26}$$

Here, λ is called similarity constant or scale factor. In general, it applies that two processes or systems are physically identical

- if there are similarity constants λ and that these are identical for the homologous values

$$\lambda = \text{const.} \tag{2.27}$$

or

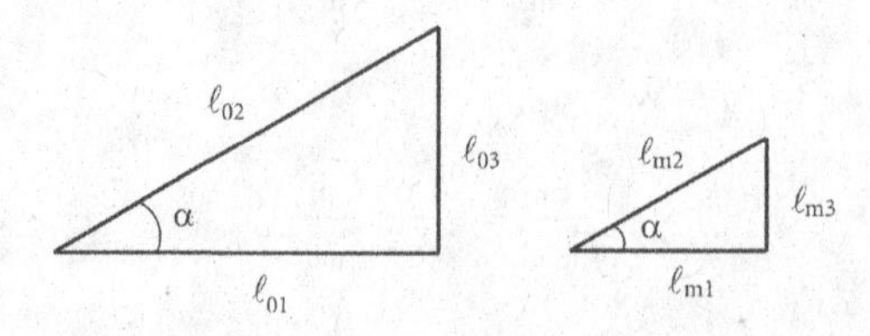

Figure 2.3 Geometric similarity of two triangles (index 0 macrosystem, index m microsystem)

- if there are basic similarity numbers π that have the same numeric value for both processes or systems

$$\pi_0 = \pi_m. \tag{2.28}$$

Example 2.3 Similarity of oscillating spring-mass-systems

We will look at a simple spring-mass-system that consists of a mass body with mass $m = \rho \cdot V_m$ (ρ density, V_m volume) and a simple compression body as the spring element (Figure 2.4).

The spring element has the compliance

$$n = \frac{u}{F_n} = \frac{\ell}{EA}, \tag{2.29}$$

with length ℓ, YOUNG's modulus E and cross-sectional area A.

In microsystem technology, such systems are used in a variety of applications, such as acceleration sensors (see Figure 1.9) and in electrostatic drives (see Figure 2.1).

For operating such oscillating systems, the ratio of the forces, the acceleration of an inertial mass m and the deformation of the compression cantilever, for instance, is important:

$$F_m = ma = \rho V \frac{d^2u}{dt^2} \tag{2.30}$$

$$F_n = \frac{E \cdot A}{\ell} \cdot u = \frac{E \cdot A \cdot \ell}{\ell^2} \cdot u = \frac{E \cdot V_n}{\ell^2} \cdot u \tag{2.31}$$

with mass and spring volumes V_m and V_n, respectively.

The following relationship results:

$$\frac{F_m}{F_n} = \frac{\rho \cdot V_m \cdot \ell^2}{E \cdot V_n \cdot u} \cdot \frac{d^2u}{dt^2} \tag{2.32}$$

or for harmonic time functions $u = \hat{u}\exp\{j\omega t\}$

$$\left|\frac{F_m}{F_n}\right| = \frac{\rho \cdot V_m \cdot \ell^2 \cdot \omega^2}{E \cdot V_n}. \tag{2.33}$$

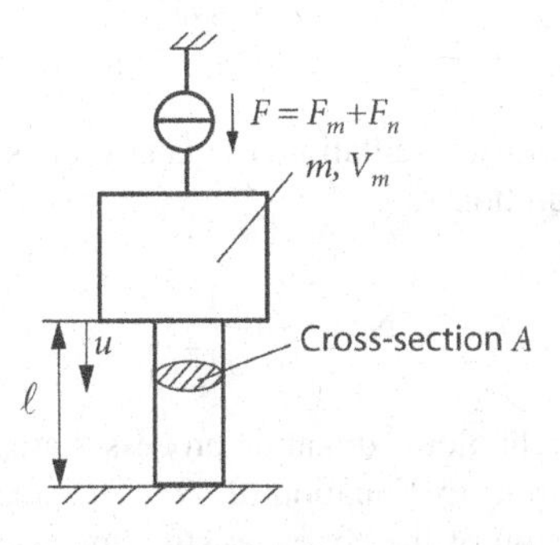

Figure 2.4 Oscillating spring-mass-system

A particular case may arise for $V_m = V_n$:

$$\left|\frac{F_m}{F_n}\right| = \frac{\rho}{E} \cdot \omega^2 \cdot \ell^2 = \text{Ca}. \tag{2.34}$$

This is the CAUCHY number Ca that describes the relationship of inertial forces and elastic forces in solid bodies [KASPAR00]. According to Equation (2.28) it is a dimensionless number describing similarity. From Equation (2.34) it results that for oscillating systems operating in a similar way and consisting of the same material (density ρ, YOUNG's modulus E), oscillation frequency ω and cantilever length ℓ are reciprocally proportional. Oscillation frequency ω thus also scales proportionally to ℓ. Microsystems have therefore high natural frequencies and a large useable dynamic range.

Example 2.4 Similarity of dynamic processes

Many microsystems have movable function elements (mass m). We will use the NEWTON number Ne to analyze how the force required for a similar microsystem decreases if all dimensions ℓ_{i0} are reduced by factor $\lambda_\ell = \ell_{im}/\ell_{i0}$ (0 macrosystem, m microsystem, i index). Acting force F accelerates mass m in accordance with NEWTON's law with

$$F_m = m \cdot a. \tag{2.35}$$

With density ρ and volume $V = \ell^3$ of the mass, the resulting ratio of acting and inertia forces is

$$\text{Ne} = \left|\frac{F}{F_0}\right| = \left|\frac{F}{\rho \cdot \ell^3 \cdot a}\right|. \tag{2.36}$$

For sinusoidal motions $u = \ell \cdot \exp\{j\omega t\}$ (ℓ is here a characteristic displacement amplitude and, in this example, correspond to the dimension of the mass), it results in

$$a = \frac{d^2u}{dt^2} = -\ell\omega^2 \cdot \exp\{j\omega t\} \tag{2.37}$$

or

$$\text{Ne} = \frac{F}{\rho \cdot \ell^4 \cdot \omega^2}. \tag{2.38}$$

correspondingly.

For non-harmonic processes, oscillation frequency ω is typically replaced by the inverse of a typical time duration t:

$$\text{Ne} = \frac{F}{\rho} \cdot \frac{t^2}{\ell^4}. \tag{2.39}$$

Therefore in microsystem technology, dynamic processes are similar if NEWTON number Ne remains constant according to Equation (2.39). That means, for example, that for constant forces and a reduction of the characteristic dimensions ℓ to one-tenth, duration of the process decreases to 1 in a 100.

If all dimensions ℓ are reduced by factor λ_ℓ and a time scale of $\lambda = t_m/t_0$ or $\lambda_t = \omega_0/\omega_m$, correspondingly, is introduced, it results from Equation (2.39)

$$\frac{\mathrm{Ne_m}}{\mathrm{Ne_0}} = \frac{F_m}{F_0} \cdot \frac{\lambda_t^2}{\lambda_\ell^4}. \qquad (2.40)$$

For physically similar systems, $\mathrm{Ne_m} = \mathrm{Ne_0}$ and therefore

$$F_m/F_0 = \lambda_\ell^4/\lambda_t^2. \qquad (2.41)$$

If processes in micro- and macrosystems are supposed to take the same time ($\lambda_t = 1$), the following results:

$$F_m/F_0 = \lambda_\ell^4. \qquad (2.42)$$

Miniaturizing systems to one-tenth of the dimensions means that only one in 10 000 of the original force F_0 is necessary.

As shown in Example 2.1, scaling can be used to translate two similar systems into each other. If both systems have the same dimensionless numbers, the characteristics of the systems coincide as to the characteristics described by these dimensionless numbers. In Example 2.3, it was the ratio of inertia and elastic force. The same two systems do not necessarily show this similarity regarding other criteria (i.e. dimensionless numbers). In Example 2.3, this could be the case of inertia and frictional forces (which were not considered).

In summary, we can draw the following conclusions:

- Dimensionless numbers do not have units and are formed of geometrical and physical values.
- Dimensionless numbers can, in general, be used for describing physical correlations. As physical laws have to be independent of any unit system, it must be possible to represent them by unitless variables, i.e. dimensionless numbers.
- Dimensionless numbers are commonly used for thermal, fluidic and acoustic systems. They are seldom used for describing mechanical systems, and for electro-dynamic systems they are typically not used at all.

[KASPAR00] uses the example of microsystems to provide a general method that uses dimension analysis to deduct dimensionless numbers.

EXERCISES

2.1 Is degree of efficiency η a dimensionless number?

2.2 Deduct the REYNOLDS' number from Equation (6.28). It describes the relationship of inertia force and viscous friction force of a moving fluid in a fluid channel.

2.3 How do the dimensions of a dust corn that is considered a cube have to scale in order for it to stick to the ceiling? The dust corn is assumed to consist of silicon. All atoms of the cube surface are singly charged (see Table 3.5); the corresponding opposite charges are found on the ceiling. For the electrostatic attraction, the minimal distance between ceiling and dust corn must not be lower than the lattice constant $a = 0.543$.

2.4 What dimensions of the oscillating cantilever-mass-system in Figure 2.4 have to be miniaturized in order to increase resonance frequency $\omega_0 = (mn)^{-1}$ (m mass, $n = \ell/EA$ elasticity) and the corresponding dynamic range?

REFERENCES

[KASPAR00] Kaspar, M. (2000) *Mikrosystementwurf*. Berlin: Springer.
[MADOU97] Madou, M. (1997) *Fundamentals of Microfabrication*. Boca Raton: CRC Press.
[PAGEL01] Pagel, L. (2001) *Mikrosysteme. Physikalische Effekte bei der Verkleinerung technischer Systeme*. Wilburgstetten: J. Schlembach Fachverlag.

3

Materials

3.1 OVERVIEW

Microelectronics deals with highly integrated, miniaturized signal and data processing components and systems. As distinct to this, microsystem technology focuses on miniaturized non-electrical functions that add sensor and actuator elements to microelectronic components. The large number of standardized products, such as processor and data storage ICs, has substantially promoted the development of highly efficient microelectronic manufacturing processes. Due to the variability of component functions and the subsequent much smaller production batches, a corresponding research and development is not feasible for microsystem technology. Microsystem technology therefore relies mainly on manufacturing processes and materials that are applied in microelectronics. This approach has two main advantages:

- Microelectronic and microsystem components can be integrated – if this is functionally and economically feasible.
- The highly efficient manufacturing procedures are based on the simultaneous processing of a large number of identical structures. It relies heavily on lithography where a structural pattern can be easily transferred from a photomask to a large number of different materials.

If we want to adapt microelectronic manufacturing procedures to microsystem technology we have, above all, to include the creation of three-dimensional structures (see Chapter 4). On the other hand, these manufacturing processes force microsystem technology to use in general the same materials as microelectronics. This is the reason why – with the exception of a few specific manufacturing processes (e.g. LIGA, Section 4.12; miniaturized classical procedures, Section 4.13) – silicon is the dominating base material:

- Silicon is a semiconductor whose conductivity can be set by doping. Conductivity or its reciprocal value, resistance ρ, can thereby be altered by more than six orders of magnitude:

  p-conducting Si: $n = 10^{14}\ \text{cm}^{-3}: \quad \rho = 2 \cdot 10^{2}\ \Omega \cdot \text{cm}$

  $n = 10^{21}\ \text{cm}^{-3}: \quad \rho = 1.5 \cdot 10^{-4}\ \Omega \cdot \text{cm}$

Introduction to Microsystem Technology: A Guide for Students Gerald Gerlach and Wolfram Dötzel

n-conducting Si: $n = 10^{14}\ \text{cm}^{-3}:\quad \rho = 5 \cdot 10^{1}\ \Omega \cdot \text{cm}$
$n = 10^{21}\ \text{cm}^{-3}:\quad \rho = 3 \cdot 10^{-4}\ \Omega \cdot \text{cm}$
(n charge carrier concentration).

- Due to thermal oxidation, silicon is an excellent insulator which is virtually defect-free. The energy gap of thermal SiO_2 is $E_g = 9$ eV, the breakdown field strength $E_D = 3 \cdot 10^6$ V/cm.
- Due to its single crystalline structure, silicon has mechanical characteristics that are largely superior to those of classical structure materials as far as elasticity, reproducibility and long-time stability is concerned.

All these are reasons why silicon, also in microsystem technology, is the most important material and will continue being that into the foreseeable future. The following section will therefore focus on single crystalline silicon.

It is very costly to produce single crystalline silicon as thin layers. Therefore, mainly polycrystalline silicon (Poly-Si) is used in thin-film technology. Thin poly-silicon layers play a dominant role in surface micromachining (see Section 4.10).

Silicon dioxide (SiO_2) and silicon nitride (Si_3N_4) are of outstanding importance as insulating layers for silicon-based microtechnology. If silicon oxynitride SiO_xN_y has the appropriate composition it can reach a thermal expansion coefficient that corresponds to that of silicon. For very thin silicon structures, it is therefore suitable as stress-free insulating and passivating material. In addition to silicon, even glass is widely used as construction material. Glass is much less expensive and shows advantageous characteristics not only for optical, but also for other applications.

Example 3.1 Piezoresistive pressure sensor

Figure 3.1 shows the typical cross section of a commercial piezoresistive pressure sensor (see Section 7.5). The pressure sensor chip consists of single crystalline (or it is often called: mono-crystalline) silicon with a pressure bending plate (or it is often called: membrane) that is locally thinned by anisotropic wet chemical etching (see Section 4.6). The etch mask used for laterally limiting the thinned area consists mainly of double layers of silicon dioxide and nitride. They are also used for passivating the sensor surface. Thermally grown SiO_2 (see Section 4.4) is used here as it is quasi free of defects, and Si_3N_4 is used because of its chemical resistance. A glass counter body is used to attach the silicon sensor chip to the chip carrier. This prevents the mechanical stresses of sensor casing or chip carrier from interfering with the deformation of the pressure plate and its pressure sensor characteristics. The reason for using glass is that its thermal expansion coefficient is similar to that of silicon (see Section 3.4). Therefore it does not cause any mechanical stress at changing temperatures.

The range of materials used in microsystem technology largely exceeds those mentioned in Table 3.1.

Silicon and glass are by far the most important substrates or construction materials.

Quartz (single crystalline SiO_2) is used for clock quartz crystals in watch industry and other piezoelectric resonators. The use of silicon carbide (SiC) and its several modifications is limited to very high temperatures (several hundred degrees Celsius). Other

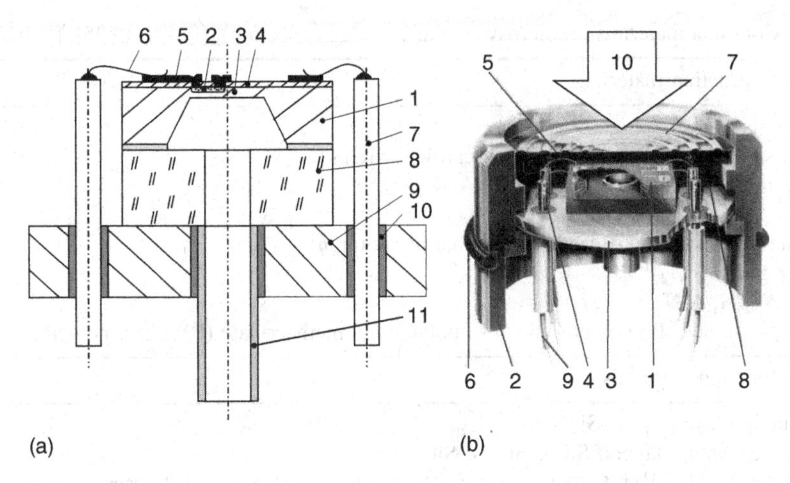

Figure 3.1 Typical cross section of a piezoresistive pressure sensor (not in correct scale). (a) Schematic construction: 1 silicon sensor chip, 2 piezoresistor (doped silicon), 3 pressure plate, 4 insulating layer (SiO_2, and Si_3N_4 if required), 5 interconnection layer (Al), 6 wire bond, 7 contact pin, 8 glass counter body, 9 IC chip carrier (metal), 10 glass channel, 11 pressure tube; (b) pressure sensor with separating membrane and oil filling: 1 measuring cell, 2 stainless steel case, 3 glass lead-through, 4 pin, 5 gold wire, 6 seal ring, 7 stainless steel membrane, 8 silicon oil, 9 flexible wire, 10 media: wet or dry gas, oil-containing chemicals, etching liquids. Reproduced by permission of Keller AG for pressure measuring technique, Winterthur, Switzerland

semiconducting materials do not play any role in microsystem technology. Due to its outstanding characteristics, silicon as such is also very important as thin film and function material. This refers to its semiconducting properties as well as to a number of sensor and actuator effects (see Chapter 7). Microsystem technology has to realize and integrate a large variety of different functions. Therefore, the range of the corresponding materials is extremely wide. Within the framework of this textbook, only the most important ones can be addressed.

Table 3.2 shows a comparison of substrate materials used in microsystem technology.

3.2 SINGLE CRYSTALLINE SILICON

Single crystalline silicon is the most important base material, both for microelectronics and microsystem technology:

- Silicon has only one single crystal modification.
- Silicon is cost-efficient and available in sufficient amounts.
- Silicon is the material that has been most extensively analyzed, both physically and chemically [HULL99].
- Cost-efficient technological procedures and advanced equipment are available for processing silicon (see Chapter 4). Lithography can be used to transfer highly complex structures to silicon or other layers that are deposited on silicon.

Table 3.1 Common materials in microsystem technology (selection), [FRÜHAUF05], [VÖLKLEIN00]

Substrate/construction materials
Single crystals:
Semiconductors: silicon (Si), silicon carbide (SiC)
Dielectrics: quartz (SiO_2), sapphire (Al_2O_3)
Glasses:
silicate glasses: Pyrex/Borofloat/Rasotherm[1]; Foturan[1]
quartz glass (SiO_2)
Ceramics: Al_2O_3, AlN
Polymers: polyamide (PA), polyimide (PI), polymethyl methacrylate (PMMA), polycarbonate (PC)
Thin films/function materials
Semiconducting layers: poly-Si, SnO_2
Insulators, passivating layers: SiO_2, Si_3N_4, SiO_xN_y
Interconnections: Al, Al/Si, Cr/Au, Ni/Cr, Si (highly doped), $InSnO_x$ (ITO, transparent)
Sacrificial layers: SiO_2, photo resist
Diffusion barriers: W, Ta
Adhesion promoters: Ti, Cr, Zr
Getter layers: borosilicate glass (BSG), phosphorosilicate glass (PSG)
Optical layers:
Wave guides: SiO_xN_y
Absorption layers: Ag, Au (porous)
Reflexion or anti-reflexion layers[2]: Al, Pt, Au, SiO_2, Si, Ge
Sensor and actuator layers:
Piezoelectrics: $PbZr_xTi_{1-x}O_3$ (PZT), ZnO, Polyvenylidenfluoride (PVDF)
Pyroelectrics: PZT, PVDF
Ferromagnetics: Ni, Fe, NiFe
Shape memory alloys: Nitinol (NiTi)
Mechanical function layers: Poly-Si

[1]Commercial names,
[2]e.g. as $\lambda/4$ layers or as ultra-thin layers (metal).

Table 3.2 Comparison of substrate materials in microsystem technology

Material	Characteristics
Silicon	semiconductor, single crystal, temperature and long-term stable, semiconductor technologies can be applied, cost-efficient, large number of usable semiconductor effects
Quartz	Electrical insulator, single crystal, temperature and long-term stable, chemical resistant, very hard, piezoelectric
Glass	amorphous material with relaxing characteristics, cost-efficient, optically transparent, chemical resistant, can be thermally adapted to silicon (borosilicate glass), in specific case photostructurable (Foturan)
Polymers	amorphous material with relaxing characteristics, cost-efficient, characteristics can be easily adapted (through polymer blend and filling materials)

- Silicon provides a large number of physical and chemical effects that are important for sensor and actuator applications.

In addition, silicon has characteristics that are preferable to those of other materials, and even other semiconductors:

- Mechanical: Under normal application temperatures, single crystalline silicon has an ideal elastic behavior. Plastic deformation and hysteresis characteristics could be so far only shown for temperatures of several hundred degrees Celsius. Proportional limit, elastic limit and tensile strength coincide. In addition, its crystalline structure provides anisotropy (property of being directionally dependent) of its characteristics and can be used for manufacturing perfectly defined shapes (see Section 4.6).
- Electrical: Doping can be used to change the conductivity of silicon by more than six orders of magnitude.
- Technological: Thermal oxidation (see Section 4.4) is a technologically simple procedure that can be used to change Si into SiO_2, which is an outstanding insulator. Thermal oxide is firmly adherent, quasi defect-free and sufficiently robust for many applications.
- Thermal: Silicon has a very good thermal conductivity.

3.2.1 Description of the Orientation of Planes and Directions in a Crystal

As a single crystal, silicon has a cubic face-centered lattice structure, such as a diamond lattice. It belongs to the crystallographic group $\bar{4}\,3m$ [PAUFLER86].

The silicon lattice originates from the overlapping of two cubic face-centered lattices with base atoms at (0, 0, 0) and at ($a/4$, $a/4$, $a/4$) with a lattice constant a of 0.5431 nm = 5.431 Å (Figure 3.2). Each Si atom is tetraedically surrounded by four other Si atoms, with a distance of 0.235 nm between two neighboring silicon atoms.

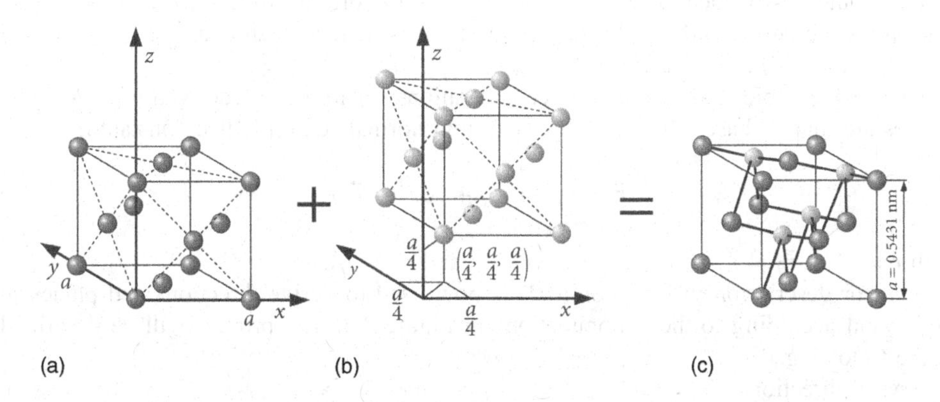

Figure 3.2 Cubic face-centered lattice structure of base atoms at (a) (0, 0, 0) and (b) ($a/4$, $a/4$, $a/4$); (c) resulting diamond lattice of silicon

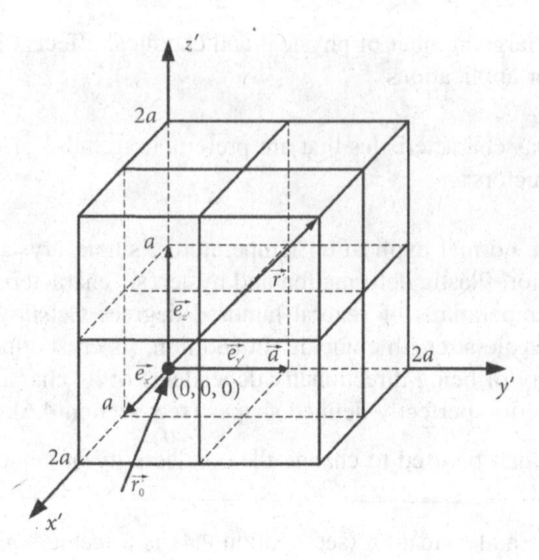

Figure 3.3 Crystal lattice of silicon in the Cartesian coordinate system

MILLER indices are used to describe planes and edges in the silicon crystal lattice. This approach is based on the Cartesian coordinate system as the silicon's diamond lattice unit cell is a cube with the same edge length a in all three coordinate directions and right angles between the cube edges. Figure 3.3 shows eight uniform cells of such a Si lattice. Each atom of the lattice can be found by periodic continuation of each corresponding atom of the cell, i.e. by using whole-numbered multiples b, c, d to displace the lattice constant into the three dimensions.

$$\vec{r}' = \vec{r}_0' + (b\vec{e}_{x'} + c\vec{e}_{y'} + d\vec{e}_{z'}). \tag{3.1}$$

In this case, $\vec{r}_0'$ and $\vec{r}'$ are the position vectors of the base atom and of the corresponding lattice atom, respectively. In the following, $\vec{r}_0'$ will be mainly used as position vector of the base atom at (0, 0, 0). However, in principle $\vec{r}_0'$ can be freely defined, as we assume an ideal, infinitely expanded crystal. This means, that the origin of the Cartesian coordinate system can be defined arbitrarily. In Figure 3.3, it is defined such that $\vec{r}_0' = (0, 0, 0)$ and $\vec{r}_0' - \vec{r}_0' =$ (b, c, d) $=$ (2a, 2a, 2a).

Due to the cubic lattice structure, considerations regarding the derivation of MILLER indices are mainly based on a coordinate system normalized on lattice constant a

$$\vec{r} = \vec{r}_0 + (b\vec{e}_x + c\vec{e}_y + d\vec{e}_z) \tag{3.2}$$

with $\vec{r} = \vec{r}'/a, \vec{r}_0 = \vec{r}_0'/a, x = x'/a, y = y'/a$ and $z = z'/a$
The vector description in Equation (3.2) is now used to derive directions and planes in the crystal according to the denominations in Figure 3.4. The process will be described in the following.

Crystal directions:

- Determine the direction components (b, c, d) of the vector of the crystal direction.

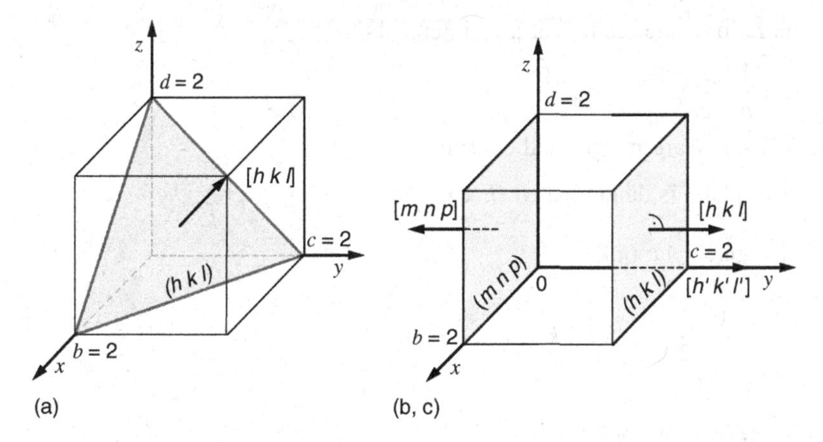

Figure 3.4 Denomination of crystal directions and crystal planes in a silicon lattice

- Form the ratio of the axis intercepts

  $$h : k : l = b : c : d \tag{3.3}$$

  with h, k and l being integers and coprime.
- The crystal direction will be denoted $[h\ k\ l]$.

Example Figure 3.4a: $[h\ k\ l]$
$b = 2, c = 2, d = 2$
$b : c : d = 2 : 2 : 2 \rightarrow h : k : l = 1 : 1 : 1$
$[h\ k\ l] = [1\ 1\ 1]$.

Example Figure 3.4b: $[h\ k\ l]$
$b = 0, c = 2, d = 0$
$b : c : d = 0 : 2 : 0 \rightarrow h : k : l = 0 : 1 : 0$
$[h\ k\ l] = [0\ 1\ 0]$
$[h'k'l] = [h\ k\ l] = [0\ 1\ 0]$.
$[h'k'l']$ and $[h\ k\ l]$ have the same direction and are therefore identical. They are not dependent on any specific point of origin.

Example Figure 3.4c: $[m\ n\ p]$
$b = 0, c = -2, d = 0$
$b : c : d = 0 : (-2) : 0 \rightarrow m : n : p = 0 : (-1) : 0$
$[m\ n\ p] = [0\ \bar{1}\ 0]$ (speak: zero, minus one, zero)

Negative values are marked with a line above the number. $[m\ n\ p]$ and $[\overline{m}\ \overline{n}\ \overline{p}\ ]$ are always pointing into opposite directions.

Crystal planes:

- Determine the axis intercepts b, c and d, where plane $(h\ k\ l)$ cuts the three coordinate axes. Displace the coordinate system, if required, in such a way that plane $(h\ k\ l)$ does not intercept the point of origin of the coordinate system.

- Form the ratio of the reciprocal axis intercepts

$$h : k : l = \frac{1}{b} : \frac{1}{c} : \frac{1}{d} \tag{3.4}$$

where h, k and l are integer and coprime.

- The crystal plane is denominated $(h\ k\ l)$.

Example Figure 3.4a: $(h\ k\ l)$
$b = 2, c = 2, d = 2$
$\frac{1}{b} : \frac{1}{c} : \frac{1}{d} = \frac{1}{2} : \frac{1}{2} : \frac{1}{2} \rightarrow h : k : l = 1 : 1 : 1$
$(h\ k\ l) = (1\ 1\ 1)$.

Example Figure 3.4b: $(h\ k\ l)$
$b = \infty, c = 2, d = \infty$
$\frac{1}{b} : \frac{1}{c} : \frac{1}{d} = \frac{1}{\infty} : \frac{1}{2} : \frac{1}{\infty} \rightarrow h : k : l = 0 : 1 : 0$
$(h\ k\ l) = (0\ 1\ 0)$.

Example Figure 3.4c: $(m\ n\ p)$
Dislocating the coordinate system in direction y, e.g., by $\Delta y = 2$. This way $(x', y', z') = (x, y - 2, z)$. Plane $(m\ n\ p)$ cuts the y' axis at $c = -2$.
$b = \infty, c = -2, d = \infty$
$\frac{1}{b} : \frac{1}{c} : \frac{1}{d} = \frac{1}{\infty} : \frac{1}{-2} : \frac{1}{\infty} \rightarrow m : n : p = 0 : (-1) : 0$
$(m\ n\ p) = (0\ \bar{1}\ 0)$.
The orientation of plane $(m\ n\ p)$ in space corresponds to plane $(h\ k\ l)$ in example Figure 3.4b. $(0\ \bar{1}\ 0)$ and $(0\ 1\ 0)$ characterize the 'lower' and 'upper' side of the crystal plane as $(h\ k\ l) = (\overline{m}\ \overline{n}\ \overline{p})$.

For cubic crystals, the following conclusions can be drawn:

- Parallel directions and planes are equivalent.
- $[h\ k\ l]$ and $[\bar{h}\ \bar{k}\ \bar{l}]$ are exact opposite directions.
- $(h\ k\ l)$ and $(\bar{h}\ \bar{k}\ \bar{l})$ describe the same plane.
- Direction $[h\ k\ l]$ is the plane normal of plane $(h\ k\ l)$ and plane $(\bar{h}\ \bar{k}\ \bar{l})$.

At first, it might appear strange that – according to Equations (3.3) and (3.4) for directions and planes – on the one hand, we use the relationship of axis intercepts and, on the other hand, that of the reciprocal axis intercepts. However, this corresponds to the transition of a – in our case normalized – reciprocal lattice, where the surface normal of the reciprocal lattice is situated vertically to the plane of original lattice [HÄNSEL96]. This way, plane $(h\ k\ l)$ and surface normal $[h\ k\ l]$ can have the same indexing.

Due to cubic symmetry it is possible to transform different crystal directions and plane into each other by using symmetry operations. This is called equivalent directions and planes (Table 3.3). This means, for instance, that coordinate axes (in positive and

Table 3.3 MILLER indices

Denomination	Explanation	Example
$[h\ k\ l]$	Crystal direction	$[1\ 1\ 1]$
$< h\ k\ l >$	Equivalent crystal direction	$<1\ 1\ 1> \in \{[1\ 1\ 1], [\bar{1}\ 1\ 1], [1\ \bar{1}\ 1], [1\ 1\ \bar{1}], [\bar{1}\ \bar{1}\ 1], [\bar{1}\ 1\ \bar{1}], [1\ \bar{1}\ \bar{1}], [\bar{1}\ \bar{1}\ \bar{1}]\}$
$(h\ k\ l)$	Crystal plane	$(1\ 0\ 0)$
$\{h\ k\ l\}$	Equivalent crystal plane	$\{1\ 0\ \bar{0}\} \in \{\{1\ 0\ 0\}, \{\bar{1}\ 0\ 0\}, \{0\ 1\ 0\}, \{0\ 1\ 0\}, \{0\ \bar{1}\ 0\}, \{0\ 0\ 1\}, \{0\ 0\ \bar{1}\}\}$

negative direction, respectively) are directional equivalents, which also means that the corresponding material characteristics in these directions are the same. Equivalent planes are characterized as $< h\ k\ l >$. In this case for instance, $<1\ 0\ 0>$ describes one or all of the following directions $[1\ 0\ 0]$, $[\bar{1}\ 0\ 0]$, $[0\ 1\ 0]$, $[0\ \bar{1}\ 0]$, $[0\ 0\ 1]$ and $[0\ 0\ \bar{1}]$. Analogously, all cubic planes of the unit cell of the Si lattice are equivalent. They are labelled $\{h\ k\ l\}$.

Angle α between two planes $(h_1\ k_1\ l_1)$ and $(h_2\ k_2\ l_2)$ or two directions $[h_1\ k_1\ l_1]$ and $[h_2\ k_2\ l_2]$, respectively, can be calculated according to the mathematical basics of vector analysis as

$$\begin{aligned} \alpha &= \arccos \angle\{(h_1\ k_1\ l_1), (h_2\ k_2\ l_2)\} = \arccos \angle\{[h_1\ k_1\ l_1], [h_2\ k_2\ l_2]\} \\ &= \arccos \frac{h_1 h_2 + k_1 k_2 + l_1 l_2}{\sqrt{(h_1^2 + k_1^2 + l_1^2)(h_2^2 + k_2^2 + l_2^2)}}. \end{aligned} \tag{3.5}$$

The angles between the three coordinate directions $\vec{x} = [1\ 0\ 0]$, $\vec{y} = [0\ 1\ 0]$, $\vec{z} = [0\ 0\ 1]$ can be calculated to

$$\begin{aligned} \alpha_1 &= \arccos \angle(\vec{x}, \vec{y}) = \arccos \frac{1 \cdot 0 + 0 \cdot 1 + 0 \cdot 0}{[(1^2 + 0^2 + 0^2)(0^2 + 1^2 + 0^2)]} = \arccos 0 = 90^\circ \\ \alpha_2 &= \arccos \angle(\vec{x}, \vec{z}) = 90^\circ \\ \alpha_3 &= \arccos \angle(\vec{y}, \vec{z}) = 90^\circ. \end{aligned}$$

This complies exactly with the requirement that the selected coordinate system has to be Cartesian. Analogously an angle of 90° results between all neighboring lateral planes of the cube (1 0 0), according to examples (b), (c) relate to Figure 3.4b,c.

Example 3.2 ***(1 1 1) side surfaces of etched cavities in (0 0 1) silicon***

When silicon is etched anisotropically, i.e. depending on crystal direction, and a {1 0 0} surface direction is used, cavities will result that have the same {1 0 0} orientation at the bottom and are delimited laterally by {1 1 1} surfaces (Figure 3.5). The reason is that {1 1 1} Si surfaces are etched off using a minimal etch rate and this way become the most resistant (see Section 4.6).

In relation to the forming edges of the etching cavity, the x, y coordinate system of the Si crystal lattice is rotated by 45°. Figure 3.5b shows a part of Figure 3.5a and provides a clearer imagination of how the two hatched planes are situated in the coordinate system.

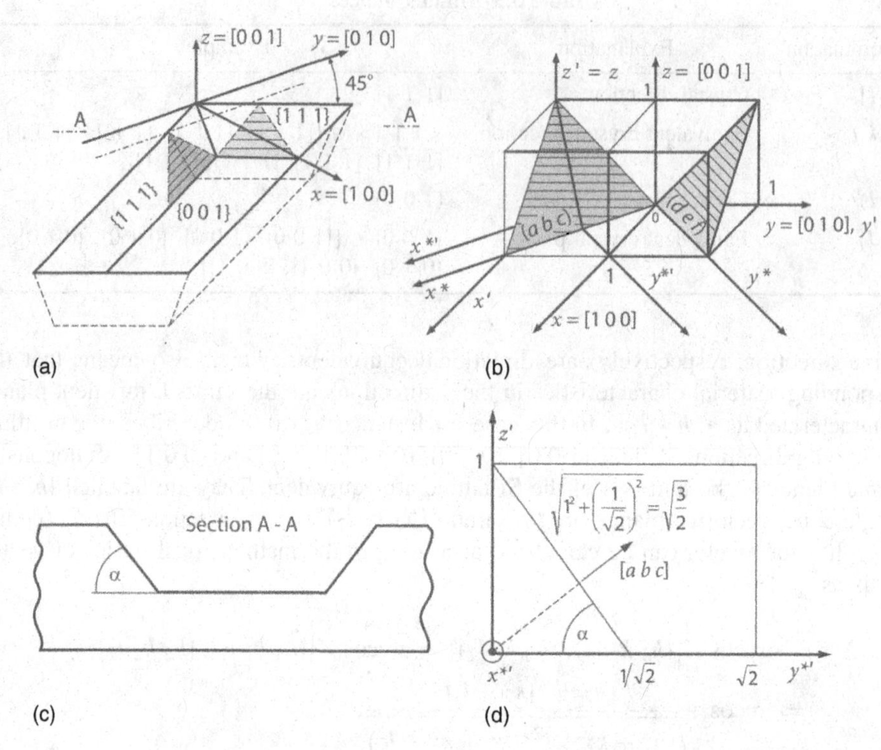

Figure 3.5 Anisotropically etched cavity in (0 0 1)-Si; (a) position in coordinate system ($x\ y\ z$); (b) detail; (c) cross-section through the etching cavity; (d) cross-section of $z' - y^{*\prime}$ plane of (b)

What specific orientation have the two {1 1 1} planes ($a\ b\ c$) *and* ($d\ e\ f$)?

The planes can be easily determined when looking at the shifted coordinate system (x', y', z'):

The intersections of ($a\ b\ c$) are situated at $x' = 1$, $y' = 1$, $z' = 1$. This way, it results that $a : b : c = \frac{1}{1} : \frac{1}{1} : \frac{1}{1}$ and $(a\ b\ c) = (1\ 1\ 1)$.

Intersections of ($d\ e\ f$) with coordinate axes are $x' = -1$, $y' = 1$, $z' = -1$. It results that $d : e : f = \frac{1}{-1} : \frac{1}{1} : \frac{1}{-1}$ and $(d\ e\ f) = (\bar{1}\ 1\ \bar{1})$ or $(1\ \bar{1}\ 1)$. The latter can be immediately verified by shifting the coordinate system to the lower right back corner.

What is the orientation of directions x^* *and* y^* *of the etching cavity edges?*

The discussion refers to coordinate system (x, y, z).

$x^* : x = 1, y = -1, z = 0$. From this results $x^* = [1\ \bar{1}\ 0]$.

$y^* : x = 1, y = 1, z = 0$. Here results $y^* = [1\ 1\ 0]$.

The etching cavity edges are situated in direction <1 1 0>.

How large is angle α *between {1 1 1} lateral faces and* (001) *bottom face?*

Using Equation (3.5) it results that:

$$\alpha = \arccos \angle\{(a\ b\ c),\ (0\ 0\ 1)\} = \arccos \angle\{(1\ \ 1\ \ 1),\ (0\ 0\ 1)\}$$

$$= \arccos \frac{1 \cdot 0 + 1 \cdot 0 + 1 \cdot 1}{\sqrt{(1^2 + 1^2 + 1^2)(0^2 + 0^2 + 1^2)}} = \arccos \frac{1}{\sqrt{3}}$$

$$\alpha = 54.74^\circ.$$

The same results for

$$\beta = \arccos \angle\{(d \text{ e } f),\ (0\ 0\ 1)\} = \arccos \angle\{(1\ \bar{1}\ 1),\ (0\ 0\ 1)\}$$

$$\beta = \arccos 1/\sqrt{3} = \alpha = 54.74^\circ.$$

As another option of deduction, it is possible to use the cross section along $z' - y^{*\prime}$ plane in Figure 3.5b (Figure 3.5d). The plane diagonal through the unit cell has a length of $y^{*\prime} = \sqrt{2}$. Plane $(a\ b\ c)$ is cut exactly in the middle at $y^{*\prime} = 1/\sqrt{2}$.

This results in $\alpha = \arctan[1/(1/\sqrt{2})] = \arctan\sqrt{2} = 54.74^\circ$ oder

$$\alpha = \arcsin(1/\sqrt{3/2}) = \arcsin\sqrt{2/3} = 54.74^\circ.$$

We will frequently encounter the angle of 54.74° in bulk micromachining with its wet chemical anisotropically etched form elements. It automatically occurs as a result of the crystallographic orientation of the {1 1 1} lateral faces during etching.

3.2.2 Surface Characteristics at Different Orientations

In silicon bulk micromachining, {1 0 0}, {1 1 0} and {1 1 1} planes are the most important ones. Table 3.4 shows the position of these planes in the unit cell. It becomes obvious that the bond density of planes with different orientations varies.

For (1 0 0) planes, there are five Si atoms at the surface. The four atoms at the corners share their four respective bonds with three neighboring cells. That means that each bond is effectively linked to this cell with only one bond. The atom in the middle

Table 3.4 Bond relationships for different silicon planes

Plane	(1 0 0)	(1 1 0)	(1 1 1)
Position in the unit cell			
Plane	a^2	$\sqrt{2}a^2$	$\frac{1}{2}\sqrt{3}a^2$
Number of atom bonds[1]	4 × 1 (corners) +1 × 2 (middle) = 6	4 × 1 (corners) +2 × 2 (edges) = 8	3 × 1 (corners) +3 × 2 (edges) = 9
Bonds per plane	$6 \cdot \frac{1}{a^2}$	$\frac{8}{\sqrt{2}} \frac{1}{a^2} = 5.66 \frac{1}{a^2}$	$9 \cdot \frac{2}{\sqrt{3}} \frac{1}{a^2} = 10.4 \frac{1}{a^2}$

[1]effective bonds of the atoms with the planes marked here.

of the plane shares two bonds with the shown unit cell. This means that the five atoms are linked by a total of six bonds to the unit cell. Assuming that all bonds have the same bonding energy, there are six bonds on plane a^2. The comparison shows that the bond density reaches its maximum for (1 1 1) Si. This is the reason why the etch rate has its minimum at {1 1 1} due to the by far highest bond density. It has to be taken into account though, that Table 3.4 can be only used for a phenomenological interpretation of etching characteristics. The calculation of the actual bond energies requires much more sophisticated models and assumptions.

3.2.3 Anisotropic Elastic Characteristics

Due to its crystal characteristics, silicon has an orientation-related elastic behavior. This is called anisotropy. Table 3.5 shows the general relation of mechanical stresses σ_j und the resulting strains ε_i.

We use VOIGT notation, where indices 1 to 3 describe normal deformation and indices 4 to 6 shear deformation. Shearing in plane 1–2 is combined with rotating the deformed plane around axis 3. It is correspondingly assigned index 6.

Table 3.5 Deduction of the general stress-strain relationships

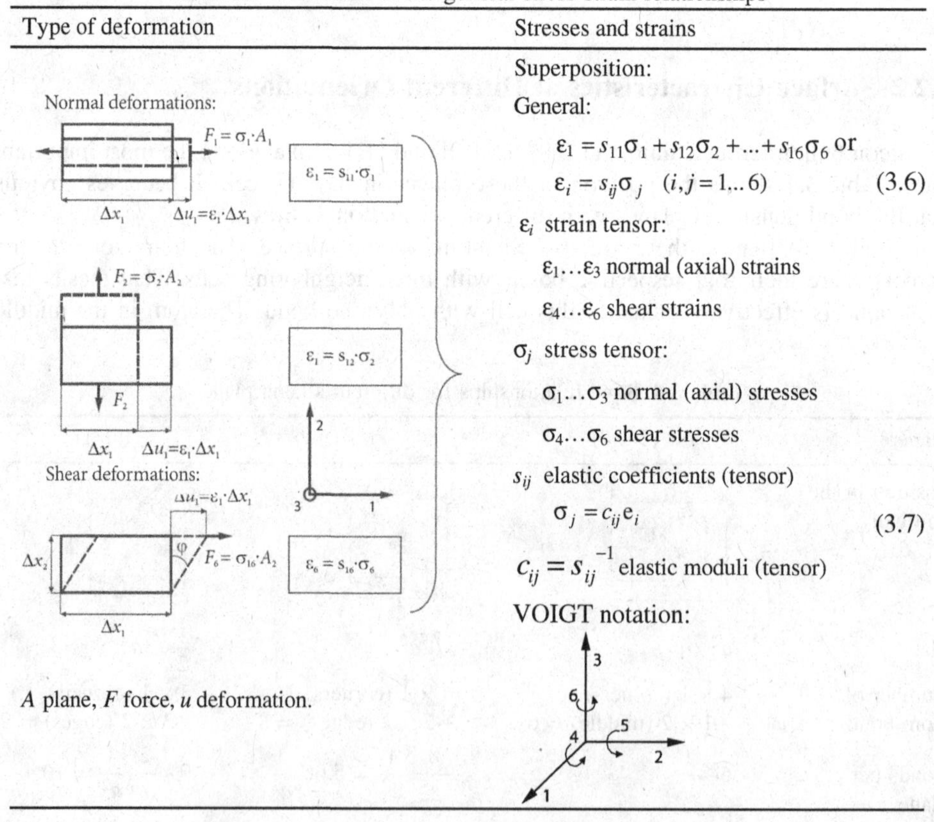

Type of deformation	Stresses and strains
Normal deformations; Shear deformations (see figure)	Superposition: General:

Superposition:

General:

$$\varepsilon_1 = s_{11}\sigma_1 + s_{12}\sigma_2 + \ldots + s_{16}\sigma_6 \text{ or}$$

$$\varepsilon_i = s_{ij}\sigma_j \quad (i,j = 1 \ldots 6) \qquad (3.6)$$

ε_i strain tensor:

$\varepsilon_1 \ldots \varepsilon_3$ normal (axial) strains

$\varepsilon_4 \ldots \varepsilon_6$ shear strains

σ_j stress tensor:

$\sigma_1 \ldots \sigma_3$ normal (axial) stresses

$\sigma_4 \ldots \sigma_6$ shear stresses

s_{ij} elastic coefficients (tensor)

$$\sigma_j = c_{ij} e_i \qquad (3.7)$$

$c_{ij} = s_{ij}^{-1}$ elastic moduli (tensor)

VOIGT notation:

A plane, *F* force, *u* deformation.

Table 3.6 Elastic coefficients s_{ij} and elastic moduli c_{ij} for silicon [BÜTTGENBACH91]

Matrix	Parameters
$s_{ij} = \begin{bmatrix} s_{11} & s_{12} & s_{12} & 0 & 0 & 0 \\ s_{12} & s_{11} & s_{12} & 0 & 0 & 0 \\ s_{12} & s_{12} & s_{11} & 0 & 0 & 0 \\ 0 & 0 & 0 & s_{44} & 0 & 0 \\ 0 & 0 & 0 & 0 & s_{44} & 0 \\ 0 & 0 & 0 & 0 & 0 & s_{44} \end{bmatrix}$	$s_{11} = 7.68 \cdot 10^{-12} \frac{m^2}{N}$ $s_{12} = 2.14 \cdot 10^{-12} \frac{m^2}{N}$ $s_{44} = 12.56 \cdot 10^{-12} \frac{m^2}{N}$
$c_{ij} = \begin{bmatrix} c_{11} & c_{12} & c_{12} & 0 & 0 & 0 \\ c_{12} & c_{11} & c_{12} & 0 & 0 & 0 \\ c_{12} & c_{12} & c_{11} & 0 & 0 & 0 \\ 0 & 0 & 0 & c_{44} & 0 & 0 \\ 0 & 0 & 0 & 0 & c_{44} & 0 \\ 0 & 0 & 0 & 0 & 0 & c_{44} \end{bmatrix}$	$c_{11} = 1.658 \cdot 10^{-11} \frac{N}{m^2}$ $c_{12} = 0.639 \cdot 10^{-11} \frac{N}{m^2}$ $c_{44} = 0.796 \cdot 10^{-11} \frac{N}{m^2}$

Normal forces F_1, F_2 and F_3 in directions 1, 2 and 3 cause a change in the deformation state Δu_1 and thereby a change in axial strain ε_1. The same applies to axial strains ε_2 and ε_3 and to shear strains ε_4 to ε_6. Shear deformations follow analogous reasoning.

Due to the symmetrical characteristics of single crystalline silicon, Equations (3.6) and (3.7) can be simplified as only three s_{ij} and c_{ij} coefficients, respectively, are independent of each other (Table 3.6).

Looking at a specific crystal orientation of silicon and assuming normal stresses and strains, exclusively, it can be considered to be quasi isotropic. This allows us to transform Equation (3.6) in the following way:

$$\begin{aligned} \varepsilon_1 &= \frac{1}{E}\sigma_1 - \frac{\nu}{E}\sigma_2 - \frac{\nu}{E}\sigma_3 \\ \varepsilon_2 &= -\frac{\nu}{E}\sigma_1 + \frac{1}{E}\sigma_2 - \frac{\nu}{E}\sigma_3 \\ \varepsilon_3 &= -\frac{\nu}{E}\sigma_1 - \frac{\nu}{E}\sigma_2 + \frac{1}{E}\sigma_3, \end{aligned} \tag{3.8}$$

$$\varepsilon_{4,5,6} = \frac{1}{G} \cdot \sigma_{4,5,6}, \tag{3.9}$$

with E being YOUNG's modulus, G the shear modulus and ν POISSON's ratio. The data for E and ν depend on the specific crystal orientation.

From Equation (3.6) to (3.8), it directly results for direction <1 0 0> of {1 0 0} Si that:

$$\frac{1}{E_{<1\,0\,0>}} = s_{11} = \frac{c_{11} + c_{12}}{(c_{11} - c_{12})(c_{11} + 2c_{12})} \tag{3.10}$$

$$\nu_{<1\,0\,0>\,\{1\,0\,0\}} = -s_{12}/s_{11} \tag{3.11}$$

$$1/G_{<1\,0\,0>\,\{1\,0\,0\}} = s_{44} = 1/c_{44}. \tag{3.12}$$

Table 3.7 Orientation-dependence of $E_{<h\,k\,l>}$ and $v_{<h\,k\,l>\{100\}}$

Planes	YOUNG's modulus	POISSON's ratio
<1 0 0>	$1.30 \cdot 10^{11}$ N/m^2	0.28
<1 1 0>	$1.69 \cdot 10^{11}$ N/m^2	0.063
<1 1 1>	$1.88 \cdot 10^{11}$ N/m^2	–

The conversion to any direction $[h\ k\ l]$ requires a coordinate transformation (Appendix B) [MESCHEDER04]:

$$\frac{1}{E_{<h\,k\,l>}} = s_{11} - s\frac{h^2k^2 + k^2l^2 + h^2l^2}{(h^2 + k^2 + l^2)^2} \tag{3.13}$$

with

$$s = 2s_{11} - 2s_{12} - s_{44} = 7.08 \cdot 10^{-12}\frac{\text{m}^2}{\text{N}}$$

for silicon.

Table 3.7 shows the correlation of YOUNG's modulus $E_{<h\,k\,l>}$ and POISSON's ratio v for several orientations in the (1 0 0) plane.

The table clearly shows that YOUNG's modulus can show variations of up to 30 %. The POISSON's ratio v depends to an even larger degree on the position of the crystal lattice. For direction <1 1 0> in {1 0 0} silicon, v decreases to almost zero. This means that there is an almost negligible small correlation between stress σ_j in one direction and deformation ε_i in the perpendicular directions.

Example 3.3 Constant volume with force F acting on a silicon body

Does the volume of a body remain constant if a uniaxial mechanical tension load is applied?

We assume a silicon deformation element and a pressure force F_1 that acts unilaterally (Figure 3.6).

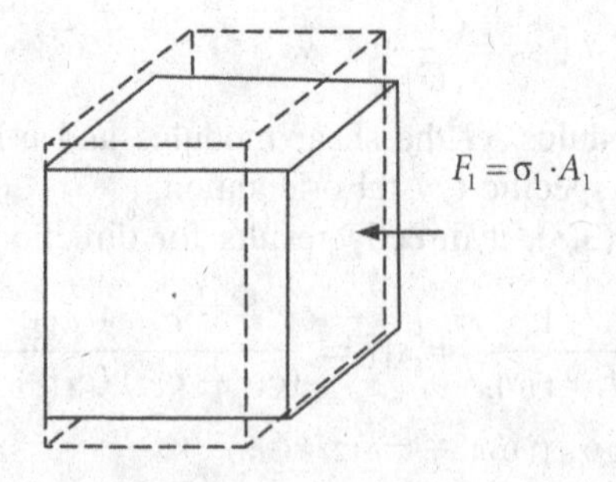

Figure 3.6 Deformation element with load F_1 (edge lengths a, b, c)

From Equation (3.8), it results for $\sigma_2 = \sigma_3 = 0$ and $\sigma_1 = F_1/A_1$

$$\varepsilon_1 = \frac{1}{E}\sigma_1 \tag{3.14}$$

$$\varepsilon_2 = -\frac{\nu}{E}\sigma_1$$

$$\varepsilon_3 = -\frac{\nu}{E}\sigma_1.$$

The resulting relative change in volume amounts to

$$\frac{\Delta V}{V} = \frac{a(1+\varepsilon_1)\, b(1+\varepsilon_2)\, c(1+\varepsilon_3) - abc}{abc} \approx \varepsilon_1 + \varepsilon_2 + \varepsilon_3 \ (\text{for } \varepsilon_i \ll 1)$$

$$\frac{\Delta V}{V} = \frac{1-2\nu}{E}\sigma_1. \tag{3.15}$$

This shows that mechanical stress causes a volume change according to Equation (3.15). The volume change is zero when POISSON's ratio $\nu = 0.5$, which is the case of rubber-like material.

3.2.4 Mechanical Strength

Silicon is a brittle material. At room temperature, it is ideal elastic. Its plastic characteristics only appear at several hundred degrees Celsius. The correlation of stress σ and strain ε is quasi proportional until the ultimate stress limit is reached (Figure 3.7).

However, studies regarding the stability of micromechanical elements under load, e.g. on the overpressure strength of piezoresistive silicon pressure sensors (see Section 7.5), have shown that it is impossible to provide a single numerical limit for tensile strength. It has become obvious that the collapse occurred at completely different loads. Therefore, the failure probability of single crystalline silicon can only be provided for specific load situations.

This effect can also be interpreted from a different perspective: silicon with a polished surface can be considered quasi smooth or defect-free, even in the microscopic range. Today, manufacturers exclude two- and three-dimensional crystallographic lattice defects.

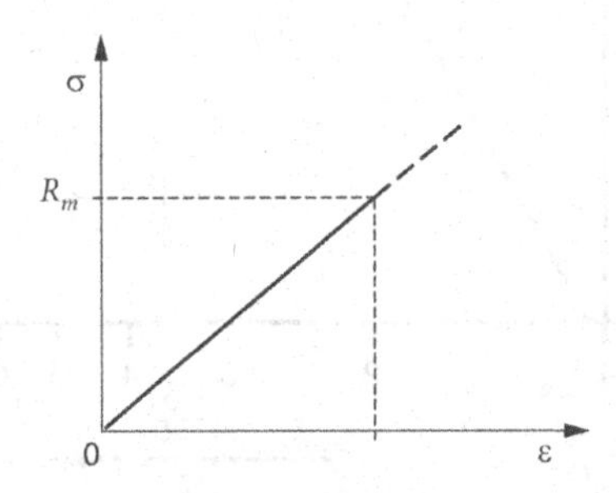

Figure 3.7 Stress-strain relationship for single-crystal silicon. Proportionality and elastic limit coincide at ultimate strength R_m

Doping atoms and vacancies are defects in the sub-nanometer range and are too small to affect the fracture behavior. Failure of micromechanical components consisting of single crystalline silicon is caused mainly by crack formation at the edges and on the surfaces of function elements. As these occur randomly, the fracture point corresponds to a failure probability distribution.

This means that even at low load, there is a certain probability that silicon will crack. Similar to other brittle materials (ceramics, hard metal), WEIBULL distribution constitutes a good representation of failure probability [KIESEWETTER92]. Failure probability $F(\sigma)$ of mechanical stress σ causing material fracture in silicon is

$$F(\sigma) = 1 - \exp\{-(\sigma/b)^k\}, \tag{3.16}$$

where k is the WEIBULL modulus or the distribution parameter. The position parameter b states at which stress σ failure probability drops to $F(\sigma) = 1 - 1/\mathrm{e} = 0.6321$. $F(\sigma)$ is a relative cumulative probability. This means that value $F(\sigma_0)$ defines the proportion of function elements that will break in all possible load situation in the range $\sigma = 0 \ldots \sigma_0$.

The correlation of Equation (3.16) is commonly presented in WEIBULL diagrams (Figure 3.8). From Equation (3.16) a linear curve results between $\log\{-ln[1 - F(\sigma)]\}$, which divides the ordinate, and $\log(\sigma/\hat{b})$ which forms the absciss. $\hat{b}$ and $\hat{k}$ are estimates of the statistical parameters b and k. The value of $\hat{b}$ can be read out directly from the curve at the value of $F(\sigma) = 63.2\,\%$. WEIBULL modulus $\hat{k}$ determines the slope and is therefore also called slope parameter.

Table 3.8 shows estimates for parameters b and k in Equation (3.16) that have been determined through experiments. The table shows that there are clear differences between

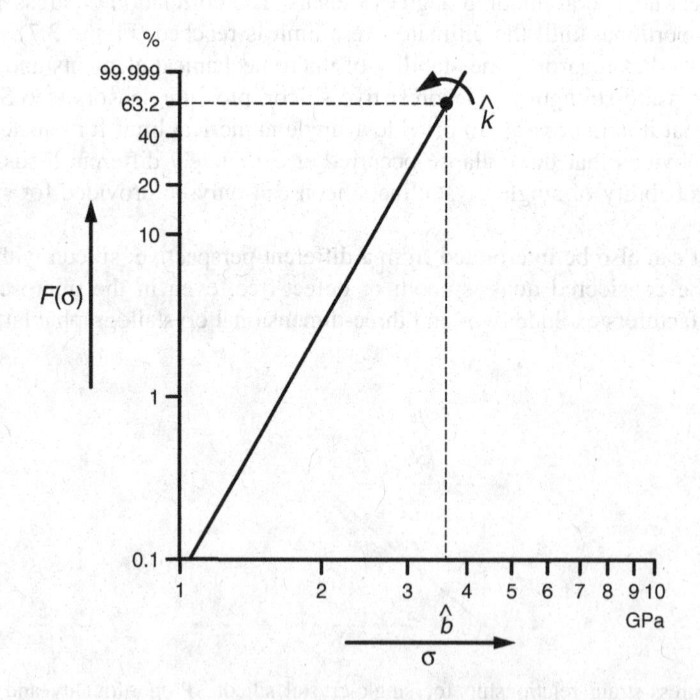

Figure 3.8 Failure probability $F(\sigma)$ in relation to mechanical stress σ

Table 3.8 Parameter of WEIBULL distribution of Equation (3.16) for CZOCHALSKI-grown (see Section 3.2.5) single crystalline silicon [KIESEWETTER92]

Wafer orientation	Temperature	$\hat{b}$ in 10^9 GPa	$\hat{k}$
{1 0 0}	22 °C	3.73[1]	6.18[1]
{1 1 1}	22 °C	4.41	
{1 0 0}	180 °C	4.00	15.37

[1]This case corresponds to Figure 3.8.

the two most commonly applied silicon orientations {1 0 0} and {1 1 1}. Usually, failure probability is intended to be as low as possible, which means that the values for $\hat{b}$ and $\hat{k}$ should be high. For those values, the curve in the diagram (Figure 3.8) has to be located as far to the right as possible. Accordingly, failure probability decreases at high temperatures.

As Table 3.8 shows, at high temperatures the position parameter $\hat{b}$ increases by a few percent, whereas WEIBULL modulus $\hat{k}$ almost doubles. Mechanical function elements of (1 1 1) silicon are less sensitive to failure through fracture than those of (1 0 0) silicon. This also corresponds to the higher resistance of (1 1 1) silicon during anisotropic etching (see Section 4.6).

***Example 3.4 Failure probability of a piezoresistive pressure sensor* [KIESEWETTER*92*]**

A piezoresistive silicon pressure sensor has a square (1 0 0) Si pressure plate. Its dimensions are 0.25 mm × 0.25 mm with a thickness of 8 μm. A nominal pressure of $p_N = 1.5$ MPa generates a maximum mechanical tensile stress of $\sigma_N = 580\ \text{N/mm}^2 = 0.58$ GPa within the pressure plate.

What is the probability that the pressure sensor fails?

According to Equation (3.16), it is valid that

$$F(\sigma_N) = 1 - \exp\left\{-\left(\frac{\sigma_N}{b}\right)^k\right\}.$$

Applying the values for (1 0 0) Si at ambient temperature given in Table 3.8 it results that:

$$F(\sigma_N = 0.58\ \text{GPa}) = 1 - \exp\left\{-\left(\frac{0.58}{3.73}\right)^{6.18}\right\} = 1 \cdot 10^{-5}.$$

Failure probability is $1 \cdot 10^{-5}$ or a failure percentage of 10 ppm, respectively.

Can the failure probability become zero?

Equation (3.16) is based on a model that shows a finite value for failure probability that is different from zero, even for a minimum mechanical stress. Therefore, often 100 % of the sensors are tested at overpressure (bursting pressure). If the applied bursting pressure corresponds to four times the nominal pressure, the mechanical tension in the pressure plate is also approximately four times as large:

$$F(\sigma_B = 4 \cdot \sigma_N = 2.32\ \text{GPa}) = 5.5\ \%.$$

The 5.5 % of sensors that fail the test also include those sensors that would have failed even at nominal pressure.

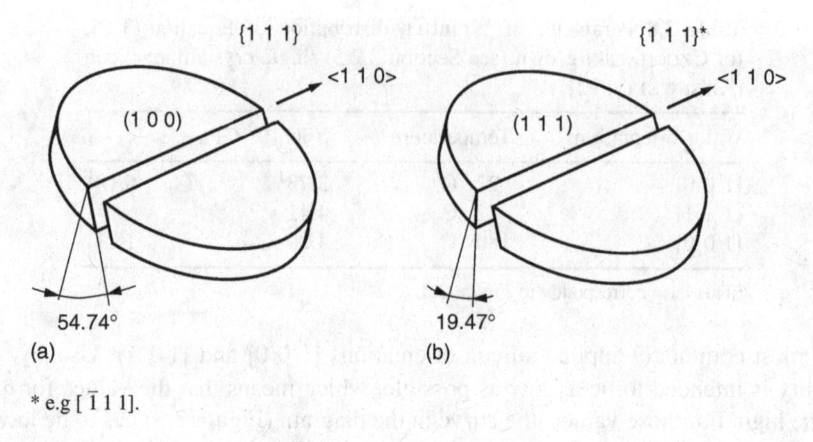

Figure 3.9 Preferred fracture planes and edges for (a) (1 0 0) and (b) (1 1 1) silicon

Sections 3.2.2 and 3.2.3 have shown that the characteristics of silicon vary for different crystal directions and planes. The result is that fracture in single crystals spreads along certain planes. For (1 0 0) and (1 1 1) silicon, mainly {1 1 1} fracture planes occur in <1 1 0> direction (preferred fracture planes and directions) (Figure 3.9).

3.2.5 Silicon Wafer

In microelectronics and microsystem technology, single crystalline silicon is used in the form of disks. Such silicon disks are called wafers. They should comply with the following requirements:

- maximum chemical purity;
- maximum crystal perfection with a precise crystal orientation;
- high doping homogeneity;
- low shape deviations with high surface perfection and a minimum roughness.

Table 3.9 comprises typical quality features of silicon wafers. The following difference between microsystem technology and highly integrated microelectronic can be observed:

- In microsystem technology, use of Si wafers with diameters in the range of 100 ... 150 mm, occasionally even 200 mm. This results from the typically low requirements regarding integration, structural size and structural precision. Therefore, older-generation equipment with smaller wafer diameter is used. Another reason is that diameter increases with higher wafer thickness. In bulk micromachining, processing depth for thin structures basically corresponds to wafer thickness and therefore, smaller wafer diameters are advantageous.

Table 3.9 Quality characteristics of silicon wafers [FRÜHAUF05][1]

Property	Characteristics
Crystallographic	• Crystal orientation on wafer plane (surface): Type of main and secondary flat • Position of a preferred direction on the wafer plane: Direction of main flat or notch[2] 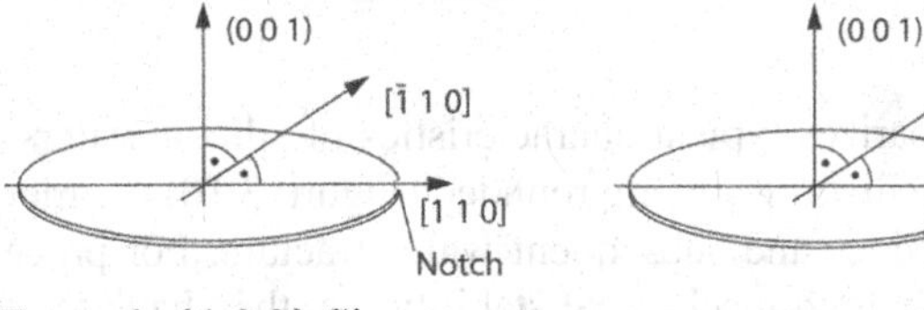 Example {1 0 0}-Si Example {1 1 1}-Si
Electrical	• Conduction type (position of secondary flat SF in relation to main flat MF) 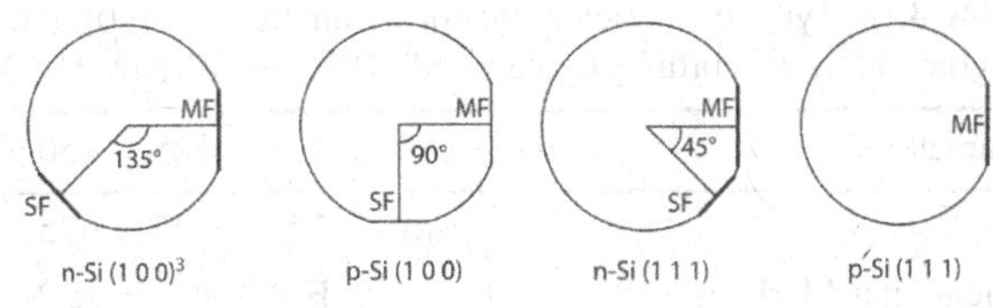 • Specific resistance ρ • Mobility μ_n, μ_p of electrons and holes • Life-time τ_n, τ_p of electrons and holes
Geometrical	• Wafer diameter *d* and thickness *h* • Total thickness variation *TTV*, Bow *B*, Warp *W*, Taper *T* 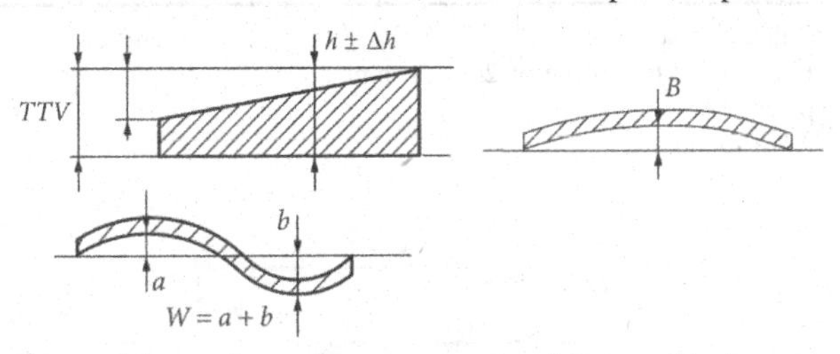 • State of the wafer surface on upper and lower side (lapped, etched or polished)

[1] According to DIN 41850 Integrated Layer Circuits: Substrates, Part 1, 1986, SEMI Standards (Semiconductor and Materials International; http://wps2a.semi.org/wps/portal) and ASTM Standards (ASTM International; http://www.astm.org)

[2] Flats apply to small disk diameters up to 150 mm, and notches to diameters from 150 mm.

[3] Previously, the angle between main flat and secondary flat was 180°.

- {1 0 0} silicon wafers are dominant in microsystem technology. With {1 0 0} oriented silicon it is possible to manufacture square and rectangular structures with an outstanding geometric definition (see Figure 3.5 and Section 4.6). When anisotropic wet chemical etching is used, the etching ratio is sufficiently large and allows for a cost-efficient processing. {1 1 1} Si – which is often used in microelectronics – is hardly important in microsystem technology. Due to its low etching ratios during anisotropic wet chemical etching, it is exclusively used when gas phase etching is applied in shaping processes (see Section 4.6).

Table 3.10 summarizes typical characteristics of silicon wafers.
The edges of the silicon wafer are rounded (Figure 3.10) in order to prevent cracks starting from the wafer edges and subsequent wafer fracture. For processing, the wafers have to have a sufficiently high mechanical stability, i.e. they have to be sufficiently thick for the corresponding diameter. For special applications, the thickness of a wafer can be reduced through chemical-mechanical polishing at the end of the processing cycle.

Table 3.10 Typical characteristics of commercial {1 0 0} and {1 1 1} oriented silicon wafers (according to standards DIN, SEMI and ASTM)

Diameter d	Unit	100	125	150	200	300
Thickness h	μm	525	625	675	725	825
Orientation Marking[1]	–	F	F	F, N	N	N
Total Thickness Variation TTV	μm	5	5	5	5	
Maximum Warp W	μm	30	30	30	30	30
Focal Plane Deviation FPD^2	μm	<1	<1	<1	0.23	0.12
Roughness R_a^3	nm	10	10	10	0.15[4]	0.1[4]
Surface orientation	–			±1°		

[1]F flat, N notch;
[2]FPD related to a plane of 20 × 20 mm²;
[3]polished;
[4]AFM roughness.

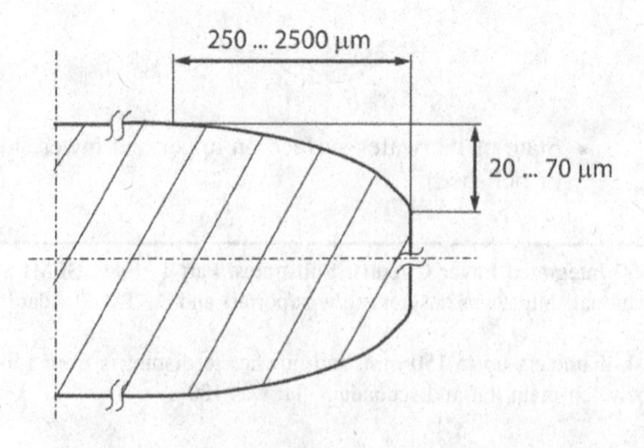

Figure 3.10 Rounded edges of silicon wafers

In addition to homogeneous Si disks, special wafers are used for specific applications:

- Wafers with epitaxial layers (usually n-epitaxy on p-Si wafers) or highly doped layers (p^+-Si) as etch-stop layer with a defined thickness for three-dimensional shaping (see Section 4.6).
- SOI (silicon on insulator) wafers with a silicon dioxide layer that is buried in the wafer at a depth of a few nano- to micrometres (as an etch stop or insulator layer, see Section 4.9).

Often, wafers are polished on both sides. This refers mainly to volume micromechanics where micromechanical function elements are shaped out of the volume of the silicon disk.

Table 3.11 represents the manufacturing of silicon wafers. Commonly, the CZOCHRALSKI technique is used for applications in microsystem technology. There, the single crystal is pulled out of the liquid silicon melt. In order to achieve maximum purity, float zone (FZ) melting can be used to extract single crystalline silicon from a cast poly-Si-stick. This method can only be used for wafer diameters of less than 200 mm and is limited to special applications in microelectronics.

3.3 GLASSES

Material which, when cooling down from melting, changes into its solid state without crystallization is called glass. Thus, glass is a generic term for a variety of materials with different compositions and characteristics. In microsystem technology, it is used for special applications as a cost-efficient substrate material to silicon. This refers, for example, to microscopic applications that require transparent carrier material or to applications with lower requirements of precision and structural size. It is possible to take advantage of the fact that certain glasses have a thermal expansion coefficient that is similar to that of silicon and that they are photostructurable.

3.3.1 General Characteristics of Glass

Among atoms, there exists a certain short-range order, but no far-range order. This corresponds to characteristics of frozen liquids (subcooled melting). Due to their amorphous state, the physical and chemical characteristics of glasses are isotropic. Amorphous materials do not have a melting temperature similar to that of crystals, but they soften above a certain temperature T_G (glass transformation temperature). Figure 3.11 shows the typical correlation between volume V of glass and temperature T.

At high temperatures, the atoms are located in the liquefied material without any order. When cooling down very slowly, the atoms can form a crystal lattice. This happens at melting temperature T_S, where volume V decreases substantially due to the fact that an order is starting to form (Figure 3.11a). The result of SiO_x melting can be single crystal SiO_2 which is also known as quartz ($T_S = 1723$ °C). If cooling down happens

Table 3.11 Manufacturing of silicon wafers [FRÜHAUF05] [NISHI00]

A. Manufacturing of high-purity, poly-crystalline silicon

- Base material: sufficiently pure quartz sand (SiO_2)
- Formation of elementary silicon by carbon reduction in an electric arc furnace
  $SiO_2 + 2C \xrightarrow{1550...1620\ ^\circ C} Si + 2CO$
  Purity of Si : (96...99) %
- Using a fluidized bed reactor to transform Si with hydrochloric gas into trichlorsilane ($SiHCl_3$)
  $Si + 3HCl \xrightarrow{300...400\ ^\circ C} SiHCl_3 + H_2$
- Purifying through multi-step distillation of $SiHCl_3$ (boiling point $T_S = 31\ ^\circ C$, clear transparent liquid, highly explosive if mixed with air)
  - Separation of impurities with other T_S
  - Critical impurities BCl_3 ($T_S = 12\ ^\circ C$), PCl_3 ($T_S = 76\ ^\circ C$) and some carbon compounds with similar boiling point: B, P and carbon compounds are hence the main impurities in high-purity silicon.
- Deposition of high-purity poly-Si in a quartz glass reactor.

Reduction:	$SiHCl_3 + H_2 \xrightarrow{1000...1100\ ^\circ C} Si + 2HCl$		
Thermal decomposition:	$SiHCl_3 \rightarrow Si + 3\ SiCl_4 + 2\ H_2$		
Content of foreign matter:	Phosphorus:	10^{13}	atoms/cm^3
	Boron:	$5 \cdot 10^{12}$	atoms/cm^3
	Heavy metals:	$<10^{13}$	atoms/cm^3
	Oxygen, carbon:	10^{16}	atoms/cm^3

B. Growing of single crystalline silicon

- Melting down crushed poly-Si in an inductively heated quartz crucible
- Dipping the seed crystal with the desired orientation (usually (1 0 0) or (1 1 1), occasionally (1 1 0)) in the molten silicon
- The crystal ingot is slowly pulled out and rotated at the same time ('bottleneck technology'). The crystal diameter is determined by the temperature of the molten silicon and the rate of pulling.
- Adding dopants (B, P, As, Sb) to the molten silicon in form of impurity atoms or through the corresponding amount of highly-doped silicon.

C. Manufacturing of silicon wafers

- Sawing off both ends of the ingot
- Diameter grinding of the crystal to create the precise diameter
- Determining crystal orientation by X-ray diffraction and grinding of flats or notches, respectively
- Partitioning of the crystal and separating of test disks in order to determine the electrical parameters (wafer cutting)
- Wafer slicing using diamond blades (internal diameter saw; up to $d = 200$ mm) or diamond wires (multi-wire saw)
- Rounding the edging through edge grind
- Double-sided lapping of the disk in order to abolish focal plane deviations
- Chemical etching of the mechanically affected surface layer (acidic etching: $HF\text{-}HNO_3\text{-}H_2O$ or basic etching: $KOH\text{-}H_2O$)
- One- or double-sided chemical-mechanical polishing (CMP) for a surface roughness in the nm-range
- Cleaning and inspection of the disk

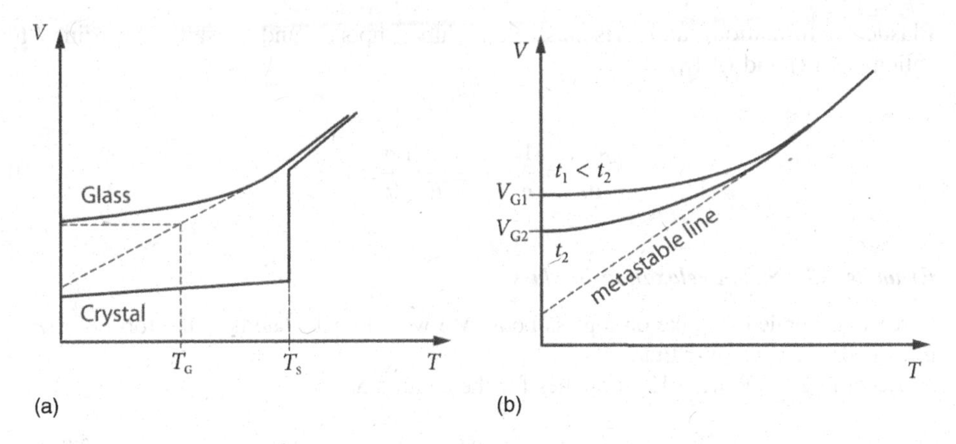

Figure 3.11 Cooling curve of glasses (explanations in the text, see also [IVERS04])

too fast, neighboring atoms (in the example of SiO_4 tetraeders) can not form a three-dimensional lattice. There will be no strict periodical far-range order, but only a short-range order. Volume decreases continuously with decreasing temperatures. Below the glass transformation temperature T_G, the slope of the $V - T$-curve (the thermal longitudinal expansion coefficient) decreases; the amorphous solid body is slowly formed. Due to the shorter-range order of glass as opposed to crystals, the volume of glass largely exceeds that of crystals with an ideal order. The resulting volume depends on how fast the glass cools down. The longer the cooling period t, the smaller the volume (Figure 3.11b). Fast heating with subsequent slow cooling to the initial temperature and vice versa result in volume changes.

3.3.2 Visco-elastic Behavior

Due to its special characteristics, glass shows a visco-elastic behavior, from a mechanical perspective. As opposed to this, an ideal solid body would have a linear elastic behavior. For the one-dimensional isotropic case, the following correlation between stress σ and strain ε applies according to Equation (3.8):

$$\varepsilon = \frac{1}{E}\sigma. \tag{3.17}$$

At the same time, glass shows as a subcooled melt the same viscous behavior as a liquid:

$$\frac{d\varepsilon}{dt} = \frac{1}{\eta}\sigma, \tag{3.18}$$

where σ and ε are mechanical stress and the corresponding strain and η is the dynamic viscosity.

Elastic deformation and viscous flux superimpose and result according to Equations (3.17) and (3.18) in

$$\frac{d\varepsilon}{dt} = \frac{1}{\eta} \cdot \sigma + \frac{1}{E} \frac{d\sigma}{dt}. \tag{3.19}$$

Example 3.5 Strain relaxation in glass

Force F is applied steplike on a glass body. We want to calculate how the form of the glass body changes over time.

According to Figure 3.12, it applies for the strain that

$$\varepsilon = u/\ell \tag{3.20}$$

or, respectively, for the speed of area A on which force F acts:

$$v = \frac{du}{dt} = \ell \cdot \frac{d\varepsilon}{dt}. \tag{3.21}$$

The transformation of Equation (3.19) for harmonic values with angular frequency ω results in the frequency range in

$$j\omega\, \underline{\varepsilon} = \frac{1}{\eta}\, \underline{\sigma} + \frac{1}{E} j\omega\underline{\sigma} = \left(\frac{1}{\eta} + j\omega\frac{1}{E}\right)\underline{\sigma}. \tag{3.22}$$

Strain ε and stress σ can be correlated to force $\underline{F}$ or deformation speed $\underline{v}$, respectively:

$$\frac{1}{\ell} \cdot \underline{v} = \left(\frac{1}{\eta A} + j\omega\frac{1}{EA}\right)\underline{F}; \quad \underline{v} = \left(\frac{\ell}{\eta A} + j\omega\frac{\ell}{EA}\right)\underline{F} \tag{3.23}$$

$$\underline{v} = (r + j\omega n)\,\underline{F}. \tag{3.24}$$

In the relationship between $\underline{v}$ and $\underline{F}$, the viscous part $(\ell/\eta A)$ acts here as frictional resistance and the elastic term as an inductance. MAXWELL's model shows this correlation (Figure 3.13). The acting force F causes an immediate elastic deformation of the glass and a long-term further increased deformation.

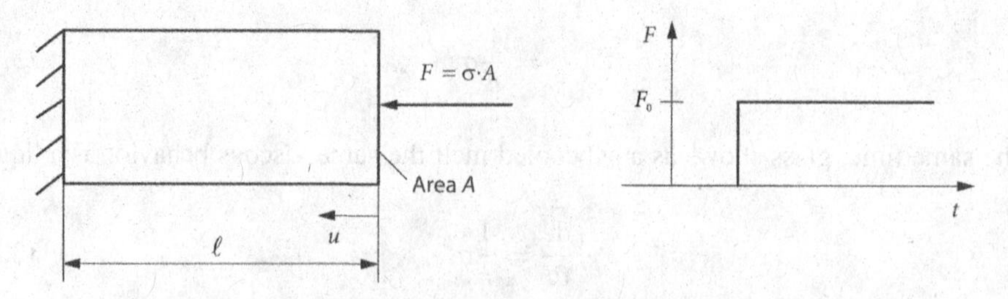

Figure 3.12 Glass body under load

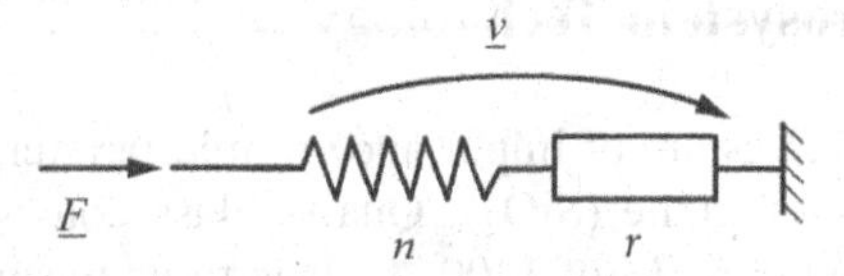

Figure 3.13 MAXWELL's model of the visco-elastic behavior of glass

In the case of a discontinuously acting force F, complex frequencies $s = \sigma + j\omega$ (σ is the real component of the complex frequency and not the mechanical stress) have to be used instead of harmonic frequency ω. Thus, Equation (3.24) becomes

$$\underline{v} = (r + sn)\,\underline{F}. \tag{3.25}$$

According to LAPLACE transform, it is valid for the input step function that

$$\underline{F} = F_0 \cdot \frac{1}{s}. \tag{3.26}$$

For deformation $\underline{u}$, it results from Equations (3.25) and (3.26) that

$$\underline{u} = \frac{1}{s}\,\underline{v} = \frac{1}{s}(r + sn)\underline{F} = \left(\frac{r}{s^2} + \frac{n}{s}\right) \cdot F_0. \tag{3.27}$$

With transform relationships $1/s^2 \leftrightarrow t$ and $1/s \leftrightarrow 1$, it follows from LAPLACE back-transform for the time-related deformation of the glass body that

$$u(t) = F_0(n + r \cdot t) = F_0\left(\frac{\ell}{EA} + \frac{\ell}{\eta A} \cdot t\right). \tag{3.28}$$

The first term describes the elastic deformation as a direct effect of the imprinted force. The deformation that immediately follows the force step increases linearly over time as the still acting force causes a flow of the glass (Figure 3.14).

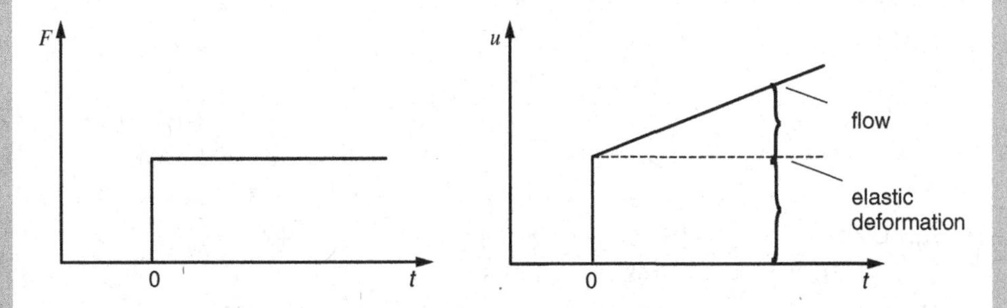

Figure 3.14 Strain relaxation of glass after a force step

3.3.3 Glasses in Microsystem Technology

In general, only silicate glasses are of importance in microsystem technology, i.e. glasses with the main component quartzite (SiO_2). Quartz glass consisting of pure SiO_2 has a glass transformation temperature T_G of 1400 °C. Due to its purity and thermal stability, it is often used in manufacturing facilities for semiconductor technology (see Table 3.11B: quartz crucible for silicon crystal pulling).

Metal ions are integrated in the lattice of the SiO_4 tetrahedral network in the quartz glass to reduce T_G in order to limit its inclination to crystallize as well as to reach specific thermal coefficients of expansion and optical refractive indices. These ions are called network-changing ions. Oxygen ions saturate the valences of the ions and this way the corresponding metal oxides are added to SiO_2 in the silicate glass.

In addition to quartz glass, glasses are used for applications in microtechnology (Table 3.12).

- Borosilicate glass (brand names: Pyrex, Borofloat, Tempax, Rasotherm). Through adding B_2O_3, Na_2O and Al_2O_3, we get a glass with a thermal coefficient of expansion α similar to that of silicon. The thermal expansion coefficient is not constant but temperature-dependent for silicon and these glasses. However, it is possible to determine a bond temperature T_{Bond} at which the temperature expansion between T_{Bond} and

Table 3.12 Composition and characteristics of important glasses in microsystem technology

Glass type	Quartz glass	Borosilicate glass	Li-Al-Silicate glass
Commercial names	–	Pyrex, Borofloat Tempax, Rasotherm	Foturan
Typical compositions			
SiO_2	100 %	81 %	75...85 %
Na_2O	–	4 %	1...2 %
Al_2O_3	–	2 %	3...6 %
B_2O_3	–	13 %	–
Others	–		3...6 % (K_2O) 0.1 % (Ag_2O) 0.3 % (Sb_2O_3)
T_G, °C	1400	820	450
α_{Glas}, $10^{-6}K^{-1}$	0.58	3.3	8.6
Thermal conductivity λ, W/mK	1.4	1.15	1.35
Density ρ, g/cm^3	2.2	2.23	2.76
Permittivity ε at 1 MHz and 20 °C	3.8	4.6	6.5

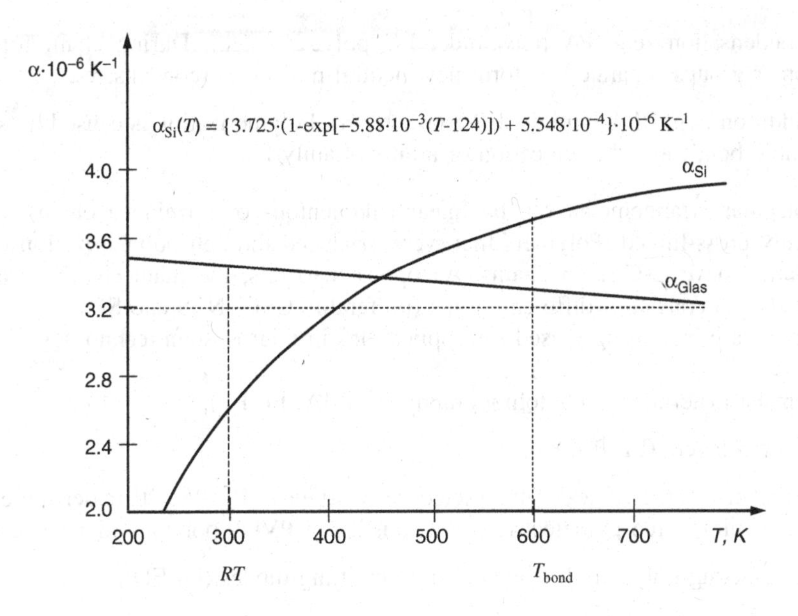

Figure 3.15 Curves of the thermal expansion coefficients of silicon and Pyrex

ambient temperature is the same for silicon and glass (Figure 3.15):

$$\int_{\mathrm{T_{Bond}}}^{\mathrm{RT}} \alpha_{\mathrm{Si}}(T)\,\mathrm{d}T = \int_{\mathrm{T_{Bond}}}^{\mathrm{RT}} \alpha_{\mathrm{Glas}}(T)\,\mathrm{d}T. \tag{3.29}$$

At high temperatures, the movable sodium ions in Pyrex glass can drift in the electric fields (ionic conduction) which makes it possible for the silicon to anodically bond with this glass (anodic bonding, see Section 4.8.2).

- Lithium aluminum silicate glass (commercial name: Foturan). This glass is photo-structurable, i.e. it is possible to use photolithography in order to etch locally limited three-dimensional shapes.

Glass substrates can be produced as flat glass using float glass processes. Applying methods of silicon manufacturing, they can be cut, lapped and polished in order to comply with specific requirements.

3.4 POLYMERS

Polymer materials (plastics) are mainly carbon organic substances that are formed by macromolecules. The macromolecules are produced synthetically from monomers through the following reactions:

- Polymerization (e.g. PE polyethylene, PP polypropylene, PVC polyvinyl chloride, PMMA polymethyl methacrylate, PTFE polytetrafluoroethylene). External effects (light, temperature) are used to break the double bonds and thus cause the formation of a chain.

- Polycondensation (e.g. PA polyamide, PC polycarbonate). During chain formation, monomer groups separate and form new neutral molecules (condensate).
- Polyaddition (e.g. PI polyimide, EP epoxy resin). Polymerization is caused by splitting of double bonds and the relocation of atoms (mainly H).

The molecular arrangements can be linear (filamentous or forming a chain), ramified or spatially cross-linked. Polymers that were produced through polymerization or polycondensation have C-C main chains. As opposed to this, the main chains of addition compounds can consist of different atoms (preferably C, O, N, S and Si).

Polymers are increasingly used for applications in microsystem technology:

- photoresist materials for photolithography (PMMA, PI, EP);
- passivation layers (PI, PTFE);
- sensor layers (e.g. changes of the dielectric constant in PI, PA, PE under the effect of humidity; piezo- and pyroelectric characteristics of PVDF polyvinylidene flouride);
- semiconducting polymers for organic light emitting diodes (OLED).

3.4.1 Thermoplastic Materials in Microsystem Technology

The majority of the polymers used in microsystem technology are thermoplastics, i.e. amorphous or partially crystal polymer materials with chain macromolecules that have either a linear or branched structure. The macromolecules form a thermostable combination based only on physical force of attraction (valence bond forces through VAN-DER-WAALS or hydrogen bonds, respectively).

(a) General characteristics

Structure and characteristics of polymer materials are primarily determined by their macromolecular structure and its recipe.

Chemical components and bond types. Polymer materials consist of macromolecular components that are combined through covalent bonds of the elements C, H, O, N, S, Cl, F, Si. There are no principal valence bonds between the macromolecules. This causes the typical plastic and visco-elastic characteristics.

Molecule configuration. According to the specific polymer type, there are different spatial arrangements of atoms and groups of atoms along the molecular chain. This can cause the formation of crystal structures.

Chain length. A larger chain length (higher degree of polymerization) causes larger elastic moduli, larger tensile strength and increasing softening temperatures. It also leads to a lower deformability, breaking elongation and solubility in solvents.

Branching, cross-linkage. The integration of poly-functional groups or the separation of double bonds with subsequent cross-linkage of the polymer chains can be used to form spatially cross-linked macromolecules.

Co-polymerization, polymer mixtures. Mixing different monomers in one macromolecule can be used to change or combine the characteristics of several polymers.

Softeners reduce the interaction of macromolecules. The chain segments become more movable; the polymer becomes softer and more flexible.

(b) Mechanical characteristics

Thermoplastic polymers show a mechanical behavior that is similar to that of glass (see Figure 3.11). They are characterized by the so called glass temperature T_G:

- At $T < T_G$, thermoplastics are solidified and hard as glass.
- At $T > T_G$, thermoplastics are in a state of a subcooled melt. Amorphous thermoplastics (PVC, PC) show a thermoplastic behavior, and partially crystal ones (PE, PP, PA) have a visco-elastic behavior. Polymers have a cristallite melting temperature. When passing this temperature, polymers change into the viscous flow range.

Macromolecules cannot evaporate; they dissolve once they pass a certain temperature (e.g. incineration).

(c) Thermoplastic polymers

Table 3.13 provides an overview of the characteristics of important thermoplastic polymers used in microsystem technology.

PMMA and PA have a very high glass temperature T_G and are therefore hard as glass at ambient temperature. In principal, they can therefore be used as mechanical form and function elements.[1] As opposed to this, PE, PTFE and PVDF show a visco-plastic behavior at ambient temperature. All these polymers have a YOUNG's modulus that is 20 to 200 times smaller than that of silicon, whereas the thermal expansion coefficient of silicon is 20 to 200 times larger than that of these polymers. Most polymers absorb water vapor from the ambient humidity, which may change their volume by a few percent.

3.4.2 Photoresists

In microelectronics and microsystem technology, the transfer of structures is usually carried out by photolithography (see Section 4.3). Light is used to change the structure and thus also the solubility of these photoresist materials regarding certain solvents. Particularly suitable resists are polymers and prepolymers that polymerise (at negative resist) or depolymerize (at positive resist) when exposed to light.

Photoresists that are used for image transfer have to fulfill a large number of requirements: adhesion resistant, defect-free, dimensional accuracy, defined edge formation, stability and resistance to developers, residue-free removal.

[1] Sager, K., Schroth, A., Nakladal, A., Gerlach, G. (1996) Humidity dependent mechanical properties of polyimide films and their use in IC-compatible humidity sensors. *Sensors and Actuators* A53, pp. 330–4.

Table 3.13 Thermoplastic polymers in microsystem technology (selection)

Polymer with structural formula	T_G [°C]	E [GPa]	α [10^{-6} K^{-1}]	Characteristics
Polyethylene (PE) $[-CH_2-CH_2-]_n$	– 68	0.2 … 1	180 … 250	Partially crystalline, visco-elastic
Polymethyl methacrylate (PMMA) $[-C(CH_3)(COOCH_3)-CH_2-]_n$	100	2.7 … 3.3	63 … 80	Amorphous, stiff, hard, scratch-resistant, transparent
Polyimide (PI)	165…400	3 … 5	50 … 63	Solid, stiff, flow-resistant and heat-resistant
Polytetrafluoroethylene (PTFE) $[-CF_2-CF_2-]_n$	– 20	7.5	160 … 200	Partially crystalline, flexible, viscous, low static friction, dielectric and non-stick layers
Polyvinylidenfluorid (PVDF) $[-CH_2-CF_2-]_n$	– 18	2.1 … 3.3	106	piezo- and pyroelectric layers

T_G glass temperature, E YOUNG's modulus, α thermal expansion coefficient

In general, photoresists consist of the following parts:

- layer-forming polymer material (resin): determines mechanical and chemical characteristics;
- photo-sensitive component/photo-initiator (sensitizer): reduces the exposure dose required for a lithographic polymer reaction;

Table 3.14 Photoresist materials

Type	Negative resist	Positive resist	
Solubility in developer	Exposed areas become insoluble.	Exposed areas become soluble	
Polymer reaction at exposure	Cross-linking of low-molecular chains (C=C-double bonds, epoxy rings) into insoluble polymers	Polymer chain parts become soluble	Separation of polymer chains into soluble low-molecular fragments (e.g. an C-S-bond)
Polymers	• Polyisoprene • Polyacrylate • Polybenzimidazole • Polyimide PI • Epoxy resin EP	• Phenol-formaldehyde (Novolak resin)	• PMMA

- solvent: determines the viscosity of the photoresist;
- additive, if required: for setting specific characteristics.

Table 3.14 provides an overview of important photoresist materials.

(a) Negative resists

The first photosensitive resists used in practice were negative resists (Kodak's Thin Film Resist KTFR, 1954). With negative resists, after exposure and developing, there remains an exact inverse structure of the image of the photomask. At the places where the photomask is transparent, the photoresist is exposed and therefore insoluble for the developer. Currently, the most commonly used negative resist is polyisoprene, solved in xylene. It contains a photo-initiator (usually diacid) which forms a radical during exposure with N_2 being separated:

$$R - N_3 \xrightarrow{h\nu} R - \overset{\bullet}{N}\bullet + N_2$$

A very reactive nitrene $R - N$ reacts to a partially cyclic polyisoprene.

The exposure requires nitrogen atmosphere in order to prevent the forming of nitroso compounds that would reduce reproduction quality. The used diacid compounds are UV-sensitive, especially in the range of 340...420 nm or 365...480 nm.

For microsystem technology, SU-8 photo resist (commercial name, developed by IBM[2], manufacturer e.g. MicroChem) [VÖLKLEIN00] has become very important. It can produce resist layers of up to 1200 μm. This resist is UV-structurable and is able to reach a length-width-ratio (aspect ratio) of 18. SU-8 resist consists of the following components:

- epoxy resin: EPON;
- organic solvent: GBL (gamma-butyrolacetone);

[2]Patent US 488 2245, 1989.

- photo-initiator: triarylsulfonium salt (10 weight %);
- developer for dissolving of unexposed areas: PGMEA (propylglycol methylether acetate).

(b) Positive resists

With positive resists, the solubility of the exposed areas increases by a manifold in relation to the applied developer (usually by a factor of 100 or more). After exposure and development, the non-transparent areas of the photoresist become the non-soluble, and thus remaining, areas of the photoresist layer.

All positive resists have the advantage that unexposed areas are not affected by the developer. It is possible to achieve a high geometric mapping precision and sharp resist edges. The most commonly used positive resists are photo resists based on Novolak (phenol formaldehyde). Diazonaphtoquinone (DNQ) acts photo-active and during UV-exposure through N_2 separation and a chemical rearrangement, it forms 3-inden carbon acid. This neutralizes in alkaline developers (TMAH tetramethylammonium hydroxide in aqueous solutions) and becomes soluble itself. Unexposed resist is hydrophobe. It does not swell in developer and therefore it is possible to achieve very thin line widths. DNQ-Novolak resists have a glass transition temperature of 70 ... 140 °C. Especially developed photo resists can be spin-coated and through-exposed at a thickness of more than 100 μm.

For polymethyl methacrylate (PMMA), exposure with X-rays leads to the separation of the polymer chains in the soluble fragments. Using high-energy synchrotron radiation, it is possible to achieve a resist thickness of up to 1 mm. Here, the inclination of the lateral edges of developed resist structures lies in the range of 0.1°. PMMA resists are used in the LIGA technology (see Section 4.12).

3.5 THIN FILMS

Thin films are used for a variety of functions in microsystem technology (see Table 3.1). In the following, we will present the characteristics of selected thin films that are of special importance to microsystem technology.

When using thin films, we always have to be aware of that they are usually not stress-free due to their manufacturing. Layers are deposited at a specific temperature T_{dep} on the substrate. Through cooling down by temperature ΔT and changes in the structure (e.g. phase transitions, grain boundary effects, lattice defects, and integration of residual gas), a free layer would change in length corresponding to a strain

$$\varepsilon_{\text{F}} = \varepsilon_{\text{F,th}} + \varepsilon_{\text{F},i} = -\overline{\alpha}_{\text{F}} \cdot \Delta T + \varepsilon_{\text{F},i} \tag{3.30}$$

$\varepsilon_{\text{F,th}} = -\overline{\alpha}_{\text{F}} \cdot \Delta T$ describes the thermal shrinkage ($\overline{\alpha}_{\text{F}}$ is the linear thermal expansion coefficient of the layer, averaged over the cooling range $T_{\text{dep}} - \Delta T \ldots T_{\text{dep}}$) and $\varepsilon_{F,i}$ the intrinsic strain due to structural changes. The substrate on which the layer was deposited only experiences a thermal shrinkage

$$\varepsilon_{\text{S}} = -\overline{\alpha}_{\text{S}} \cdot \Delta T, \tag{3.31}$$

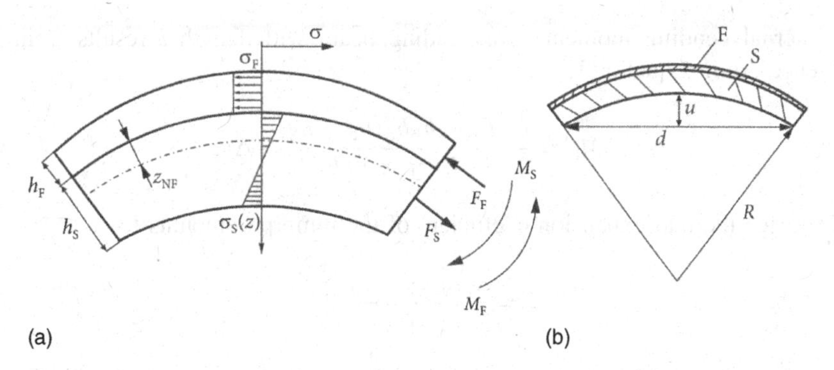

Figure 3.16 Cross-section of the compound thin film/substrate after cooling down from deposition temperature to ambient temperature: (a) mechanical stress in the bimorph; (b) geometric relationship on a coated circular plate; index F: layer (film), index S: substrate

resulting in strain difference

$$\Delta\varepsilon = \varepsilon_F - \varepsilon_S = (\overline{\alpha}_S - \overline{\alpha}_F)\Delta T + \varepsilon_{F,i} \tag{3.32}$$

which leads to a bowing of the two-layer compound (Figure 3.16).

The stress distribution in the bimorph element layer/substrate results from the clamping conditions, where forces F_F and F_S as well as moments M_S and M_F resulting from σ_F und σ_S over the cross section, balance exterior forces and moments. If thickness h_F of the thin film is substantially smaller than thickness h_S of the substrate, it can be assumed that there is an approximate constant stress σ_F in the entire layer. Equation (3.32) then results in

$$\sigma_F = \sigma_{F,\text{th}} + \sigma_{F,i} \sim \Delta\varepsilon. \tag{3.33}$$

That means that film stress σ_F consists of a thermal part which is caused by the difference of the thermal expansion coefficients, and an intrinsic part caused by structural changes.

This bimorph thin film/substrate can be described using the classical bending theory, but taking into consideration the following modifications[3]:

- Stiffness EI of a bending beam with width b results as average stiffness $\overline{EI}$ with

$$\frac{1}{\overline{EI}} = \frac{1}{b}\,\frac{12(E_F h_F + E_S h_S)}{E_F^2 h_F^4 + E_S^2 h_S^4 + E_F E_S h_F h_S (4h_F^2 + 6h_F h_S + 4h_S^2)}. \tag{3.34}$$

- The neutral axis (regarding the bending) is located at distance z_{NF} from the boundary area between layer and substrate:

$$x_{NF} = \frac{1}{2}\,\frac{E_F h_F^2 - E_S h_S^2}{E_F h_F + E_S h_S}. \tag{3.35}$$

[3]Gerlach, G. (1990) Mechanische Störeinflüsse an integrierten piezoresistiven Drucksensoren kleiner Nenndrücke. Dissertation B, TU Dresden.

- The internal bending moment of a bending beam with length l results from strain difference $\Delta\varepsilon$ in Equation (3.32):

$$M_\varepsilon = \frac{\ell}{2}\,\frac{E_F E_S h_F h_S (h_F + h_S)}{E_F h_F + E_S h_S}\cdot\Delta\varepsilon. \qquad (3.36)$$

- The average thermal expansion coefficient of the bimorph amounts to

$$\overline{\alpha} = \frac{\alpha_F h_F E_F + \alpha_S h_S E_S}{h_F E_F + h_S E_S}. \qquad (3.37)$$

For plate structures, analogous correlations can be used. Layer stresses in thin films can be experimentally determined from deflection u of a coated disk (e.g. Si wafer) with diameter d, which is bent into the section of a spherical surface with bending radius R (Figure 3.16) [TU92] [VÖLKLEIN00]:

$$\sigma_F = \frac{1}{6}\,\frac{E_S}{1-\nu_S}\,\frac{h_S^2}{h_F}\,\frac{1}{R} \qquad (3.38)$$

with

$$R = \frac{d^2}{8u} + \frac{u}{2} \approx \frac{d^2}{8u} \quad (\text{für } d \gg u). \qquad (3.39)$$

Equation (3.38) is also known as STONEY's equation. The bent coated wafer usually is measured with a surface profilometer or using interference optics for wafer with three-point suspension. Often, wafers are pre-bent when still uncoated (pre-bending radius R_0). In that case, $1/R$ has to be replaced by $(1/R - 1/R_0)$ in Equation (3.38). Analogous to Equation (3.39), R_0 can be determined through deflection u_0.

Example 3.6 Bowing of a 100 mm silicon wafer oxydized on one side

A standard silicon wafer with a diameter of $d = 100$ mm is thermally oxydized on one side (thickness $h_{SiO_2} = 1$ µm). According to [FRÜHAUF05], there is an internal stress in the silicon dioxide layer of $\sigma_{F,i} = -0.3$ GPa.

Using the values in Table 3.15, bowing u of a wafer that was previously completely plane according to Equation (3.38) and (3.39) will amount to:

$$u \approx \frac{1}{8}\frac{d^2}{R} = \frac{1}{8}d^2\cdot 6\cdot\sigma_F\cdot\frac{1-\nu_S}{E_S}\frac{h_F}{h_S^2}.$$

Using the values $d = 100$ mm, $h_S = 525$ µm (for typical characteristics of silicon wafers, see Table 3.10), $E_S = E_{Si} = 169$ GPa, $\nu_S = \nu_{Si} = 0.063$ (Table 3.15) and $h_F = h_{SiO_2} = 1$ µm, it results that $u = -45$ µm.

The negative sign means that the Si wafer bows upwards due to intrinsic tensions in SiO_2. The maximum bowing in the wafer centre is 45 µm. In addition to the internal stress, the difference in the thermal expansion coefficient would lead to a superimposed bowing. Due to $\alpha_{SiO_2} < \alpha_{Si}$ and $\Delta T < 0$ (cooling), it results $(\alpha_{Si} - \alpha_{SiO_2})\cdot\Delta T < 0$. This means the bowing occurs in the opposite direction of the intrinsic bowing.

Table 3.15 Selected mechanical and thermal characteristics of materials used in microsystem technology (according to [FRÜHAUF05] [PETERSEN82]

Material	YOUNG's modulus E in GPa	POISSON's ratio ν	Density ρ in g cm^{-3}	Linear thermal expansion coefficient $\alpha_{l,T}$ in 10^{-6} K^{-1}	Thermal conductivity λ in Wm^{-1}K^{-1}
Si (single crystal, (1 0 0) <1 1 0>)	169	0.063	2.3	2.3	157
Poly-Si	160	0.22	2.3	2.6	24
SiO_2	73	0.17	2.2	0.5	2.1
Si_3N_4	290	0.3	2.8	3.0	18
Al	70	0.34	2.7	23	236
Polyimide (PI)	3	0.3	1.1	50	0.3
Steel (unalloyed)	210	0.3	7.9	19	97

3.5.1 Silicon Dioxide, Silicon Nitride

Silicon dioxide SiO_2 and silicon nitride Si_3N_4 are dielectric and insulating. Therefore, they are used in microelectronics as insulating and passivating layers for electronic structures. Silicon dioxide can be produced using thermal oxidation directly on the surface of the silicon wafer (see Section 4.4.3), which requires a comparatively low technological effort. Thermal SiO_2 forms an amorphous, homogeneous, closed layer which is quasi defect-free. Due to simple manufacturing and outstanding layer characteristics of thermal SiO_2, silicon has become the dominating material in microelectronics.

Due to manufacturing, the transition from Si to SiO_2 shows a very small transition range with a nonstoichiometric composition SiO_x ($x = 0$ to 2) which affects the electronic characteristics of Si at the interface.[4] In addition to thermal oxidation, SiO_2 can also be produced by deposition on the silicon surface. These layers have not such a good homogeneity and a higher defect rate. SiO_2 is used in microsystem technology also for other functions:

- Sacrificial layer for the production of movable function elements in surface micromachining (see Section 4.10). This process uses the fact that SiO_2 has a different chemical resistance to specific chemical etchants. This way it can be selectively separated from a compound Si wafer/SiO_2/Si thin film.
- Etch resist for a three-dimensional structuring of silicon, e.g. for anisotropic, wet chemical etching of structures in silicon wafers (see Section 4.6.2). Regarding the etching solutions used there, SiO_2 has a substantially higher resistance than Si.

[4]Balk, P. (ed.) (1988) *The Si-SiO$_2$ System*. Amsterdam: Elsevier.

In many cases, instead of simple SiO_2 layers

- SiO_2/Si_3N_4 double layers or
- silicon oxide nitride (SiO_xN_y) layers are used.

Those first mentioned combine the advantages of SiO_2 (outstanding homogeneity and no defects) with those of Si_3N_4 (extraordinary chemical resistance). The linear thermal expansion coefficient of SiO_2 is smaller ($\alpha_{l,T} = 0.3 \ldots 0.5 \cdot 10^{-6}\ K^{-1}$), that of Si_3N_4 is larger ($2.5 \ldots 3.0 \cdot 10^{-6}\ K^{-1}$) than the coefficient of Si (see Table 3.15). If an appropriate layer thickness ratio is used, it is thus possible to create a mechanical, almost stress-free film stack on silicon. It has to be taken into account, though, that $\alpha_{l,T}$ of the layers depends considerably on the manufacturing conditions. It is possible to use silicon oxide nitride SiO_xN_y and achieve the same thermal expansion coefficient as for silicon. This requires an appropriate composition which depends on layer deposition conditions. In accordance with the expansion coefficient, the average refraction index of SiO_xN_y lies between that of SiO_2 $n = 1.44 \ldots 1.48$) and Si_3N_4 (1.9...2.5) [FRÜHAUF05].

Example 3.7 Stress compensation of SiO_2 and Si_3N_4 layers

The stress of a 1 μm thick silicon dioxide layer on a silicon plate is to be mechanically compensated using a silicon nitride layer. Both layers are separated at the same temperature. The stress is compensated once the thermal expansion of the double layer SiO_2/Si_3N_4 equals that of silicon. According to Equation. (3.37), it follows that:

$$\overline{\alpha}_{SiO_2+Si_3N_4} = \frac{\alpha_{SiO_2} h_{SiO_2} E_{SiO_2} + \alpha_{Si_3N_4} h_{Si_3N_4} E_{Si_3N_4}}{h_{SiO_2} E_{SiO_2} + h_{Si_3N_4} E_{Si_3N_4}} \stackrel{!}{=} \alpha_{Si}.$$

For optimum thickness $h_{Si_3N_4}$ of the silicon nitride layer and using the values in Table 3.15, it results that

$$h_{Si_3N_4} = h_{SiO_2} \cdot \frac{E_{SiO_2}}{E_{Si_3N_4}} \cdot \frac{\alpha_{Si} - \alpha_{SiO_2}}{\alpha_{Si_3N_4} - \alpha_{Si}} = 0.65\ \mu m.$$

For practical application, it has to be taken into account that SiO_2 and Si_3N_4 layers are usually deposited at different temperatures. In addition, the thermal expansion coefficient is often correlated to temperature which means that average values have to be used for the calculation.

3.5.2 Conducting Layers

For electrically conducting layers, mainly metals are used. Aluminum layers dominate as they can be produced cost-efficiently through evaporation.

In order to prevent that there are SCHOTTKY contacts instead of OHM's contacts, Al is usually alloyed with 0.5...1% Si (AlSi 0.5...1). This also prevents the indiffusion of Si atoms from the silicon material.

The adhesion of the conducting interconnections on the insulating layers, such as SiO_2, can be improved through adhesive layers with ions that can be incorporated into the amorphous network of the sublayers. For Al layers, often titanium is used.

If, after conductor deposition, there are other technological processes with high temperatures, it is necessary to use layer systems with conductor materials that prevent the diffusion of conductor metals into the semiconductor. This is often achieved through diffusion inhibiting intermediate layers. Common layer stacks consisting of adhesion layer/diffusion barrier/conducting metal are Cr/Au or Ti/Pt/Au.

3.5.3 Polysilicon Layers

Surface micromachining (see Section 4.10) mainly uses form and function elements consisting of thin silicon layers. A disadvantage is that thin Si layers cannot be deposited as single crystals, but only with a fine-crystalline structure (poly- or micro-crystalline, respectively). This is called poly-crystalline silicon or 'polysilicon'. A subsequent recrystallization requires a substantial technological effort and is therefore not commonly used in practical applications. The structure of poly-Si is heavily dependent on the deposition process. Certain deposition conditions often result in the formation of a certain texture, i.e. a specific distribution of the crystal orientation. The characteristics result from averaging the material parameters that depend on crystal orientation over the distribution function of crystallite. Here, the grain boundary also has to be taken into account.[5] Material characteristics that depend only little on crystal orientation (thermal expansion coefficient, thermal conductivity) are similar for both poly-Si and single crystalline silicon. Characteristics that heavily depend on crystal orientation (YOUNG's modulus and POISSON's ratio, see Table 3.7; piezoresistive coefficients) adopt values for poly-Si that lie between the extreme values of single crystalline silicon (Table 3.15, Appendix B).

3.6 COMPARISON OF MATERIAL CHARACTERISTICS

Table 3.15 summarizes important mechanical and structural characteristics of materials that are most commonly used in microsystem technology. The following conclusions can be drawn:

- Regarding YOUNG's modulus, silicon as structural material can be compared to steel that dominates mechanical engineering. Silicon has a smaller density and a better thermal conductivity, though. In addition, it has outstanding linear elastic characteristics and therefore lends itself particularly for micromechanical form and function elements.
- Polycrystalline and single crystalline silicon have the same density and a similar YOUNG's modulus. The anisotropic characteristics of single crystalline Si becomes obvious through the extremely small value of POISSON's ratio υ. Thermal conductivity of poly-Si is substantially reduced by the grain boundaries.
- Silicon dioxide (SiO_2) and silicon nitride (Si_3N_4) have a density similar to that of Si and a thermal expansion that is substantially smaller than that of metals and plastics. Due to the high layer deposition temperatures, the small difference between the thermal

[5] Schubert, D., Jenschke, W., Uhlig, T., Schmidt, F.-M. (1987) Piezoresistive properties of polycrystalline and crystalline silicon films. *Sensors and Actuators* 11, pp. 145–55.

expansion coefficients of SiO_2 or Si_3N_4, respectively, and Si is sufficient to produce mechanical deformations.

- Thin metal layers, e.g. consisting of Al or AlSi1, respectively, have a large thermal expansion and conductivity.
- Plastics, such as polyimide, have a very low YOUNG's modulus, which means that small forces can result in large deformations. In addition, they are good thermal insulators.

EXERCISES

3.1 Is it possible, on principle, for {1 0 0}, {1 1 0} or {1 1 1} silicon wafers to form inclined planes {1 1 1} that are situated vertically on the wafer surface?

3.2 During anisotropic deep etching of silicon with alkaline etching solutions, mainly slow-etching {1 1 1} lateral planes are formed that delimit the cavities. Show for the wafer orientation in exercise 3.1 what an etching cavity with boundaries of all possible six {1 1 1} planes would look like.

3.3 Calculate YOUNG's modulus for silicon in direction <1 1 1>.

3.4 Calculate the failure probability of a piezoresistive pressure sensor with a maximum tensile stress of 0.058 GPa; 0.58 GPa; 5.8 Gpa in the pressure plate.

3.5 Determine the change over time of the stress in the glass body in Figure 3.12. It is exposed to a deformation step of $\underline{u} = u_0 \cdot 1/s$.

3.6 Calculate layer stress in a 5 μm thick polyimide layer of a standard 150 mm Si wafer. Bowing is 10 μm.

REFERENCES

[BÜTTGENBACH91] Büttgenbach, J. (1991) *Mikromechanik*. Stuttgart: B.G. Teubner Verlag.

[FRÜHAUF05] Frühauf, J. (2005) *Werkstoffe der Mikrotechnik*. Leipzig: Fachbuchverlag.

[HÄNSEL96] Hänsel, H., Neumann, W. (1996) *Physik. Band IV: Moleküle und Festkörper*. Heidelberg: Spektrum Akademischer Verlag.

[HULL99] Hull, R. (Hrsg.) (1999) *Properties of Crystalline Silicon*. EMIS Datareviews Series No. 20. London: INSPEC.

[IVERS04] Ivers-Tiffée, E.; von Münch, W. (2004) *Werkstoffe der Elektrotechnik*. 9. Aufl. Stuttgart: B.G. Teubner.

[KIESEWETTER92] Kiesewetter, L., Houdean, D., Löper, G., Zhang, J.-M. (1991) Wie belastbar ist Silizium in mikromechanischen Strukturen? *F&M Feinwerktechnik, Mikrotechnik, Meßtechnik* 100, 6, pp. 249–54.

[MESCHEDER04] Mescheder, U. (2004) *Mikrosystemtechnik*. 2. Aufl. Stuttgart, Leipzig: B.G. Teubner.

[NISHI00] Nishi, Y.; Doering, R. (2000) *Handbook of Semiconductor Manufacturing Technology*. New York, Basel: Marcel Dekker.

[PAUFLER86] Paufler, P. (1986) *Physikalische Kristallografie*. Berlin: Akademie-Verlag.

[PETERSEN82] Petersen, K. (1982) Silicon as a mechanical material. *Proceedings of the IEEE* 70, pp. 420–57.

[TU92] Tu, K.-N.; Mayer, J.W.; Feldmann, L.C. (1992) *Electronic Thin Films Science for Electrical Engineers and Materials Scientists*. New York: Macmillan Publishing Co.

[VÖLKLEIN00] Völklein, F.; Zetterer, T. (2000) *Einführung in die Mikrosystemtechnik*. Wiesbaden, Vieweg-Verlag.

4

Microfabrication

4.1 OVERVIEW

Microsystem technology is characterized by the fact that extremely miniaturized components with integrated electrical, mechanical and other functions are produced using three-dimensional processing of materials from semiconductor technology. It applies the principal technological methods of microelectronics. Due to the programmability of the components (processors, memory circuits), microelectronics usually works with much larger batches and it is therefore easier to finance technological development. As opposed to components in microelectronics, microsystem technology does not only use the near-surface range of the silicon wafer, but also the depth of the wafer for manufacturing microsystems. The goal is therefore to develop – based on traditional semiconductor technology – technologies that lead to three-dimensional shapes.

Due to the integration of electronic, mechanical and other functions, the production of microsystems will also require the application of traditional microelectronic processing steps which are completed by three-dimensional shaping. We will illustrate this with the production of piezoresistive pressure sensors.

Example 4.1 Manufacturing of piezoresistive pressure sensors

The goal is to manufacture the piezoresistive absolute pressure sensor shown in Figure 4.1. The pressure sensor should have the following characteristics (see Section 7.2.5):

- rectangular pressure membrane of n-silicon;
- piezoresistors as p-doped silicon ranges;
- metal interconnects and bonding pads that are electrically insulated from the Si chip.

The following conclusions can be drawn for the manufacturing process:

- The dimensions of the pressure sensor chip lie in the range of a few square millimeters. Therefore during manufacturing, many sensor elements are simultaneously produced on one wafer and singled only at the end.

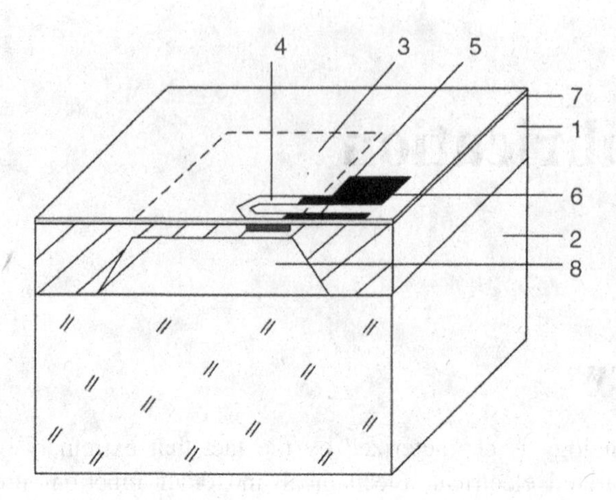

Figure 4.1 Piezoresistive absolute pressure sensor (cross-section not to scale): 1 silicon sensor chip, 2 glass substrate, 3 silicon pressure membrane, 4 piezoresistor, 5 interconnect, 6 bonding pad, 7 insulation layer (SiO_2), 8 etch groove

- The piezoresistor is doped on the upper side of the pressure membrane, in the thinned part of the sensor chip. It is doped starting from the front face, whereas the pressure membrane is thinned from the back. This double-sided processing requires a precise alignment of front and backside through double-sided lithography (see Section 4.3). A displacement of front and back would result, for instance, in the resistor not being located at that edge of pressure membrane where the largest mechanical stresses occur, but instead in the thicker chip area. This would mean that the sensitivity of the pressure sensor is reduced (see Example 7.8).
- The thinning of the pressure membranes results in a reduced mechanical stability of the silicon wafer in general and, specifically, of the thinned areas. The anisotropic deep etching that produces the desired rectangular form of the membrane (see Section 4.6) is therefore only carried out at the end of manufacturing process.
- The silicon wafer that contains the sensor chips is entirely bonded to a glass wafer (see Section 4.8) before the chips are separated.

Figure 4.2 shows a possible sequence of the manufacturing process.[1]

The original material is a silicon wafer with specific characteristics and the desired crystallographic orientation. For piezoresistive pressure sensors, mainly n-conducting silicon with a (1 0 0)-oriented surface is used. Both the edges of the membrane and the longitudinal axis of the piezoresistor are oriented in direction <1 0 0>. Due to this crystallographic orientation, anisotropic etching (see Section 4.6) can be used to shape rectangular membranes. For p-conducting piezoresistors that are doped into n-Si, it allows a high sensitivity of the sensor (see Section 7.2.5).

[1]As opposed to the general rules for Technical Representation, microsystems are usually represented with the direct plane of section and other visible unit edges are not shown.

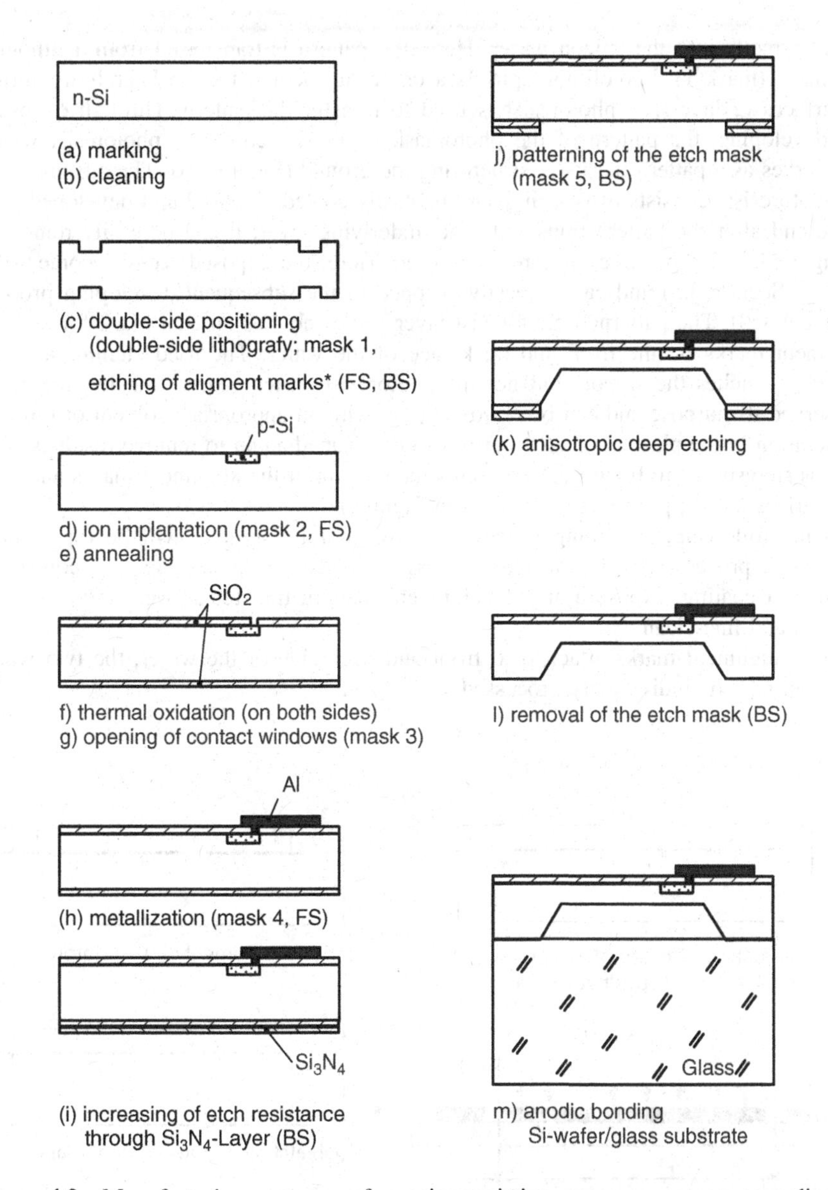

Figure 4.2 Manufacturing process of a piezoresistive pressure sensor according to Figure 4.1. FS front side, BS back side
* The following figures do not include the alignment marks. Alignment mark production is discussed in Figure 4.3

The Si wafer is first marked, i.e. it receives a continuous number, and then cleaned in order to prevent particles or other contaminations from interfering with the generation of structures (Figure 4.2b). After that, double-sided lithography is be used to imprint alignment marks on both faces of the wafer in order to be able to achieve a precise and unique alignment of front and back (Figure 4.2c). In this case, the alignment marks are etched directly into the silicon wafer. Here, the pattern is transferred from a lithography mask (mask 1). Photolithography is a universal method for the lateral structuring of surfaces or layers. A photoresist is used to transfer the pattern. Through exposure and developing, the pattern of the photomask is transferred to the photoresist which then serves as a pattern mask for structuring the ground (Figure 4.3). The advantage of the photoresists consists of that they can be easily coated, exposed and developed and, at the end after the pattern transfer to the underlying layer, it can be easily removed. In Figure 4.3c2, a positive photoresist is used. Here, the exposed areas become soluble (see Section 3.5) and can be easily stripped in the subsequent developing process (Figure 4.3c3). The patterned photoresist layer serves now as an etch mask for etching alignment marks on the front and back face of the wafer. The used etching solution selectively etches the silicon, but not the photoresist. After pattern transfer, the resist has served its purpose and can be removed (e.g. with an appropriate solvent or through incineration in the plasma). The wafer has to be cleaned again to remove residues. The process step shown in Figure 4.2c presents the etching of the alignment marks and thus summarizes the six process steps shown in Figure 4.3.

In the following, patterning processes using photolithography will be represented as a single process step. In the real process sequence, the patterning step comprises photoresist coating, exposure and development, pattern transfer as well as photoresist removal and fine-cleaning.

With alignment marks attached to front and back side of the wafer, the two wafer sides can now be individually processed.

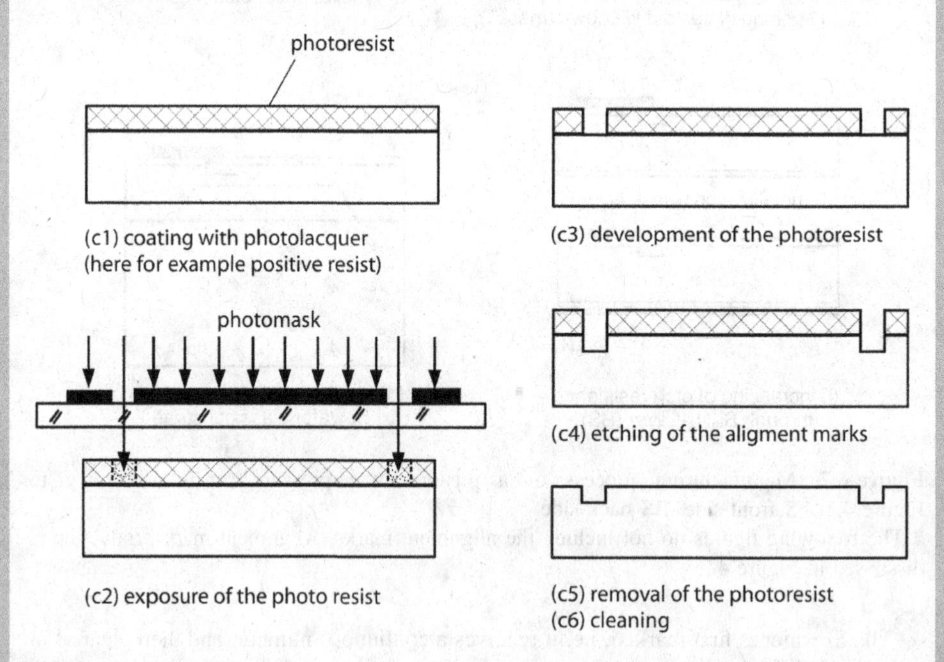

Figure 4.3 Photolithographic process for generating alignment marks according to Figure 4.2c

Ion implantation is used to create p-conducting piezoresistors in n-conducting silicon wafers (Figure 4.2d, see Section 4.7). The implantation mask is again a photoresist layer patterned by a photomask. In order to anneal defects caused by collisions during ion implantation, the wafer subsequently undergoes a heat treatment. Now, the wafer undergoes thermal oxidation (Figure 4.2f). The result is a homogeneous, quasi defect-free, electrically insulating SiO_2 layer on both sides (see Section 3.6). On the front face, this layer is used as an insulation layer between interconnects for contacting the piezoresistors and the silicon chip. On the back face, it is used as an etch mask for thinning the membrane in the Si wafer. In the contact area of piezoresistor and interconnect, the front face oxide has to be opened (Figure 4.2g). This requires another photolithographic process as well as an SiO_2 etching. The next step is the full-area metal layer deposition and its structuring (Figure 4.2h). As mentioned above, anisotropic wet chemical etching (e.g. KOH etching, see Section 4.6) is used to thin the membrane starting from the back face of the wafer. The thickness of the thermal oxide layer on the back face is not sufficient for common wafer thickness in the range of 500...700 μm (see Example 4.6). Therefore, the etch mask has to be reinforced. Usually, silicon nitride Si_3N_4 is used as it is affected by the etching solutions to a much smaller degree than SiO_2 (Figure 4.2i). The structured SiO_2/Si_3N_4 double layer now serves as etch mask for the anisotropic etching process (Figure 4.2k). Anodic bonding is used to bond the silicon wafer with a glass substrate (see Section 4.8). For this, the SiO_2/Si_3N_4 etch mask on the back of the wafer has to be removed again (Figure 4.2l). After the bonding process (Figure 4.2m), the sensor elements are singled and are transformed into complete sensors using bonding and packaging technology (see Chapter 5).

The manufacturing of piezoresistive pressure sensors in Example 4.1 comprises 13 different steps, five of which are lithography steps which themselves consist of several substeps. The manufacturing of complex microsystems may include a total of hundreds of different process steps, especially when integrating microelectronic devices.

The lithography steps are particularly important for the yield of the manufacturing processes of microsystems as especially during the pattern transfer process, functional defects can be caused. Therefore, a minimum of lithography steps would have the advantage of reduced technological requirements and higher manufacturing yield. But as Example 4.1 showed, the manufacturing process of microsystems consists of a large number of diverse process steps (Table 4.1) and each of them has to be carried out reliably and successfully, and none of them must interfere negatively with another step. Many of the procedures are standard processes of microelectronics; others are especially designed for the production of three-dimensional micromechanical function and form elements.

The latter techniques particularly comprise special techniques such as:

- two-sided alignment for adjusting mechanical and function elements on front and back side of a silicon wafer;
- deep-etching for generating three-dimensional structures;
- generation of insulating layers in order to geometrically limit three-dimensional etchings;
- bonding techniques for structuring micro-mechanical function elements.

Table 4.1 Microfabrication techniques

Wafer processes
• Cleaning, lithography • Film deposition techniques • Growth: Oxidation, epitaxy Deposition: CVD (Chemical Vapor Deposition), PVD (Physical Vapor Deposition: sputtering, evaporation) • Layer-removal techniques (etching) • Doping techniques (ion implantation, diffusion) • Bonding techniques (wafer bonding)[a] • Techniques of three-dimensional shaping
Assembly processes
• Separation techniques • Bonding techniques (die bonding)[a] • Bonding techniques • Encapsulating, coating • Final test

[a] In general, for connecting silicon wafers (wafer bonding) and diced-up chips (die bonding), the same interconnection techniques are used.

4.2 CLEANLINESS DURING PRODUCTION

Similar to the production of integrated circuits, the manufacturing of microsystems and their components constitutes a complex technical process consisting of many individual steps. Any mistake during an individual process can affect function or form elements and result in failure of the entire microsystem. In that context, the photolithographic pattern-transfer processes are particularly critical moments as any mistake during that step can lead directly to failing functionality. Surface quality is particularly important. On the one hand, it is affected by surface morphology (e.g. roughness, flatness) which is determined by the wafer production (see Table 3.10). On the other hand, the surface must not be impacted by too many particles or chemical contamination.

In order to reach a high yield, there are two prerequisites [ALBERS05]:

- Minimum contamination due to particles: Similar to the production of microelectronic components, microsystem components are manufactured in clean room where the air contains a limited number of particles that must not exceed a specific size (Figure 4.4).
- Minimum chemical contamination: Special cleaning procedures are used to achieve that.

4.2.1 Clean Room Technology

Clean rooms ensure clean ambient conditions for the production of microsystems. They have the following functions:

- generating filtered, i.e. low-particle, air;

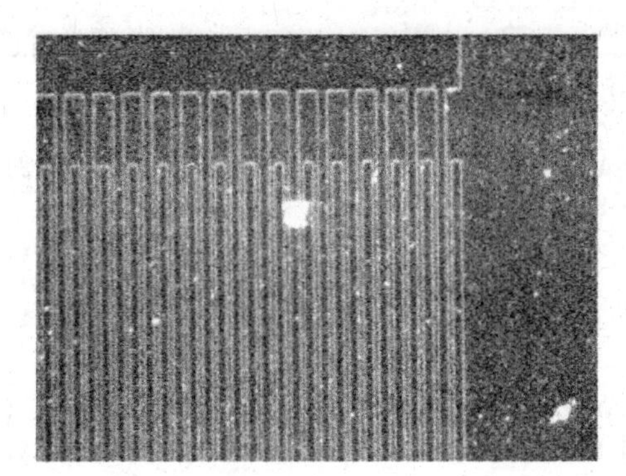

Figure 4.4 Examples of how a surface is affected by particles, (a) short-circuit, (b) disruption. Reproduced by permission from the Dresden University of Technology, Germany

Table 4.2 Emission of particles with a particle diameter of more than 0.5 μm for different activities (per minute)

Activity	Normal cloths	One-piece protective cloths incl. head, mouth and nose cover
Motionless sitting	$3 \cdot 10^5$	$7 \cdot 10^3$
Head movement	$6 \cdot 10^5$	$1 \cdot 10^4$
Body movement	$1 \cdot 10^6$	$3 \cdot 10^4$
Slow walking	$3 \cdot 10^6$	$5 \cdot 10^4$
Fast walking	$6 \cdot 10^6$	$1 \cdot 10^5$

- providing sufficiently clean media (process gas, compressed air, cooling water, power supply, vacuum);
- disposal (toxical exhaust air, sewage water).

Experience shows that particles exceeding 1/10 of the minimum structural dimensions can lead to failure in components. Microsystem technology works with typical minimum lateral structural sizes of 5 μm. Therefore, maximum particle size is 0.5 μm. Table 4.2 gives an overview of particle emissions related to human activity.

(a) Yield

When manufacturing microsystem components, the yield is to a large extent affected by particles. Yield Y is defined as

$$Y = \frac{\text{Number of good components per wafer}}{\text{Total number of components per wafer}}. \tag{4.1}$$

The major part of yield loss is caused by randomly distributed defects due to lithographical patterning. Defect density D is the number of defects per wafer. When producing numerous identical components on a wafer, component area A and number N of the critical mask level are decisive for total yield Y [NISHI00]:

$$Y = (1 + A \cdot D)^{-N} \tag{4.2}$$

and with $A \cdot D \ll 1$

$$Y = (1 - A \cdot D)^{N}. \tag{4.3}$$

Equation (4.2) assumes that the probability of one or more defects per chip follows a POISSON probability distribution and that all mask levels have the same defect density. The latter, however, applies only to a limited degree.

Example 4.2 Yield for the production of piezoresistive sensors

The pressure sensor in Figure 4.1 is assumed to have a chip area of $A = 4 \times 4\ \text{mm}^2$. According to Figure 4.2, manufacturing comprises $N = 5$ lithographical steps. Defect density is assumed to be $D = 10^{-3}\ \text{mm}^{-2}$ (one defect per $10\ \text{cm}^2$). Applying Equation (4.2), the yield is

$$Y = (1 + AD)^{-N} = (1 + 16\ \text{mm}^2 \cdot 10^{-3}\ \text{mm}^{-2})^{-5} = 92\,\%.$$

Due to $A \cdot D = 1.6 \cdot 10^{-2} \ll 1$, Equation (4.3) results in the same approximate result.

If we want to integrate the electronic circuitry on the same sensor chip, additional mask levels are required. For a CMOS process, this would mean up to seven mask levels [GARDNER01]. In that case, yield is reduced to

$$Y_{\text{total}} = (1 + AD)^{-N_{\text{total}}} = (1 + 16\ \text{mm}^2 \cdot 10^{-3}\ \text{mm}^{-2})^{-(5+7)} = 83\,\%.$$

This is the reason why microsystem chip and electronic circuitry for signal processing are often manufactured separately and then are interconnected through packaging techniques (see Chapter 5).

(b) Clean room

The following measures are necessary to reduce particles:

- operation of a clean room that is basically free of particles;
- minimized introduction or generation of particles due to equipment, devices, operating staff and supply media;
- continuous monitoring of particle density in the clean room.

ISO 14644-1 or US Federal Standard 209 (Table 4.3) classify the air purity of clean rooms.

The production of very-large-scale integrated circuits requires class 1 clean rooms or better whereas clean room classes of 10 to 10 000 are sufficient for the manufacturing of microsystem technology (Table 4.4).

Table 4.3 Definition of clean room classes according to ISO 14644-1 and to US Federal Standard FED STD 209 D and E; the measures given is the maximum number of particles per cubic meter of air

Class			Particle size		
ISO	FED 209 D*	FED 209 E	0.1 μm	0.3 μm	0.5 μm
ISO 1			10		
ISO 2			100	10	4
ISO 3	1	M1.5	1000	102	35
ISO 4	10	M2.5	10000	1020	352
ISO 5	100	M3.5	100000	10200	3520
ISO 6	1000	M4.5	1000000	102000	35200
ISO 7	10000	M5.5			352000
ISO 8	100000	M6.5			3520000
ISO 9					35200000

* The number for the clean room classes according to US FED STD 209D results from the number of admitted particles with a size of 0.5 μm per cubic foot (1 $ft^3 = 2.84 \cdot 10^{-2} m^3$). 35200 particles/$m^3$ correspond to e.g. 1000 particles/ft^3.

Table 4.4 Clean room classes required for various process steps of microsystem technology

Process step	Clean room class
Photolithography	10... 100
Deposition processes	100... 1000
Wet chemical processes (etching, cleaning)	100... 1000
Plasma processes (plasma cleaning, etching)	1000... 10000
Analytics and measuring equipment	1000... 10000
Service areas (media supply, vacuum pump equipment, electronics equipment, exhaust gas disposal)	1000... 10000

There are different options for creating clean rooms.

- The most common method so far has been to try and keep critical areas with a high air quality very small and separate the working area into areas with different degrees of air purity. This approach often uses comblike structures. Figure 4.5a shows the separation into a white area for wafer and chip processing and a gray area (service area, see Table 4.4). White areas have a slight overpressure (10...20 Pa) in relation to the gray areas in order to reduce particle introduction from the outside. The same relation exists between gray areas and outer areas. Laminar flow boxes have proved to be successful in providing such particle absence even in rooms with substantially poorer air purity.
- More recent clean rooms use large, pillar-free rooms with laminar flows between ceiling and floor (ballroom layout). Here, fan units are integrated into the ceiling; the floor has perforations and allows the laminar flow to flow in (Figure 4.5b). Under the floor, there are installation lines and underneath, there are service rooms. Air quality areas

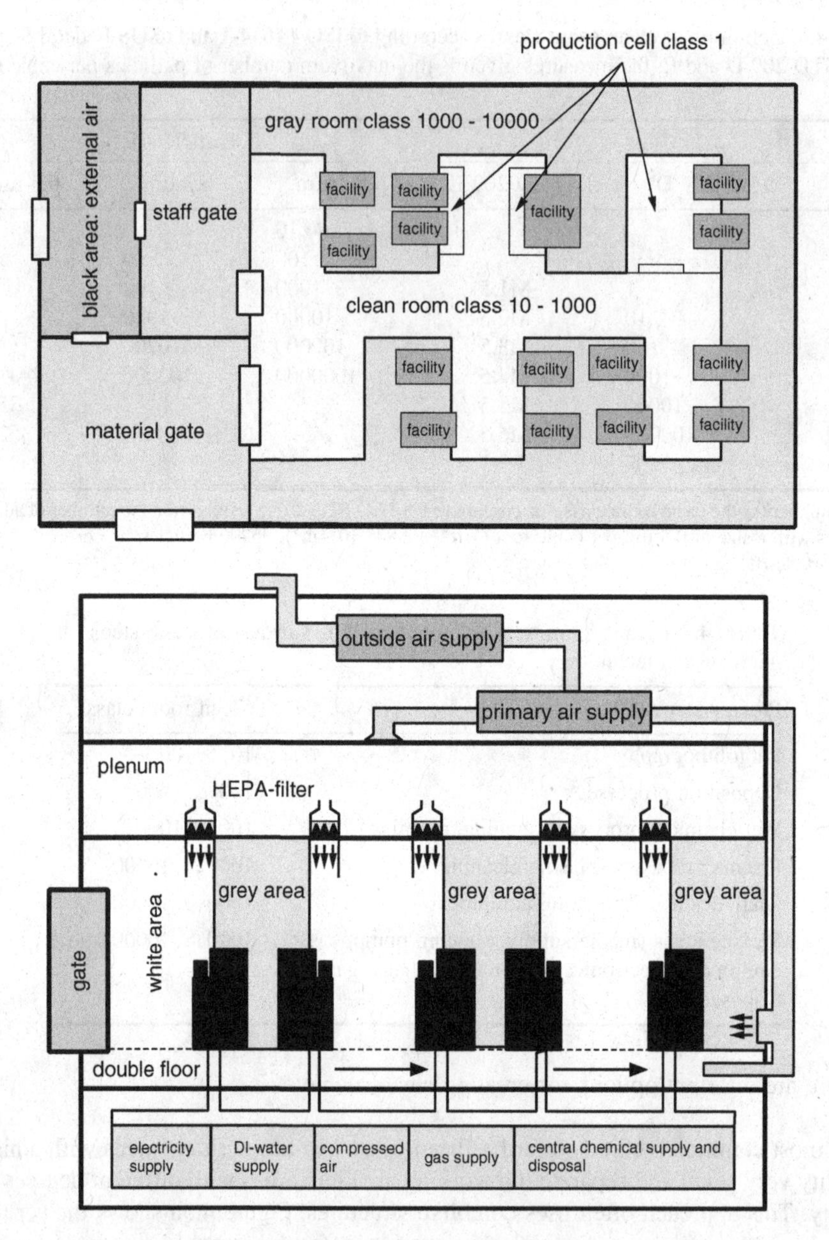

Figure 4.5 Typical structure of clean rooms for the manufacturing of microsystem technology: (a) comb structure; b) ballroom structure (according to [ALBERS05])

are determined by the number and density of air filter units installed in the ceiling. With a sufficient laminar flow, it is possible to create areas complying with class 1 specifications of clean rooms, without having to create any spatial separation. This concept has the advantage that the clean room can be used in a very flexible way due to the variable arrangement of air filters in the ceiling.

All clean room concepts require clean room cloths and mouth cover in order to reduce particle introduction by operating staff (see Table 4.2).

4.2.2 Wafer Cleaning

High yields require to remove any contamination on the wafers prior to each process step. Contaminations are caused by

- particles from the air;
- contaminations from process media; and
- contaminations from previous processing steps.

The quality of the clean room deals with the first contamination cause. Contamination by process media is reduced by using ultra pure gases with a 99.999 % (often called 5N) or 99.9999 % (6N) degree of purity.

The most important contaminations caused by the manufacturing process are again particles, organic and metal substances as well as natural oxide layers on the silicon surface. Table 4.5 shows common solvents for the removal of such substances. They contain alkaline solutions and acids, hydrogen peroxide H_2O_2 as a strong oxidation medium as well as de-ionized water.

A complex fine-cleaning regime has to consist of a sequence of cleaning steps with different solvents for the contaminations that have to be removed. Between each cleaning bath, the wafers are rinsed with de-ionized water. At the end, the wafers are dried in a centrifuge.

A well-proven standard fine-cleaning regime is the cleaning procedure named after W. KERN[2] or former company RCA, respectively (Table 4.6). The cleaning regime is

Table 4.5 Common solutions for wet chemical cleaning [SCHADE91]

Solvents	Typical process parameters	Removable substances
$NH_4OH/H_2O_2/H_2O$	70...80 °C, 5...10 min	Organic contaminations (e.g. residual resist, grease, wax), metals (Cu, Ag, Ni, Co, Cd)
HNO_3 (70 %)/H_2O	80...100 °C, 7...10 min	Organic and inorganic substances
HF/Acetone	1 min	Oxidation products
$HCl/H_2O_2/H_2O$	80 °C, 10 min	Metals (Au, Cu, Cr, Fe, Na^+)
HF (10...48 %)/H_2O	23 °C, 5...30 s	Natural SiO_2

NH_4OH ammonium base; H_2O_2 hydrogen peroxide; HNO_3 nitric acid; HCl hydrochloric acid; HF hydrofluoric acid

[2] Kern W., Puotinen D.A. (1970) Cleaning solutions based on hydrogen peroxide for use in silicon semiconductor technology, *RCA Review* 31, pp. 187–206.

Table 4.6 RCA cleaning procedure [KERN93]

Step	Solvents	Temperature	Duration	Effect
1	$H_2SO_4 + H_2O_2$ (4:1)[a]	120 °C	10 min	Organic substances
2	DI–H_2O	AT	1 min	Rinsing
3	NH_4OH (29 %) + H_2O_2(30 %) + DI–H_2O (1:1:5)[b]	75...80 °C	10 min	Particles
4	DI–H_2O	AT	1 min	Rinsing
5	HCl (37 %) + H_2O_2(30 %) + DI–H_2O (1:1:6)[c]	75...80 °C	10 min	Metal ions
6	DI–H_2O	AT	1 min	Rinsing
7	HF + H_2O (1:50)	AT	15 s	Oxide
8	DI–H_2O	AT	1 min	Rinsing

[a] Piranha solution; [b] RCA SC1 solution;[c] RCA SC2 solution; SC Standard Clean; DI de-ionized; AT Ambient temperature.

monitored by controlling the specific resistance of the ultra pure water bath (specific resistance $\rho > 18\ \text{M}\Omega \cdot \text{cm}$).

4.3 LITHOGRAPHY

4.3.1 Principle

The lithographic process is used to pattern layers and substrates by selective material removal. Figure 4.3 has provided an example of the pattern transfer from a mask to a photoresist layer which is then used as a mask for patterning the underlying layer. Depending on whether the photoresist layer is a positive or negative resist (see Section 3.5) the final pattern generated in the layer either corresponds to the pattern of the mask or is complementary to it (Figure 4.6).

The most important step of the pattern transfer is the exposure as during this step the pattern reaches the wafer for the first time. In lithography, the following methods of exposure are used:

- electromagnetic radiation: light (preferably in the UV range), X-ray;
- particle beams: ion or electron beam.

Due to its wavelength, UV light is of limited value regarding the fine details of the reachable structural dimensions. However, it can expose large areas with only little equipment requirements. This way, complete wafers can be efficiently and productively exposed. It is therefore even in very-large-scale integrated microelectronics the most important exposure procedure.

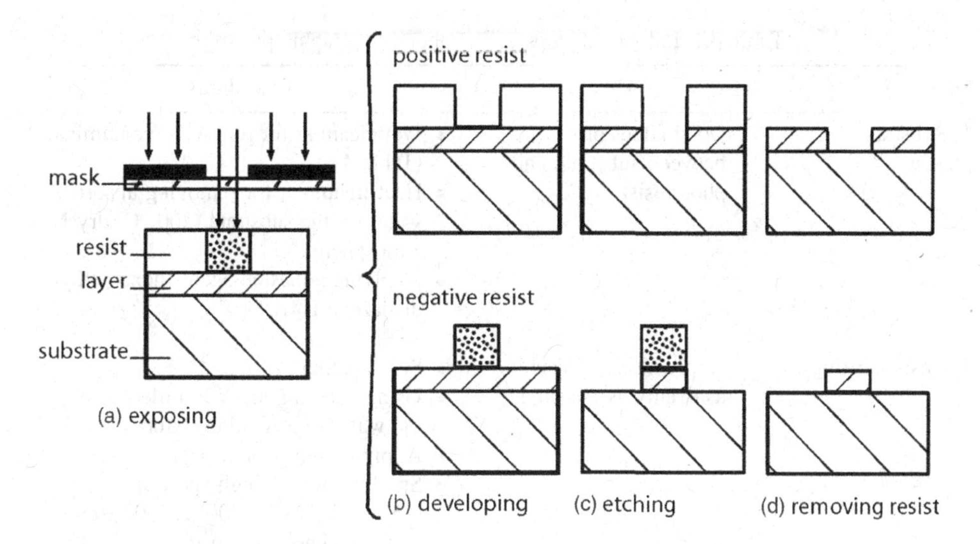

Figure 4.6 Pattern transfer using lithography

Due to its much smaller wavelength, X-ray allows pattern resolutions down to the nanometer range. It requires sophisticated instrumentation, though. Currently, it is only used for exposing very thick resist layers that require high-energy radiation (e.g. LIGA technique see Section 4.12).

Electron beam techniques also allow very small pattern sizes, but requires a vacuum. Their application focuses on the production of photomasks which have to fulfill higher requirements regarding the precision of the pattern transfer.

4.3.2 Lithographic Process

Figure 4.6 shows an abbreviated version of the photolithographic process which consists of numerous steps. Table 4.7 provides an overview of these steps.

Lithography can be carried out as contact, proximity or projection exposure.

Contact lithography means that the mask is pressed against the photoresist on the wafer surface. This improves pattern resolution as compared to proximity lithography where there is a slight gap of 10 . . . 40 μm between substrate and mask.

The in-contact technique has the following advantages and disadvantages:

- very precise pattern transfer down to the μm range (for UV exposure);
- low durability of the mask due to mechanical wear;
- higher defect density due to particles between mask and resist layer;
- pattern distortion due to possible bending of mask and substrate.

It is a cost-efficient procedure for small batches. However, for large series the disadvantages prevail due to the low lifetime of the masks.

Table 4.7 Individual steps of the photolithographic process

Step	Aim	Procedure
1. Substrate priming	Good adhesion between substrate and photoresist	• Wet cleaning for removing contamination (Table 4.6) • Heat treatment for removing absorbed water on the substrate (200 °C, dry N_2 atmosphere) • Applying adhesion agent (hexamethyle disilazane HMDS)
2. Resist coating	Applying the resists as evenly as possible	• Spin coating • Using vacuum chuck in order to mount the wafer to a rotating fixture • Applying the photoresists • Spinning with a high speed n: (15…30 min at 2000…5000 r/min) resist thickness $d \sim n^{1/2}$ • Spray deposition (for three-dimensional topography of the substrate surface)
3. Baking (softbake, prebake)	Generation of a thin resist film, maximum removal of solvents from resist	• In a convection oven (30 min, 70…90 °C) • On a hot metal plate (hotplate, 1 min, 100…120 °C)
4. Aligning and exposure	Transferring the mask pattern to the resist	• Adjusting the mask aligner • In-contact, proximity or projection printing
5. Post-exposure bake	Reducing the effect of an inhomogeneous exposure dose due to standing waves in the resist	• Thermal treatment (30…60 s, 110…120 °C, hotplate)
6. Development	Formation of the pattern in the photoresist	Removing the soluble areas of the photoresist (wet chemical etching, reaction process depends on the resist material)
7. Hardbake	Evaporating residual solvent, improving resist adhesion on the wafer for the following etching or implantation processes	Thermal treatment in the convection oven (ca. 30 min) or on a hotplate (1 min): • Positive resist: 100…120 °C • Negative resist: 120…150 °C
8. Inspection	Checking the quality of the resist mask	Using incident light microscope for evaluation of undesired residual resist, possible vertical resist edges, position of resist pattern

Table 4.7

Step	Aim	Procedure
9. Process step	Transferring the resist pattern to the underlying film or substrate	By • etching, • doping or • lift-off of film or substrate
10. Removing the resist	Removing the resist pattern after pattern transfer by stripping (resist stripping)	By • wet oxidation in acid or alkaline solutions (e.g. CrO/H_2SO_4 chromosulphuric acid at 70 °C) • solvents (e.g. in acetone, chloroform, partially assisted by ultrasound) or • plasma incineration (oxygen plasma, 0.1 ... 500 Pa in HF plasma reactor)

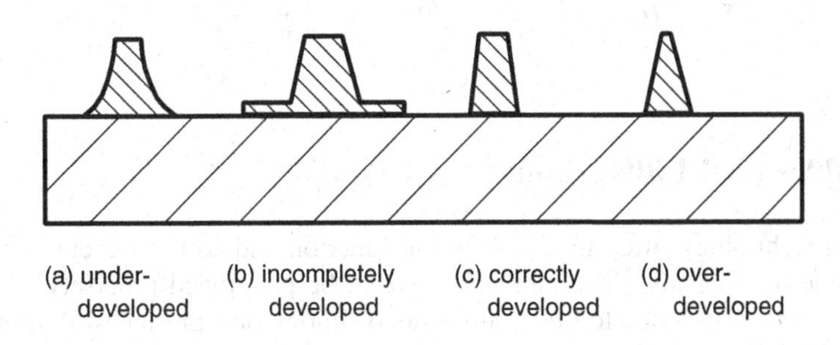

Figure 4.7 Cross-sections of resist patterns after exposure and development

During projection exposure, only parts of the wafer are exposed by projecting the mask patterns in a miniaturized way on the wafer. It is mainly used for highly integrated microelectronic components, and only rarely in microsystem technology. It is important for the optical exposure to remember that light has wave characteristics und is bent and dispersed at the edges of the mask. Thermal treatment during post-exposure bake (PEB) prevents the formation of standing waves in the photoresist, which can lead to undulated sidewalls. Optimum conditions of the resist development result in basically vertical resist edges (Figure 4.7).

4.3.3 Minimum Structurable Line Width

It was mentioned above that during exposure there are scattering effects due to the gap between mask and photoresist. At proximity lithography, scattered light reaches even unexposed resist areas due to the proximity gap. For in-contact exposure, the thickness of the photoresist is decisive for scattering. The diffraction-related minimum structurable feature size L is

$$L = 1.5\sqrt{\lambda \cdot (G + d/2)}. \tag{4.4}$$

with wavelength λ of light, proximity distance G and resist thickness d. For in-contact exposure $G = 0$.

Microsystem technology mainly uses mercury high-pressure lamps at wave lengths of 436 nm (g-line) or 365 mn (i-line); less frequently excimer laser (248 nm for KrF laser, 193 nm for ArF laser) is used.

Example 4.3 Pattern resolution during exposure

A photoresist layer with a thickness of 1 μm is exposed with a wave length of $\lambda =$ 436 nm. The minimum line width that can be reached is

$$L = 1.5\sqrt{\lambda \cdot d/2} = 1.5\sqrt{0.436\ \mu\text{m} \cdot 0.5\ \mu\text{m}} = 0.70\ \mu\text{m}.$$

If exposure takes place at the bottom of a 500-μm-deep etch groove in the patterned silicon wafer, for example, the effect is a proximity distance G which results in a minimum structurable line width of only

$$L = 1.5\sqrt{\lambda \cdot (G + d/2)} = 1.5\sqrt{0.436\ \mu\text{m} \cdot 500.5\ \mu\text{m}} = 22.2\ \mu\text{m}$$

4.3.4 Double-sided Lithography

Microsystem technology often places different function and form elements on the upper and lower side of the wafer. The elements have to be positioned precisely in relation to each other. Commonly, one step of double-sided lithography produces alignment marks on front and back side (see Figure 4.2c). Subsequently, both sides can be processed individually. Figure 4.8 shows the principle of double-sided lithography. The alignment precision between front and back side lies in the range of micrometers.

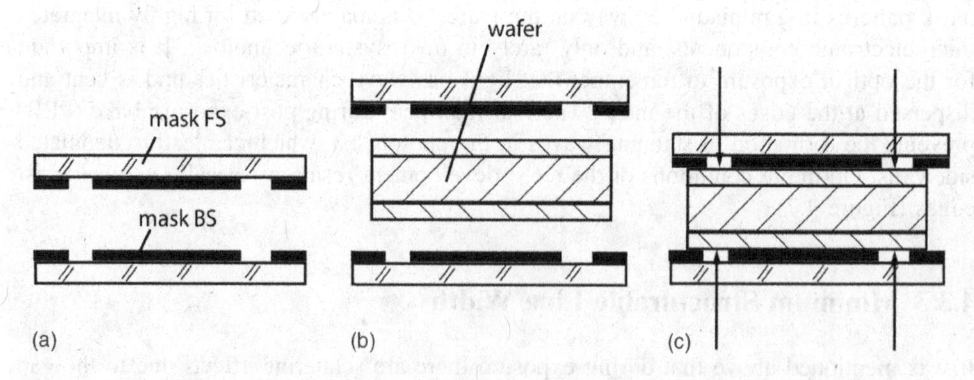

Figure 4.8 Process steps in double-sided lithography: (a) taper error compensation, alignment of both masks for FS (front side) and BS (back side); (b) the masks part in base position, the transport sledge keeps the wafers – that have resist on both sides – at the alignment distance; the wafer is aligned in relation to the lower mask; (c) the wafer is put on the lower mask, the upper mask moves to the pre-selected exposure distance; exposure

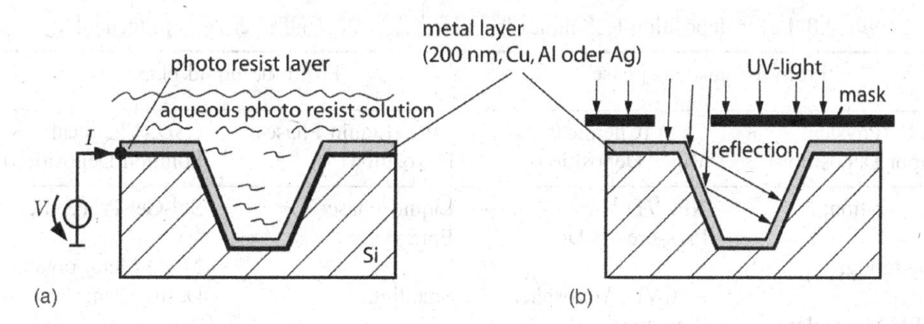

Figure 4.9 Photolithography in structures with deep profiles: (a) electro-deposition of photoresist; (b) exposure

4.3.5 Lithography in Structures with Deep Profiles

In microsystem technology, there are often three-dimensional patterns with complicated forms (e.g. deep etchings). Pattern transfer in such structures with deep profiles is complicated. This does not only refer to the conformal resist deposition in the structure, but also to their exposure. In addition, it is only possible to produce very wide lines in such deep etchings (see Example 4.2).

Figure 4.9 shows an option of how lithography can be used for producing conducting paths and other function layers in deep-etched surfaces.[3] As the resist cannot be applied conformally to the deep etch grooves by spin-coating or spraying, electro-deposition is used for depositing the resist. A water-soluble positive resist is used for the photoresist layer (PEPR 2400, Shipley Ltd.). It grows onto a thin metal layer that has been previously applied and serves as an anode. The thickness of the photoresist is self-limiting as current density, and subsequently, deposition rate decreases with increasing resist thickness. Reflexions on sloped walls may cause problems as they result in undesirable ghost images on parts that are not supposed to be exposed. TM-polarized light can diminish this effect.

4.4 THIN-FILM FORMATION

In microsystem technology, thin films have a variety of functions (see Sections 3.1 and 3.6). Mostly, they are amorphous, micro- or poly-crystalline. They have other characteristics than bulk materials, with manufacturing techniques having a large impact.

4.4.1 Overview

In principle, two different approaches can be used for forming thin films.

1. Layer formation through surface transformation. Chemical transformation of the surface is used to form the thin film. During thermal oxidation, for example, oxygen

[3]Linder, S., Baltes, H., Gnaedinger, F., Doering, E. (1996) Photolithography in anisotropically etched grooves. In: *The Ninth Annual International Workshop on Micro Electro Mechanical Systems. Proceedings. IEEE Robotics and Automation Society*, pp. 38–43.

Table 4.8 Layer deposition techniques in microelectronics and microsystem technology

From the gaseous phase		From the liquid phase	
PVD (Physical Vapor Deposition)	CVD (Chemical Vapor Deposition)	LPD (Liquid Phase Deposition)	CSD (Chemical Solution Deposition)
Evaporation;	NPCVD Normal Pressure CVD;	Liquid Phase Epitaxy;	Sol-Gel-Procedure;
Sputtering;	APCVD Atmospheric Pressure CVD;	Smelting;	MOD Metal-organic Destruction;
MBE Molecular-Beam Epitaxy;	LPCVD Low Pressure CVD;	LANGMUIR-BLODGETT-Technique;	ECD Electro-chemical Deposition (Electroplating);
Laser Ablation;			
PLD Pulsed Laser Deposition	PECVD Plasma Enhanced CVD;	Spin-on-methods (SOG Spin-on-Glass, SOD Spin-on-Dielectric)	Electroless Plating
	MOCVD Metal-organic CVD;		
	MOMBE Metal-organic Molecular Beam Epitaxy		

causes silicon on the surface to transform to SiO_2. A growing SiO_2 layer reduces the thickness of the silicon (see Section 4.4.2). The chemical reaction for the layer formation takes place at the boundary to the underlying substrate. The contributing parts of the reaction have to diffuse through the layer resulting in reduced growth rate or speed, respectively, with increasing layer thickness.

2. Layer deposition: Particles are generated, transported to the surface and deposited. Condensation of the particles leads to layer growth. Growth rate is independent of film thickness.

For the first group, only thermal oxidation is relevant for microelectronics and microsystem technology. As opposed to this, there is a large number of layer deposition techniques that are applied for manufacturing microsystems (Table 4.8). Most widely used are those procedures where deposition takes place during the gaseous phase (evaporation, sputtering, CVD Chemical Vapor Deposition).

4.4.2 Layer Conformity

Commonly, a uniform layer thickness over the surface contour is desirable (Figure 4.10). In microsystem technology, there are often edges due to the three-dimensional structure.

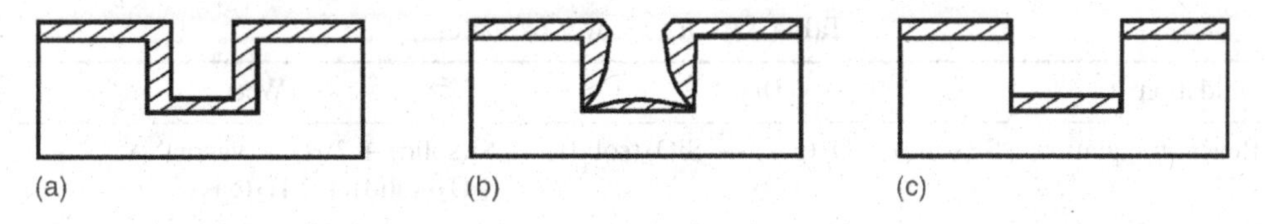

Figure 4.10 Edge coating of layers on surfaces with deep structures: (a) conformal coating; (b), (c) nonconformal coating

If layer thickness is the same in all areas, this is called layer conformity (Figure 4.10a). Conform layers require the following:

- uniform transport of particles over the entire surface (e.g. for thermal oxidation, oxygen diffusion through the SiO_2 layer that already has been formed);
- a growth rate during layer formation that is determined by reaction rate (e.g. oxidation of silicon or chemical reactions during CVD procedures).

If the growth rate is restricted by the rate the particles diffuse to the surface and if diffusion channels to the surface vary, the result will be a deterioration of conformity (Figure 4.10b). This is also the case for a directed transport of particles, as for PVD procedures, for instance. In the extreme, there will be no particle deposition at all on the side walls (Figure 4.10c).

4.4.3 Thermal Oxidation

During thermal oxidation, silicon reacts chemically to silicon oxide. The growing SiO_2 layer utilizes silicon and the silicon thickness decreases (Figure 4.11). However, the thickness of the Si/SiO_2 stack increases. This effect is used in surface micromachining to plug narrow channels (Figure 4.49).

During dry oxidation, there is a reaction with oxygen (Table 4.9). As the O_2 molecule is comparatively large, but anyway has to diffuse through the SiO_2 layer already formed at the boundary with the silicon, the drift and subsequent growth rate are very low. Dry oxidation is used for very thin oxides with high quality requirements (e.g. gate oxide in CMOS technology). During wet oxidation, water vapor is used for the chemical reaction.

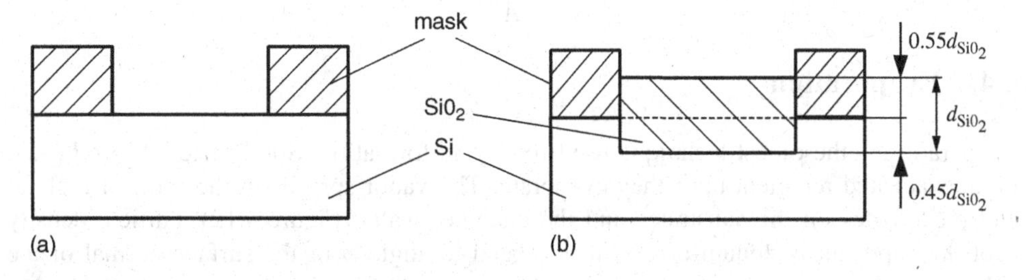

Figure 4.11 Utilization of silicon during thermal oxidation: (a) prior to and (b) after oxidation

Table 4.9 Dry vs. wet oxidation

Oxidation	Dry	Wet
Reaction equation	Si(solid) + O_2(gas) → SiO_2(solid)	Si(solid) + 2 H_2O (vapor) → SiO_2(solid) + 2 H_2(gas)
Film thickness	1 μm: 1200 °C, 1400 min 0.5 μm: 1200 °C, 85 min 1100 °C, 650 min 1000 °C, 880 min	1 μm: 1200 °C, 85 min 0.5 μm: 1200 °C, 21 min 1100 °C, 30 min 1000 °C, 52 min
Characteristics	Smallest defect density Best dielectric characteristics	High growth rate Higher defect density
Growth coefficient on (1 0 0)-Si	1200 °C: B = 720 nm^2/mm; B/A = 29 nm/min 1100 °C: B = 390 nm^2/mm, B/A = 5.3 nm/ min	1200 °C: B = 12000 nm^2/mm; B/A = 240 nm/min 1100 °C: B = 8500 nm^2/mm, B/A = 77 nm/min

The considerably smaller water molecule results in layers with a factor 3...6 higher thickness in the same oxidation time.

According to the DEAL-GROVE-model [SCHADE91], the linear-parabolic correlation between thickness d_{SiO_2} of the oxidation layer and oxidation time t is

$$d_{SiO_2}(t) = \frac{A}{2}\left(\sqrt{1 + (2B/A^2)(t + \tau)} - 1\right) \tag{4.5}$$

with

$$\tau = (d_0^2 + Ad_0)/B. \tag{4.6}$$

Here, B/A is the linear and B the parabolic constant as well as d_0 the initial oxide thickness. The following approximation is valid for long oxidation times:

$$d_{SiO_2}^2 = B \cdot t \tag{4.7}$$

and for short oxidation periods (up to $d_{SiO_2} \approx 200 \ldots 300$ nm)

$$d_{SiO_2} = \frac{B}{A}(t + \tau). \tag{4.8}$$

4.4.4 Evaporation

Evaporation is the oldest technique used for layers formation. Solid materials are heated in an evacuated recipient until they evaporate. The vapor spreads in the form of a cloud and precipitates on the substrates and the chamber walls (Figure 4.12). Particle density n of the vapor jet is nonuniform as it is related to angle φ of the surface normal of the ablating surface:

$$n(\varphi) = n(\varphi = 0) \cdot \cos^4 \varphi. \tag{4.9}$$

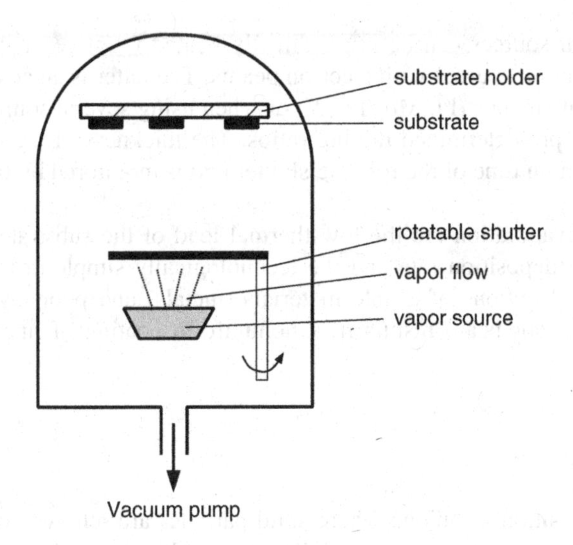

Figure 4.12 Schematic representation of an evaporation chamber

The pressure in the evaporation chamber has to be set in a way that the atoms reach the substrate without impingement. In order to achieve that mean free path length L has to exceed the distance between evaporation source and substrates:

$$L = \frac{1}{\sqrt{2}} \frac{kT}{\pi \cdot d_A^2 \cdot p}. \tag{4.10}$$

Here, k is BOLTZMANN constant, T the absolute temperature, d_A the diameter of the atom or molecule of the vaporized species and p the pressure in the evaporation chamber. A distance of 50 cm between source and substrate requires a vacuum of $10^{-3} \ldots 10^{-4}$ Pa. Evaporation rate Φ_A (number of atoms or molecules per area and time unit) is both for solid and liquid surfaces [OHRING92], [CAMPBELL01]

$$\Phi_A = \frac{\alpha_A \cdot p_S(T)}{\sqrt{2\pi kT/m_A}}, \tag{4.11}$$

where α_A is the evaporation coefficient describing the surface effects ($0 \leq \alpha_A \leq 1$), m_A the molecule mass and p_S the saturation vapor pressure, which is exponentially correlated to $1/T$ and thus follows ARRHENIUS equation. For aluminum, it is according to [OHRING92]

$$\lg\frac{p_S}{\text{Pa}} = 15.993\frac{1}{T/\text{K}} - 10.284 - 0.999\lg\frac{T}{\text{K}} - 3.52 \cdot 10^{-6}\frac{T}{\text{K}}, \tag{4.12}$$

with both terms to the right being only small correcting terms. The exponential relation to temperature leads to two important conclusions:

- A highly-constant growth rate requires a very constant temperature in the device.
- Pure metals can easily be evaporated. For alloys, the component with the higher vapor pressure evaporates preferably resulting in a varying composition of the layer.

Various evaporation sources can be used: directly heated resistive or inductively heated crucibles or evaporation sources with electron beams. The latter is an advantageous option for hard-to-evaporate metals (Pt, Mo, Ta, W) or when using several sources for depositing alloys with exactly pre-determined mixing ratios. The thickness of the deposited layer is set through the opening time of the rotating shutter und is monitored by the layer thickness sensor.

Advantages of evaporation are the low thermal load of the substrates (<100 °C), the comparatively high deposition rate and the technologically simple procedure. Disadvantages are the limited options of usable materials (metals) and poor layer adhesion. The orientation of the vapor beam results in a nonuniform coating of uneven or structured surfaces.

4.4.5 Sputtering

Sputtering is a deposition technique where solid particles are removed out of the cathode surface (target) by ion bombardment. The ejected particles use the kinetic energy to move to the substrate, condense there and form a layer.

Sputtering has two important advantages in relation to evaporation:

- Due to the impulse during the collision with the target, the solid particles have a much higher kinetic energy (1 ... 20 eV) than the molecules at evaporation (0.1 eV). Therefore they adhere much better to the substrate.
- Sputtering makes it possible to deposit many different material classes as thin layers: metals, alloys, semiconductors, ceramics, glasses etc. The composition of the layer is, apart from evaporation processes in the deposited layer, determined by the composition of the target.

Figure 4.13 shows a schematic of a sputtering reactor. In its high vacuum ($10^{-2} \ldots 10$ Pa), the recipient contains the spatially extended target and the substrates with a distance of

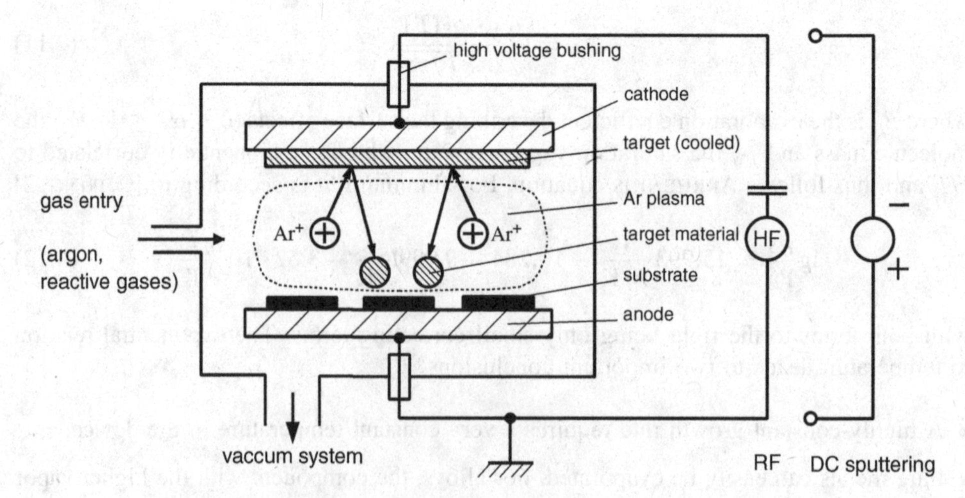

Figure 4.13 Schematic of a sputtering equipment

only a few centimeters from each other. Between the electrodes burns a plasma that ionizes the noble gas atoms (mostly argon due to the desired atomic mass). The electric field accelerates the argon ions towards the target. This impulse leads to a series of collisions ejecting target material that has a certain kinetic energy. The particles migrate to the substrate surface where they condense and contribute to the layer-formation process.

The sputtering process is determined by the sputtering yield. It determines how many atoms of target material per gas ion collision are ejected from the target. For argon, the sputtering yield amounts to 0.1 . . . 3 [OHRING92]. The ion bombardment heats the target to a degree that it has to be cooled.

Due to the ion bombardment, the targets are ablated over the time. Therefore, they have to be replaced when about 50 % of the material are sputtered off.

The sputtered layers on the substrate have not the same purity as those resulting from evaporation as gas atoms from the plasma are incorporated in the layer. Due to the collision process, target particles hit the substrate from different directions. The deposited layers have therefore a better conformity than those resulting from evaporation. We can distinguish several sputtering variants:

- DC sputtering: There is DC between target and anode. This variant can only be applied to conducting target materials. For insulating materials, there would be a charging that disrupts the plasma discharging process.
- RF sputtering: For insulators, the gas discharging is excited by high-frequency. Here, the freely available radio frequency of 13.56 MHz is used. In the plasma, only the electrodes, but not the Ar ions can follow the alternating field. That means that the electrodes receive a charge. The capacitor produces a DC-like disruption which results in a different charging of the electrodes (self-biasing).
- Bias sputtering: Substrate carrier and target are fed by separated HF generators. This allows precise control of the sputtering process. When reverting the bias voltage, substrate material is sputtered instead of target material (sputter etching). This can be used to clean the substrate prior to the actual sputtering process.
- Magnetron sputtering: Magnetic fields in the target area can increase the ionization rate in the plasma. This results in higher depositing rates at lower operating voltages and lower power conversion. Almost all sputtering machines have magnetrons.
- Reactive sputtering: Reactive gases can be conducted to the substrate surface und cause a reaction of the target material. Reaction products will be deposited. The applications most commonly used are oxides (SiO_2, Al_2O_3, SnO_2, InO_2, Ta_2O_5), nitrides (Si_3N_4, TiN, AlN) and oxinitride (SiO_xN_y).

4.4.6 Chemical Vapor Deposition

CVD processes are based on chemical reactions of gaseous reagents (precursors) on the substrate surface, where a solid reaction product contributes to layer growth, whereas all other reaction products are gaseous and are exhausted by the vacuum system (Figure 4.14).

Reaction

$$AX(gas) + BY(gas) \rightarrow AB(solid) + XY(gas) \tag{4.13}$$

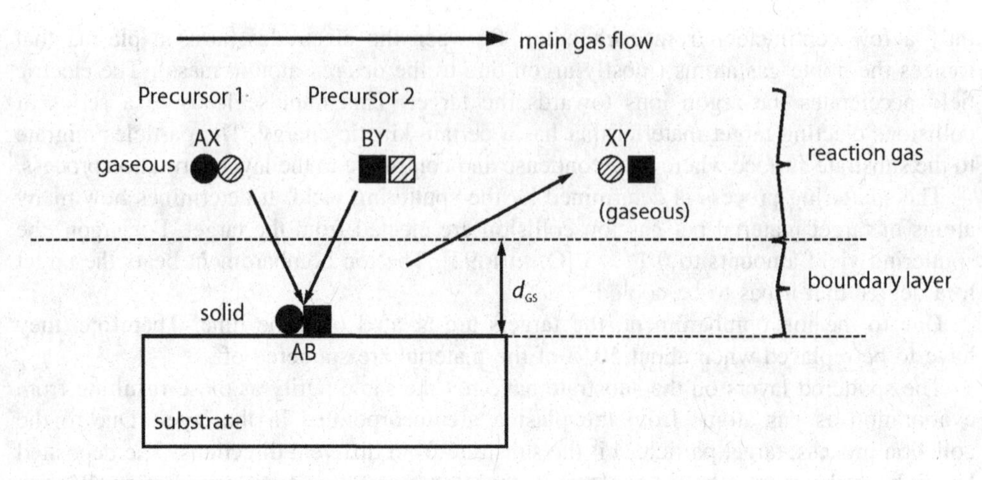

Figure 4.14 Basic principle of CVD processes

requires an activating energy which can be thermally provided by a plasma (PECVD Plasma Enhanced CVD) or something else. The reaction process consists of the following steps:

- diffusion of reactive particles AX and BY to the substrate surface;
- surface reaction (adsorption of reactive particles AX and BY, chemical reaction, surface diffusion of solid reaction products AB with adsorption to the solid body, desorption of reaction products XY);
- diffusion of particles XY from the boundary layer into the gas.

Particle flow density S_{Diff} of the particles diffusing to the substrate surface is [SCHADE91]

$$S_{\mathrm{Diff}} = \frac{D}{kT} \cdot \frac{p_{GS} - p_S}{d_{GS}} \tag{4.14}$$

with

$$D = D_0 \cdot \frac{T^m}{p}. \tag{4.15}$$

Here, D is the diffusion coefficient, p_{GS} and p_{S} the mean partial pressure of gaseous particles AX at the interface between boundary layer and substrate surface, d_{GS} the thickness of the boundary layer, D_0 a constant and $m = 1.5\ldots2$. The proportion of the particle current density on the substrate surface that contributes to layer formation is

$$S_{\mathrm{react}} = k_S \cdot p_S = k_{S0} \cdot \exp\left\{-\frac{E_{\mathrm{a}}}{kT_S}\right\} \cdot p_S. \tag{4.16}$$

Here, k_S is the specific reaction constant, E_{a} the activation energy and T_{S} the substrate temperature.

Depending on which of the two partial processes is dominating, the CVD process has different features:

- $S_{diff} \gg S_{react}$: The process is diffusion-restricted, i.e. determined by S_{diff}. The deposition rate is only slightly dependent on the temperature. The thickness resulting in the boundary layer is heavily dependent on the reactor geometry, though. Diffusion restriction is favorable for epitaxial growth.
- $S_{react} \gg S_{diff}$: The process is reaction-restricted and determined by Equation (4.16). This is favorable for polycrystalline or amorphous growth. Due to the exponential term in Equation (4.16), the deposition rate depends heavily on temperature.

Three variants are common in chemical vapor deposition:

- Normal-pressure CVD (APCVD, NPCVD): At normal pressure, layer deposition is restricted by the diffusion process of the reaction species to the substrate surface. Therefore, APCVD is suitable for epitaxy processes. This way, epitaxial silicon layers can be deposited at 1000...1300 °C during the silane (SiH_4) process:

$$SiH_4(gas) \rightarrow Si(solid) + 2H_2(gas)$$

- Low-pressure CVD (LPVCD): LPCVD processes occur at pressures of 10... 100 Pa. They are determined by the reaction, as with low pressure the diffusion coefficient increases by several orders of magnitude. The wafers can be placed close to each other, which substantially decrease the required gas quantity (Figure 4.15). This way, the deposition of dielectrics and poly-Si becomes cost-efficient:

$$SiH_4 + O_2 \rightarrow SiO_2 \text{ (solid)} + 2H_2O \text{ (LTO or low temperature oxydation)}$$
$$3\ SiH_2Cl_2 + 4\ NH_3 \rightarrow Si_3N_4 \text{ (solid)} + 6\ HCl + 6\ H_2$$
$$SiH_4 \rightarrow Si \text{ (solid)} + 2\ H_2.$$

Process temperatures lie in the range of 400...900 °C, and deposition rates at 2...10 nm/min.

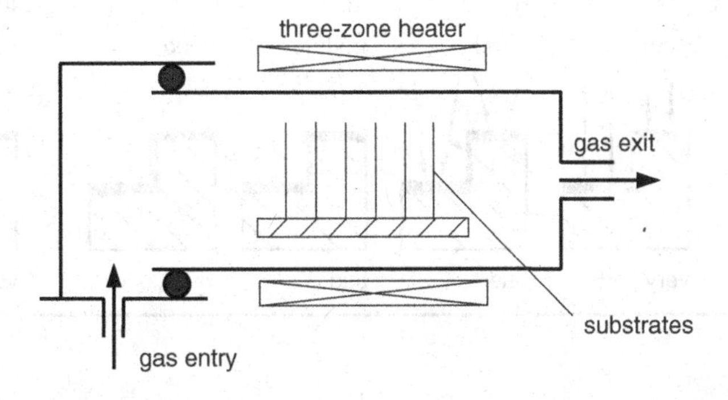

Figure 4.15 Schematic of an LPCVD reactor

- Plasma-enhanced CVD (PECVD): For PECVD, the activation energy for the chemical vapor reaction according to Equation (4.16) is not provided by thermal energy $k \cdot T$, but by gas discharge. This allows for much lower process temperatures of 200 ... 400 °C. PECVD reactors resemble sputtering reactors (Figure 4.13), however, they do not require a target. In addition, the electrically insulated substrate carrier acts as a cathode, whereas the anode is electrically connected to the chamber itself. Typical PECVD processes are:

$$SiH_4 + 2\,N_2O \rightarrow SiO_2(\text{solid}) + 2\,N_2 + 2\,H_2$$
$$SiH_4 + NH_3 \rightarrow Si_xN_yH_z(\text{solid}) + H_2.$$ [4]

4.4.7 Comparison

Table 4.10 presents the essential characteristics of the different deposition techniques.

PVD and CVD techniques provide a broad range of processes for a large variety of goals:

- PVD and PECVD processes are particularly suitable in situations where low process temperatures are required.
- Vapor deposition and LPCVD processes do not require very sophisticated devices and have a high productivity.
- As LPCVD and PECVD processes are chemical processes, they show an excellent conformity, i.e. edge coating.

Table 4.10 Comparison of different deposition techniques

Technique	PVD		CVD		
	Vapor deposition	Sputtering	APCVD	LPCVD	PECVD
Temperature	ca. 100 °C	<300 °C	>1000 °C	400 ... 1000 °C	200 ... 400 °C
Pore density	high	middle	middle	very good	very good
Contaminations	low	high	high	low	low
Adhesion	poor	good	middle	good	good
Conformity	very bad	bad	bad	very good	very good

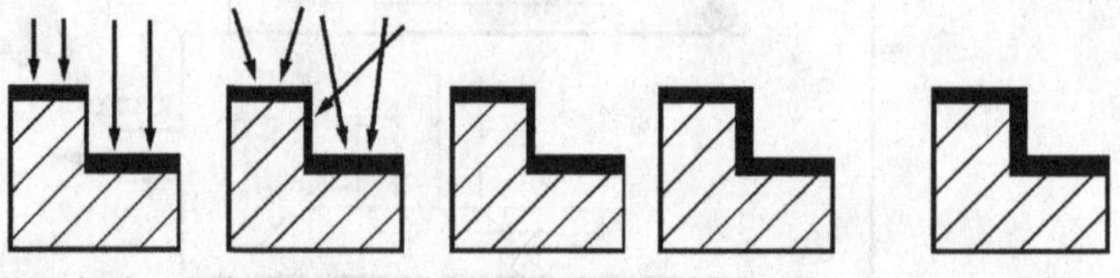

[4]These silicon nitride films contain hydrogen.

4.5 LAYER PATTERNING

In microsystem technology, thin films are applied to the entire surface and subsequently are structured using photolithography. Dry and wet etching is used during this process. The pattern of a masking layer consisting of photoresist or another etch resist is transferred to this layer. Similar to the etching of thin layers, bulk material can also be etched. Due to the specifics of etching deep three-dimensional structures, deep-etching will be discussed separately in Section 4.6.

4.5.1 Basics

Figure 4.16 shows a cross-section of a structure that has been masked with etch resist before and after the etching process. The layer is covered with an etch mask consisting of an etch resist material. During the etching process, the layer is removed with an etch rate of

$$R = \text{etching depth/etching time} \tag{4.17}$$

In addition to the vertical removal of the layer (vertical etch rate R_V), there is also an undercutting removal (etch rate R_L) and often also a removal of the etch resist (etch rate R_{V2}) (Figure 4.16b).

The selectivity of the etchant is the ratio of the etch rates of the two different materials:

$$S = R_{V1}/R_{V2}. \tag{4.18}$$

Usually, a very large selectivity between layer to be etched and etch resist, but also between layer and substrate, is desirable. In the latter, a low selectivity results in overetching into the substrate. The larger the selectivity, the lower is the overetching. A sufficiently high selectivity for vertical etching ($R_{V\text{substrat}} \ll R_{V1}$) is called etch stop.

An etchant as shown in Figure 4.16b with an identical etch rate in all spatial directions of a material is called isotropic. Etchants with an etching rate that is dependent on the direction is called anisotropic. Anisotropy can be defined as follows:

$$A = 1 - R_L/R_V. \tag{4.19}$$

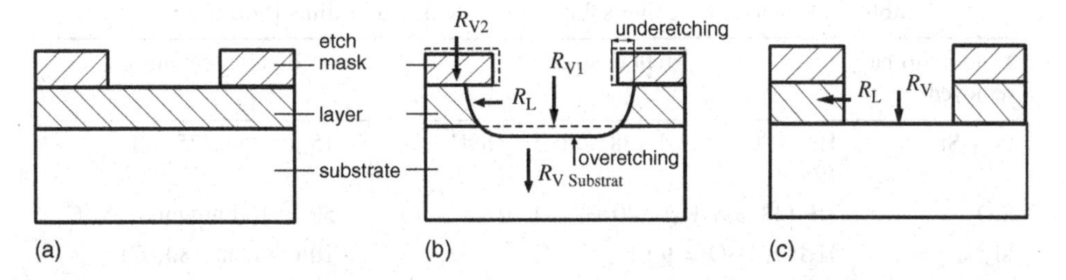

Figure 4.16 Etching structure (a) before and (b), (c) after etching; (b) isotropic, (c) anisotropic etching process

If in Figure 4.16b lateral and vertical etch rates are identical, then $A = 0$: The etchant is isotropic, it does not show any anisotropy. In the case presented in Figure 4.16c, where $R_L = 0$, it results that $A = 1$: The etching process is anisotropic.

Anisotropy can occur

- if there is a directional effect of the etching impact, e.g. for ion etching; or
- if the material to be etched is crystalline and there are bonding energies in different crystal orientations that have to be dissolved by the etching process.

4.5.2 Wet Etching

Wet chemical etching processes are technologically simple and do not require sophisticated equipment. There is no contamination by redeposition of disposals from previous etching processes or radiation damage due to hard UV radiation in plasma, which occurs during dry etching. However, wet etching is limited regarding pattern size. Devices required for wet etching are etching baths, cleaning and rinsing baths as well as spin-drying. Etching solutions typically contain oxidants, complexing agents for dissolving the oxides and reaction-reducing components for an even etch rate. CP4 solution, for instance, as a polishing etchant for silicon, contains 38.5 percent by volume of concentrated nitric acid HNO_2 (oxidant), 19.2 % by volume of hydrofluoric acid HF (complexing agent) and 42.3 % by volume of acetic acid CH_3COOH (etch rate adjustment).

Table 4.11 shows a few selected etching solutions for patterning thin layers. Silicon nitride is chemically less active and has therefore only a low wet chemical etcheability. It is, therefore, usually dry-etched.

4.5.3 Dry Etching

In principle, etching processes in the gaseous phase use two main mechanisms:

- chemical reaction with the material to be removed with reactive species in plasma (reactive or plasma etching);
- physical impact processes due to particle bombardment (ion or ion beam etching).

Table 4.11 Selected etching solutions for etching thin films [SCHADE91]

Material to be removed	Etching solution	Etching conditions
Poly-Si	HNO_3 (72 %)/HF (38 %)/CH_3COOH (98 %) = 2:1:1	15 μm/min (25 °C)
SiO_2	HF (48 %)/NH_4F (40 %) = 1 : 21.5 ... 15	50 ... 100 nm/min (25 °C)
Si_3N_4	H_3PO_4/H_2O = 9 : 1	10 nm/min (180 °C)
Al	H_3PO_4O (85 %)/HNO_3 (65 %)/H_2O = 20 : 1 : 5	220 nm/min (40 °C)

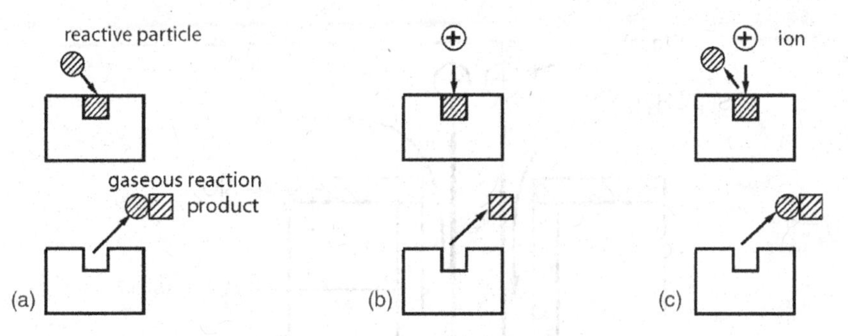

Figure 4.17 Principle of dry etching techniques: (a) plasma etching; (b) ion etching; (c) reactive ion etching (RIE)

The named mechanisms can be used individually or in combination (Figure 4.17).

Plasma etching: Similar to plasma-enhanced CVD, ions or radicals that are formed in the plasma of the etching gas react with the surface. The composition of the etching gas has to guarantee that only volatile reaction products with the surface are formed (Figure 4.17) and can be exhausted later. As there is a variety of different simultaneous reactions in the plasma, no single reaction equation can be given for the plasma etching process. For processes that are determined by the reaction, the etching removal is isotropic and has a high removal rate. Depending on the used etching gas, a high selectivity can be achieved.

Ion etching: Similar to sputtering, for ion etching an ion gas or an ion beam is generated and accelerated in the electric field towards the area to be removed. The material is removed by physical impact processes. As the ions are directed towards the surface, there results a higher anisotropy ($A \rightarrow 1$). With the respective process procedure, steep sidewalls can be created. However, the values of the bond energy of the different materials are very similar, which means a lower selectivity of the etching process in relation to various etch resist materials (Table 4.12). The substrate to be etched has to be cooled because it will be heated by the ion bombardment.

Reactive Ion Etching (RIE): Combines the advantages of plasma and ion etch. The ion bombardment enhances the anisotropy of the etching process and the use of reactive gases increases selectivity and etch rate.

Table 4.12 Etch rate for ion beam etch (Ar^+; 0.5 keV; ion flow 1 mA/cm^2)

Material	Etch rate R in nm/min
Si	20...40
SiO_2	30...40
Au	140
Al	30
Photo resist AZ 1350	20...30

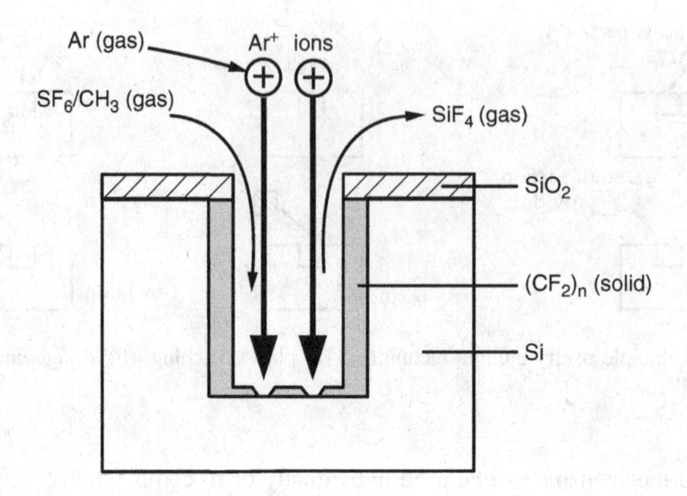

Figure 4.18 Principle of DRIE process

Deep Reactive Ion Etching (DRIE): Reactive ion etching can be modified to create very deep etchings with almost vertical sidewalls. In order to prevent undercutting ($A \to 1$), the used etching gas chemically passivates the etch groove. Etching only takes place where the ion beam breaks up the passivating layer (Figure 4.18). The following processes take place:[5]

- A SF_6/Ar gas mixture flows into the reactor chamber. In the plasma, SF_6 forms F^* radicals that etch silicon:

$$\mathrm{Si(solid)} + 4\ \mathrm{F^*(gas)} \to \mathrm{SiF_4(gas)}.$$

- In the plasma, argon is ionized to Ar^+ and is accelerated towards the substrate due to the negative bias ($-5 \ldots -30$ V) of the substrate. It hits the surface almost vertically and enhances the etching process.
- After the etching process has started, CHF_3 gas is added. In the plasma, CHF_3 forms CF_2 that polymerizes on all surfaces, both side and bottom surfaces, to teflon-like $(-CF_2-)_n$. This polymer layer prevents further etching of such surfaces.
- The ion bombardment removes the polymer layer on the bottom of the etch groove and the etching process can continue with F^* radicals. The ion bombardment does not affect the polymer layer on the sidewalls.

There is an alternation of deposition of polymer layer, polymer removal on the bottom surface and etching removal on the bottom surface, which means that the etching process continues vertically, but not laterally. Aspect ratios (ratio of depth to width) of up to

[5]Lärmer, F., Schilp, P. (1994) *Verfahren zum anisotropen Ätzen von Silizium*. (Techniques for the anisotropic etching of silicon) Patent DE 4.241.045.

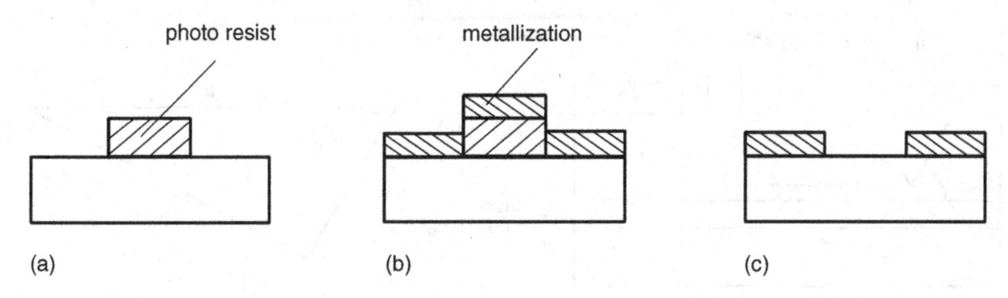

Figure 4.19 Lift-off process: (a) deposition and patterning of the photoresist; (b) metallization of the entire surface; (c) lift-off of the photoresist layer

30 can be achieved with this technique. The sidewalls have angles of $(90 \pm 2)^{\circ}$. The company that developed this process was Bosch and the process is therefore often called the 'Bosch process'. In the meantime, a large number of variations of this production technique has been developed [ELWENSPOEK98].

4.5.4 Lift-off Process

The lift-off technique is an adhesion mask technique where on the substrate initially the mask – consisting of patterned photoresist – is produced. Then the entire layer is deposited on the mask. In the parts where the layer is thinner than the photoresist and the sidewalls of the resist mask are not coated by layer material, the photoresist can be removed together with the layer on the photoresist (Lift-off, Figure 4.19). Due to the requirements (low film conformity, limited thermal loading capability of the previously applied photoresist), the lift-off-process is mainly suitable for patterning thin metal layers on very smooth surfaces.

4.6 ANISOTROPIC WET CHEMICAL DEEP ETCHING

Anisotropic wet chemical deep etching is of outstanding importance to microsystem technology. It can be used to carry out mechanically movable and deformable components with high geometric precision in silicon [GERLACH05]. In addition to surface micromachining, it continues to be the technique that has the largest economic significance.

4.6.1 Principle

Figure 4.20 shows the cross-section of an etch groove during anisotropic wet etching. The silicon wafer with orientation $\{a\ b\ c\}$ is coated with an etch mask consisting of an appropriate resist material. Through the opening in the etch mask, the Si substrate is removed by the etching solution. The etchant acts anisotropically and removes areas with different crystallographic orientation with different etching rates. As a result, the etch

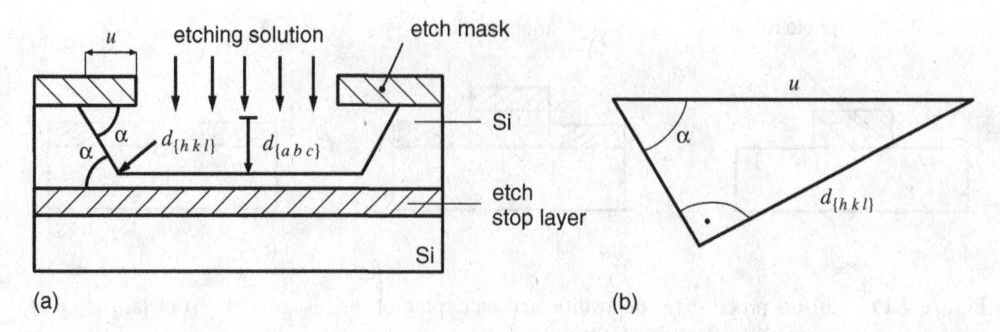

Figure 4.20 Principle of anisotropic wet chemical deep etching: (a) cross-section through the etching in a silicon wafer; (b) geometric relationship to undercutting u

groove is delimited by several surfaces. The geometric relation between the resulting layers leads to the determined geometric form. The bottom area of the etching has the same orientation $\{a\ b\ c\}$ as the surface. Depending on edge orientation and patterning of the etch mask, determined areas $\{h\ k\ l\}$ will be formed on the sidewalls of the etch groove.

According to Equation (4.17), etch rates for bottom area and sidewalls are

$$R_{\{a\,b\,c\}} = d_{\{a\,b\,c\}}/t_{\text{etch}} \tag{4.20}$$

and

$$R_{\{h\,k\,l\}} = d_{\{h\,k\,l\}}/t_{\text{etch}}. \tag{4.21}$$

For $R_{\{a\,b\,c\}} > R_{\{h\,k\,l\}}$ the anisotropy results in

$$A = 1 - R_{\{h\,k\,l\}}/R_{\{a\,b\,c\}}. \qquad (4.21)$$

The lateral boundary of the etch groove is determined by the form of the etch mask. The depth of the etch groove is determined by etching rate $R_{\{a\,b\,c\}}$ (see Section 4.6.3) and etching time t_{etch} or by etch stop layers integrated in the silicon bulk, respectively (see Section 4.6.4).

The lateral removal rate of the sidewalls results in an undercutting u under the etch mask. It is the projection of etching removal $d_{\{h\,k\,l\}}$ onto surface $\{a\ b\ c\}$, which is determined by angle α of the lateral areas (Figure 4.20b):

$$u = d_{\{h\,k\,l\}}/\sin\alpha. \tag{4.22}$$

4.6.2 Anisotropic Etching Solutions

Alkaline and amine-containing etching solutions have an anisotropic etching effect in silicon. For acid etching baths, no relation between etching rate and crystal orientation could be shown.

Table 4.13 Characteristics of the most common etching solutions [VÖLKLEIN05]

Description	KOH	TMAH	EDP
Etching solution[1]	Potassium hydroxide KOH (24 %)/water H_2O, 85 °C	Tetramethyl-ammonium-hydroxide TMAH (8 %)/water, 90 °C	Ethylendiamin NH_2-$(CH_2)_2$-NH_2, Pyrocatechol $C_6H_4(OH)_2$
$R_{\{1\,0\,0\}}$ in µm/h	100	55	75
$\frac{R_{\{1\,0\,0\}}}{R_{\{1\,1\,1\}}}$	400 600[2]	40	35
Masking layer (Etch rate)	SiO_2 thermal. (0.43 µm/min); PECVD-SiO_2 (0.7 µm/min); Si_3N_4 (70 µm/min); Au/Cr	SiO_2 ($R_{SiO_2} = 10^{-4}$ $R_{\{1\,0\,0\}}$), Si_3N_4	SiO_2 (12...30 nm/h), Si_3N_4 (6 nm/h), Au, Cr, Ag, Cu
$\frac{R_{\{1\,0\,0\}}(p^{++}\text{-Si})}{R_{\{1\,0\,0\}}}$	1:20 (boron, $> 10^{20}$ cm^{-3})	1:40 (boron, $> 4 \cdot 10^{20}$ cm^{-3})	1:50 (boron, $> 5 \cdot 10^{19}$ cm^{-3})
Remarks	IC compatible, strong formation of H_2 gas, easy handling	IC compatible, easy handling	toxic, fast ageing, to be used excluding O_2

[1]Percent refers to mass percent.
[2]$R_{\{1\,1\,0\}}/R_{\{1\,0\,0\}}$.

Table 4.13 presents characteristics of common etchants. The selection of suitable etching solutions is determined by the following criteria:

- easy handling (incl. toxicity);
- etch rate;
- etch stop efficiency;
- selectivity, especially regarding masking materials;
- IC compatibility;
- topology of the etching surfaces.

KOH and TMAH etch solutions are of high technical importance. Both are easy to handle and can be integrated in circuit production processes. KOH has a higher etch rate as TMAH, whereas TMAH has a better selectivity in relation to SiO_2. In addition, KOH – as opposed to TMAH – attacks aluminum interconnects with a higher etch rate.

EDP etching solutions require high temperatures of more than 100 °C for technically attractive etch rates. They are also toxic. Due to their characteristics, they put higher demands regarding devices and safety than KOH und TMAH. They have the advantage, though, that they substantially reduce the etch rate for a significantly lower doping dose. This can be used as etch stop for layers that are highly doped with boron (see Section 4.6.4). Today, most applications use electrochemical etch stop instead of highly doped p^{++} Si. Therefore, EDP etching solutions have become less important during the past years.

4.6.3 Etch Rates

Table 3.5 has shown that silicon atoms on (1 0 0), (1 1 0) and (1 1 1) surfaces bond in varying degrees and that therefore the extraction of individual atoms requires different energies. This results in different etch rates of this etch surface, i.e. anisotropy. However, this effect alone cannot completely explain the large differences in etch rate and there are several other models that try to explain this phenomenon [MADOU97]. One example is a model that includes the fact that the bond between Si surface atoms and the underlying atoms have different energy levels.[6] In addition, [ELWENSPOEK98] characterizes the nuclear roughness of different surfaces as nucleation barrier that reduces the etch rate by several dimensions. {1 1 1} planes have the lowest nuclear roughness of all planes.

The etching process itself consists of an oxidation of Si atoms on the surface

$$Si + 4OH^- \rightarrow Si(OH)_4 + 4e^-, \tag{4.23a}$$

and the reduction of hydrogen in the water and the solution process of the silicate complex into the hexahydrosilicate complex:

$$Si(OH)_4 + 4e^- + 4H_2O \rightarrow Si(OH)_6{}^{2-} + 2OH^- + 2H_2 \tag{4.23b}$$

Equations (4.21a, b) summarize the total reaction equation:

$$Si + 4H_2O + 2OH^- \rightarrow Si(OH)_6{}^{2-} + 2H_2. \tag{4.23c}$$

Reaction Equations (4.23a), (4.23b) and (4.23c) show that:

- The etching mechanism requires OH^- groups, which are supplied by the basic (alkaline and amine-containing) etching solutions.
- The etching process requires water, i.e. aqueous solutions.
- The reaction generates molecular hydrogen that rises as bubbles in the etching solution. The etching process can be visually controlled by the formation these gas bubbles.

[6]Seidel, H., Csepregi, L., Heuberger, A., Baumgartel, H. (1990) Anisotropic etching of crystalline silicon in alkaline solutions. Part I. Orientation dependence and behavior of passivation layers. *Journal of the Electrochemical Society* 137, pp. 3612–3626. Part II. Influence of dopents, pp. 3626–32.

- The first partial step according to Equation (4.23a) results in electrons that enter the conduction band of silicon and are used up again later in Equation (4.23b). An appropriate bias voltage to the silicon can be used to extract the electrons out of the silicon, thus disrupting the etching process (electrochemical etch stop, see Section (4.6.4).

If the etch rate is determined by reaction kinetics, it will – similar to particle flow density for CVD processes (see Equation (4.16)) – depend on the temperature and concentration N_i of the involved chemical species (in Equation (4.16), this corresponds to particle pressure on the reaction surface):

$$R = k_S \cdot \prod_i N_i^{n_i} = k_{SO} \cdot \exp\left\{-\frac{E_a}{kT}\right\} \cdot \prod_i N_i^{n_i}. \tag{4.24}$$

Here, k_S is the reaction rate constant, k_{SO} the corresponding constant without correlation to temperature, N_i the concentration of the involved chemical species, n_i the order of reaction for N_i, E_a the activation energy of the chemical etching reaction, k the BOLTZMANN constant, and T the absolute temperature. The exponential term describes that only particles with thermal energy kT, exceeding activating energy E_a, can participate in the reaction. The probability corresponds to a BOLTZMANN distribution. Equation (4.24) describes an ARRHENIUS' law

$$\lg \frac{R}{k_{SO}} \propto -E_a \cdot \frac{1}{T}. \tag{4.25}$$

Example 4.4 Etch rate for KOH etching of silicon

When etching silicon in an aqueous potassium hydroxide solution, in Equation (4.23a) there results one Si atom and four OH ions and in Equation (4.23b) four water atoms. Including all involved species, Equation (4.23c) will be as follows:

$$Si + 4H_2O + 4OH^- + 4e^- \rightarrow Si(OH)_6{}^{2-} + 2OH^- + 4e^- + 2H_2O$$

It results for Equation (4.23)[6]:

$$R = k_{SO} \cdot \exp\left\{-\frac{E_a}{kT}\right\} \cdot N_{Si}^{\alpha} \cdot N_{H_2O}^{4} \cdot N_{KOH}^{1/4}. \tag{4.26}$$

The concentration of silicon at the interface Si/etching solution is constant and determined by the crystallographic orientation. The relation of the etching rate to the concentration of KOH in the etch solution is shown in Figure 4.21. It corresponds to Equation (4.26) and shows the maximum of the product $N_{H_2O}^4 \cdot N_{KOH}^{1/4}$ at an approximately 20 % KOH solution.

Figure 4.22 presents the ARRHENIUS relation between etch rate and temperature. It corresponds to Equation (4.25). The following slope of the curve over $1/T$ can be determined from the activation energy (Table 4.14). The behavior shown in Figure 4.21 and Figure 4.22 is analog to that of TMAH and EDP.

Table 4.14 illustrates the direct relation between etch rate and amount of activation energy E_a.

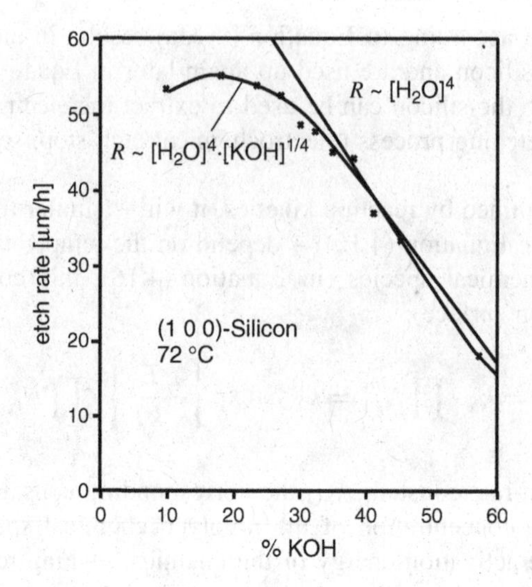

Figure 4.21 Relation of Si (1 0 0) etch rate to the concentration of the KOH etching solution [SEIDEL86]

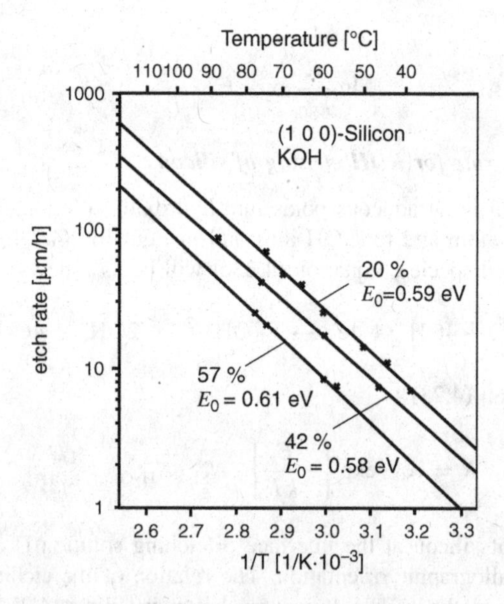

Figure 4.22 ARRHENIUS diagram of the temperature-dependence of the etch rate of KOH

Table 4.14 Activation energy E_a for different Si orientations and etching solutions [SEIDEL90]

Etching solution	Si (1 0 0)	Si (1 1 0)	Si (1 1 1)	SiO_2	Si_3N_4
KOH (20 %)	0.59 eV	0.61 eV	0.67 eV	0.81 eV	0.76 eV
EDP	0.40 eV	0.33 eV	0.53 eV		

4.6.4 Etch Stop Techniques

We talk about an etch stop if the etch rate decreases from value R_1 to value $R_2 \ll R_1$. A decrease of $R_2 : R_1 \leq 1 : 20$ is of technological interest in order to achieve sufficiently deep etch grooves. A decrease of $R_2 = 0$ is desirable; it cannot be practically achieved, though, in many cases.

The anisotropic or selective characteristics of etching solutions can be used to achieve an etch stop.

- Anisotropic etch stop: When producing etchings with only concave angles, sidewalls with very low etch rates will be generated. It will be mostly {1 1 1} lateral planes. As the etching process continues, the bottom area with the orientation determined by the surface orientation will disappear. This means that the etch groove will be surrounded only by slow-etching {1 1 1} planes (Figure 4.23). The deep etch rate, e.g. $R_{\{1\,0\,0\}}$, is reduced to $R_{\{1\,1\,1\}}$ (see Table 4.13, Column 4). The etch stop rate is determined by the ratio $R_{\{1\,1\,1\}}/R_{\{1\,0\,0\}}$.
- Selective etch stop on insulating layers: The implantation of O^+ ions under the surface of Si wafers and the reaction to SiO_2 can be used to carry out SiO_2 layers buried in the silicon (SIMOX technique, see Section 4.9). The implantation of N^+ ions will result in Si_3N_4 layers. Both act as etch stop due to $R_{SiO_2, Si_3N_4} \ll R_{\{1\,0\,0\}}$ (Table 4.13, Column 5).
- Selective etch stop on highly doped p^{++} silicon: High doping density of boron in silicon results in a dramatic decrease of the etch rate in anisotropic etching solutions. In p^{++}-Si, the holes recombine with the electrons generated during the etching (see Equation (4.22a)). This means that they are not available any longer in the etching process according to Equation (4.23b). The reaction rate is largely reduced. This etch stop mechanism requires boron concentrations of $5 \cdot 10^{19}$ cm^{-3} (EDP) or even much larger (KOH, TMAH) (see Table 4.13, Column 6). Due to the clearly lower required doping concentration for EDP, EDP used to be the preferred anisotropic etchant due to this etch stop technique. The electrochemical etch stop does not require such high doping concentration any longer. Therefore, EDP is rarely used today.

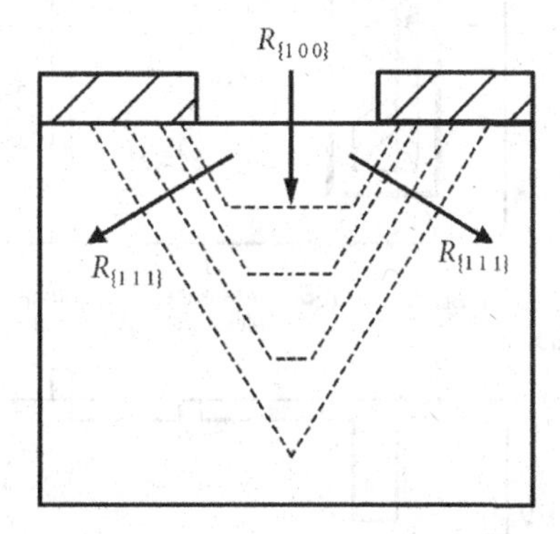

Figure 4.23 Anisotropic etch stop

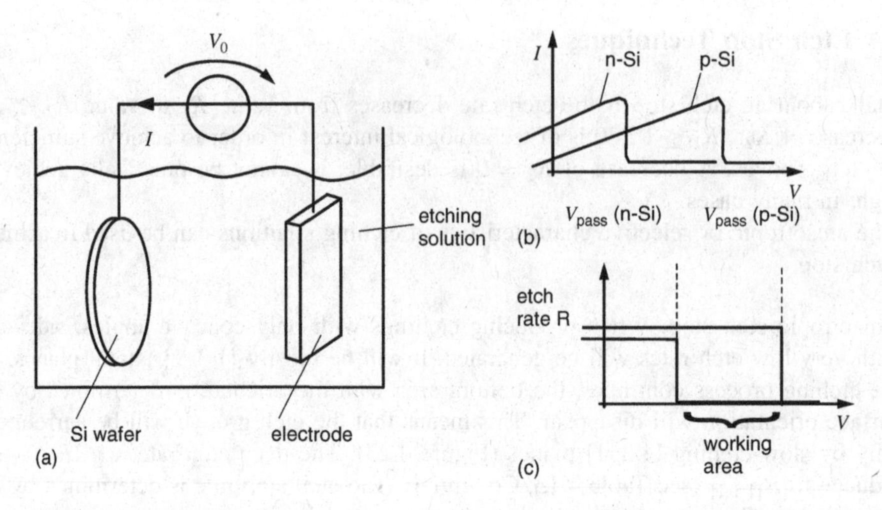

Figure 4.24 Electrochemical etch stop on silicon: (a) setup; (b) current-voltage characteristic; (c) corresponding etch rate-voltage characteristic; V_{pass} passivation voltage

- Electrochemical etch stop: By applying a sufficiently large positive voltage to the Si wafer, it is possible to achieve a reduction of electrons similar to that of p^{++}-Si (Figure 4.24). The required voltage varies for n- and p-conducting silicon. Due to the higher voltages for p-Si, the passivating voltage for pn-junctions can be set in a way that p-Si is still etched whereas n-Si is passivated. The electrochemical etch stop is therefore mainly used for pn-junctions (Figure 4.25). In addition to the counter-electrode, a potentiostat with reference electrode is necessary. In order to achieve a

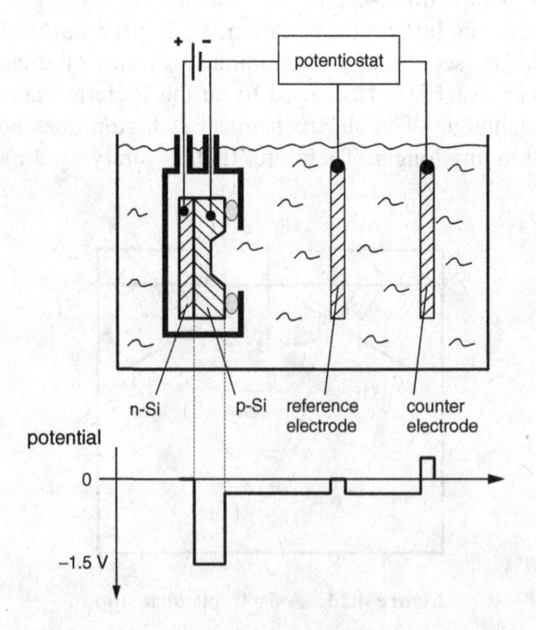

Figure 4.25 Four-electrode arrangement for the electrochemical etch stop at pn-transitions

high process reliability, the p- and n-Si areas are often separately biased. The goal is to create reproducible etching conditions.

4.6.5 Etch Figures

Anisotropic etching solutions have different etch rates in the various crystallographic directions of silicon. Depending on the anisotropic characteristics of the etchant, the orientation of the silicon wafer as well as form and direction of the etch mask, specific patterns will result in etchings.

(a) Etch rate diagram

Etch rates of different crystallographic planes are represented either in tables (see Table 4.13) or as polar diagrams (Figure 4.26). The latter are determined by experiments with specific etch masks (wagon wheel or etch rosette, Figure 4.27). They consist of radial etch mask wedges with a uniform wedge angle γ (e.g. 1°).

It is possible to calculate undercutting u and subsequently undercutting etch rate R_u from shortening a, which is easily optically measured:

$$\tan \gamma/2 \approx \gamma/2 = u/a, \tag{4.27}$$

$$u = \gamma \cdot a/2 \tag{4.28}$$

and

$$R_\mathrm{u} = \gamma \cdot a/(2 \cdot t_\mathrm{etch}). \tag{4.29}$$

If the orientation of lateral plane $\{h\ k\ l\}$ delimiting the undercutting is known as well as the angular relationship with the surface (see Equation (3.5)), etch rate $R_{\{h\ k\ l\}}$ can be calculated from undercutting etch rate R_u.

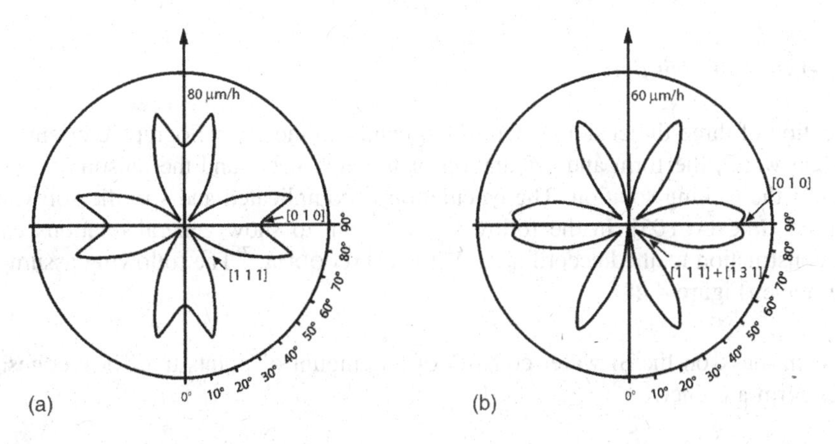

Figure 4.26 Relation of crystal orientation and undercutting etch rate R_u on a Si wafer: (a) (1 0 0)-wafer and (b) (1 1 1)-wafer [SEIDEL86]

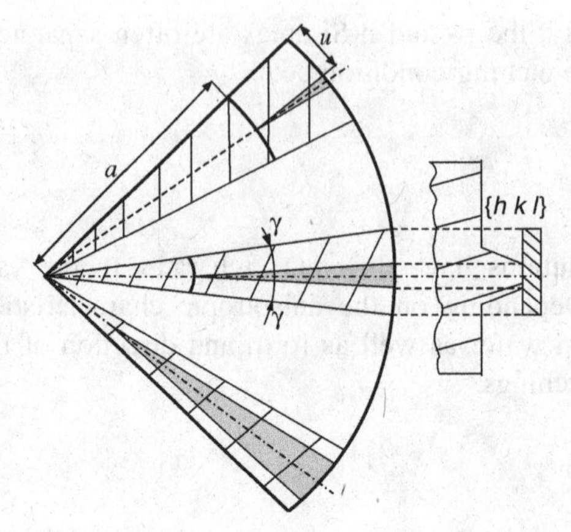

Figure 4.27 Etch rosette for determining the angle-related undercutting etch rate R_u (hatched: etch mask, dark hatched: area not undercut)

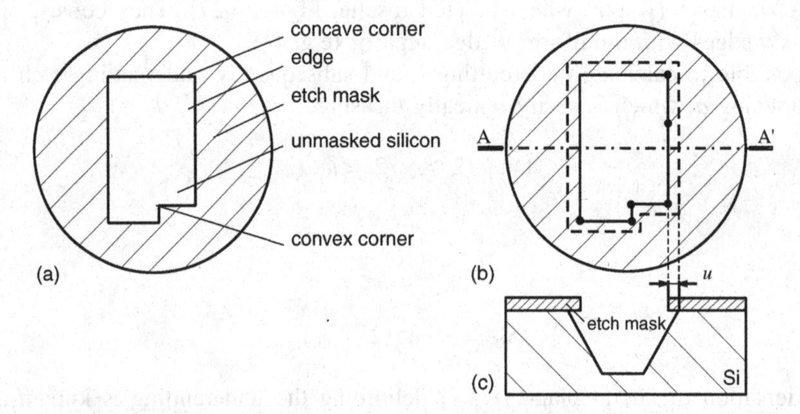

Figure 4.28 Construction of the etching process according to WULFF-JACCODINE: (a) un-etched; (b) after etching time t_{etch}; (c) cross-section after etching

(b) Generation of etch figures

The creation of three-dimensional etchings depends on the crystallographic orientation of the silicon wafer, the form and orientation of the etch mask and the anisotropic characteristics of the etching solution. The calculation is complicated and specific software has to be used [FRÜHAUF05]. In the following, we want to show general relations using a simple construction method according to WULFF-JACCODINE.[7] The following assumptions will be made (Figure 4.28):

- The etch mask on the Si wafer consists of an amount of points that form edges. Two edges form a corner.

[7] Jaccodine, J. (1962) Use of modified free energy theorems to predict equilibrium growing and etching shape. *Journal of Applied Physics* 33, pp. 2643–7.

- Several edges form closed stretches that form internal and external boundaries of the etch mask.
- Corners and edges can be concave ($<180°$) or convex ($>180°$).
- During the etching process, the intersection edges of lateral boundary planes and etch mask shift or rotate away from the original mask edges due to undercutting.

Concave and convex corners and edges show different etching behavior:

- Concave corners: Sidewalls with the lowest etch rate are formed.
- Convex corners: Sidewalls with angle-related local maximum etch rate are formed. Different etching solutions generate different plane with maximum etch rate (Table 4.15).
- Straight mask edges: The points of the mask edge form a line and are neither concave nor convex. The shifting of the edge represents the progressing etching-off of a sidewall.

The construction of an etch figure results from the original position of the etch mask as well as minimum and maximum values of undercutting etch rate R_u resulting from the undercutting etch diagram. The following procedure is applied (Figure 4.29):

- Straight edges are shifted by the undercutting etch rate towards this direction (Figure 4.29c). It corresponds to the progress of the sidewalls projected on to the surface (etch mask).
- On concave edges, the direction with the minimum undercutting etch rate has to be determined; and the etch fronts vertical to this direction are shifted by undercutting distance $u = R_u \cdot t_{etch}$ (Figure 4.29d).
- The same procedure is applied for convex corners using the maximum value of the undercutting rate in this case (Figure 4.29e).
- Total undercutting is the sum of the etch fronts that have progressed the furthest (Figure 4.29f).
- During the etching process, new convex or concave corners and edges can be formed (e.g. during corner undercutting). These new corners have to be treated in the same way as described above.

Table 4.15 Fast-etching planes for different etching solutions [Elwenspoek98]

Etching solution	Fast-etching planes
KOH (40 °C)	{1 1 0}
KOH/propanol/H_2O (30:25:45)	{2 1 2}, {3 3 1}
KOH (80 %)	{1 3 0}
KOH (15...50 %, 60...100 °C)	{4 1 1}
EDP (acc. to composition)	{2 1 2}, {3 3 1}

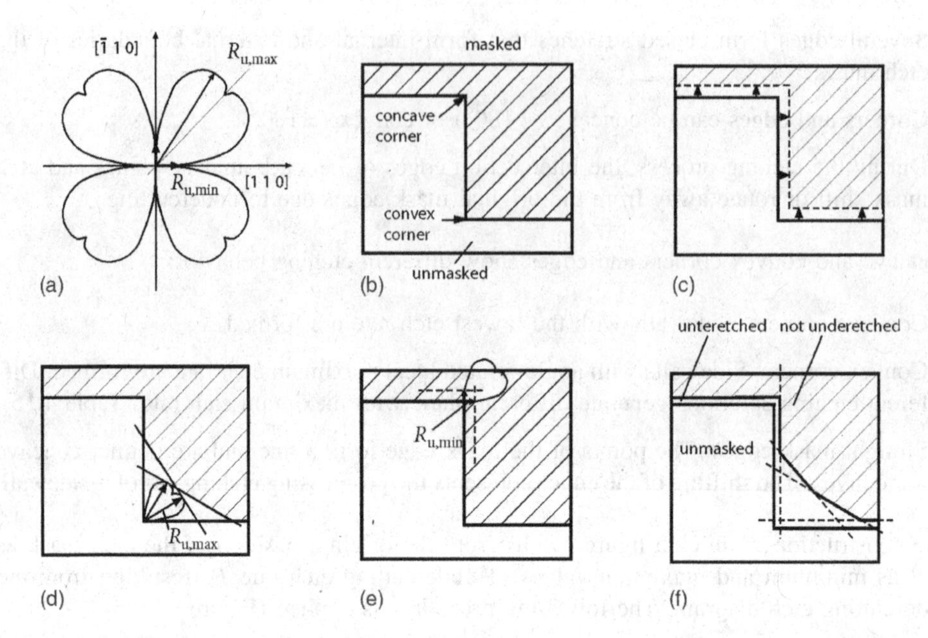

Figure 4.29 Undercutting of an etch mask with concave and convex corner: (a) orientation-related undercutting etch rate R_u; (b) etch mask on silicon; c) undercutting of mask edge; (d), (e) undercutting of the concave and convex corner; (f) total undercutting (hatched: undercut)

Example 4.5 Undercutting for a rectangular etch mask with angular displacement

We want to look at how an etch mask with a rectangular opening is undercut if the etch mask edges are not orientated, as seen in Figure 4.29, precisely in <1 1 0> direction, but have an angle of rotation φ (Figure 4.30).

We will individually describe the construction of the undercutting of the various edges and corners (Figure 4.30a). As there are only concave corners, no fast-etching areas will be formed:

- The lateral boundary planes that are parallel to the four mask edges progress at etch rate R_{edge}.
- Starting from the concave corners, planes with minimum etch rates R_{corner} will be formed.
- If the planes with the smaller etch rate are located closer to the mask edge, they will delimit the etch groove as they are the planes with the highest chemical resistance (Figure 4.30b).

When the etching process progresses the sidewalls parallel to the mask edges will become smaller and smaller and finally disappear completely (Figure 4.30c). The final etching is again delimited by sidewalls that are orientated along the <1 1 0> edges. On the surface of the Si wafer, it completely encloses the mask opening.

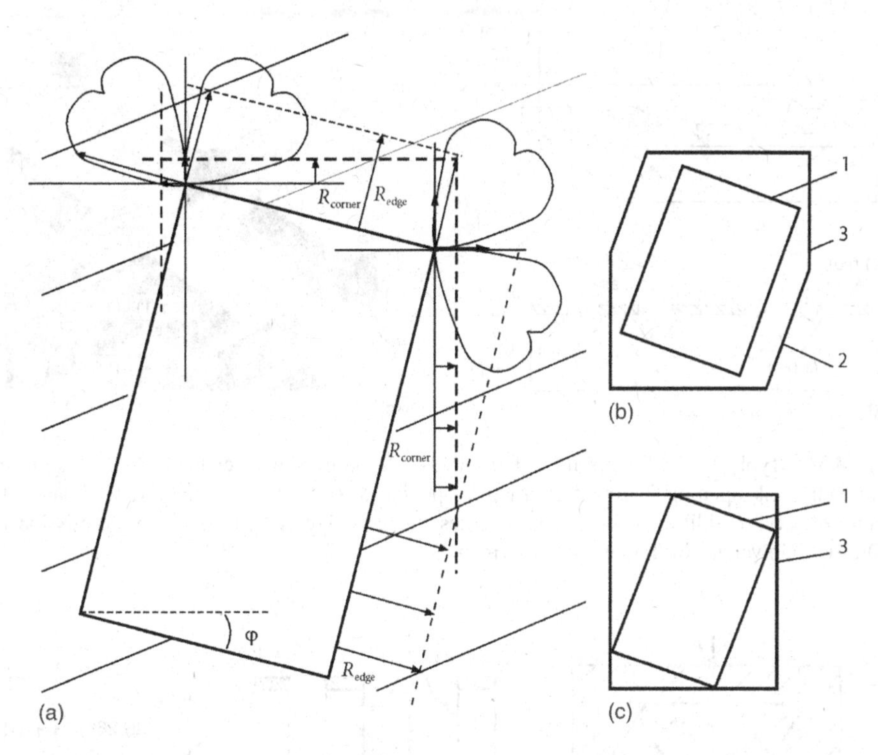

Figure 4.30 Undercutting of a rectangular etch mask with angle of rotation φ: (a) construction at time t_1; (b) undercutting front at time t_1; (c) undercutting front after extended etching time ($t_1 \gg t_2$); 1 Etch mask edges, 2 Undercutting front with faster etch rate, 3 Undercutting front with minimum etch rate

As {1 1 1} silicon planes in general have the lowest etch rate, anisotropically etched etch figures are often laterally delimited by {1 1 1} areas. Figure 4.31 shows typical etch figures in (1 0 0) Si. Along <1 1 0> orientations, {1 1 1} sidewalls are formed at an angle of 54.74° to the (1 0 0) surface. In Figure 4.31a, the etch mask opening has become large enough for the {1 1 1} sidewall to form. However, the (1 0 0) etch groove planes still continue to exist. In Figure 4.31b, the etch mask opening is so small that the etched bottom plane completely touches the {1 1 1} sidewalls. The etching that previously had the form of a truncated pyramid has now become a complete pyramid-shaped etching. Figure 4.31c shows that such an inverse pyramid is even formed for round etch mask openings. The technique shown in Figures 4.29 and 4.30 can even be used for complex three-dimensional structures. For this, we need to know the orientation and size of the etch rates of fast- and slow-etching areas as well as the deep etch rate of the surface (Figure 4.33).

For <1 1 0>-oriented mask edges it is possible to generate vertical sidewalls in (1 0 0)-Si. As those are also {1 0 0}-oriented, deep etch rate and undercutting rate are identical.

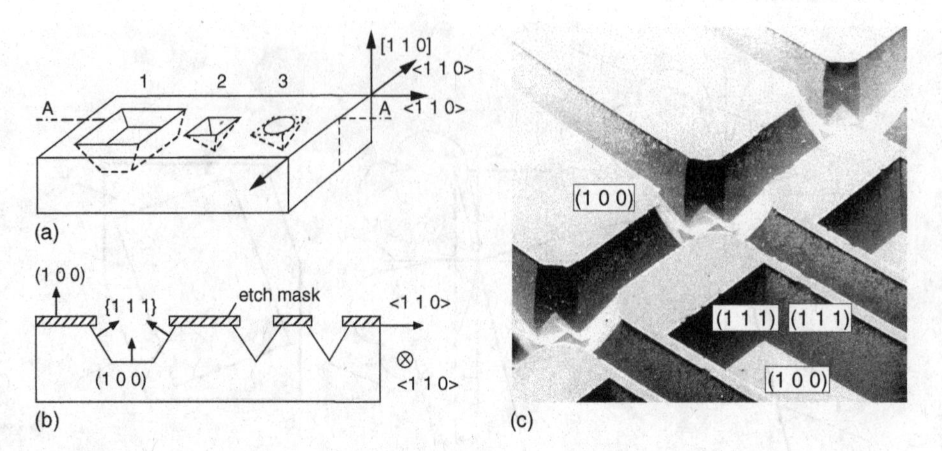

Figure 4.31 Typical etch figures in (1 1 0) silicon: (a) large square etch mask opening 1, small square etch mask opening 2, round etch mask opening 3; (b) cross-section AA; (c) image of the structure of 1 with additional separation trenches. Image 4.31c is reproduced by permission from the Dresden University of Technology, Germany

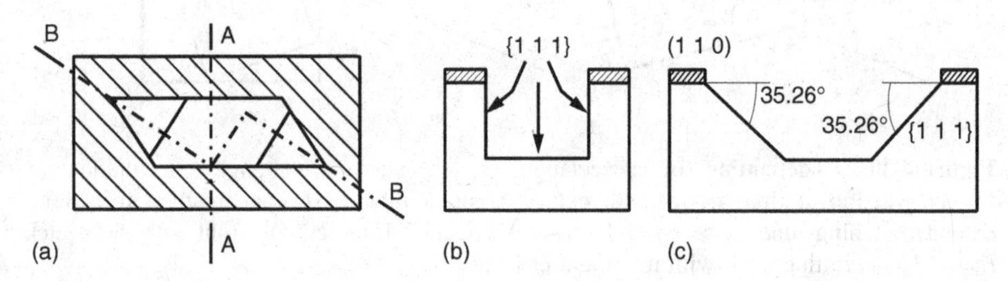

Figure 4.32 Vertical and inclined {1 1 1} sidewalls in (1 1 0) silicon: (a) top view; (b) cross-section AA; (c) cross-section BB

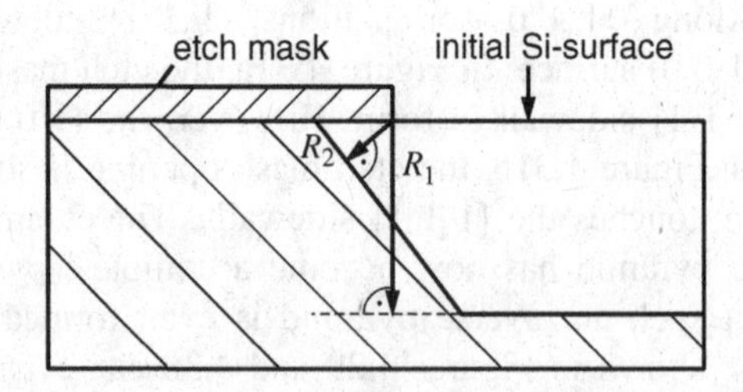

Figure 4.33 Construction of the developing form of an etching in silicon bulk: R_1 etch rate of Si surface, R_2 Amount and orientation of the etch rate minimum

In (1 1 0)-Si, there will be also vertical groove boundaries for the respective etch mask orientation. These boundaries are {1 1 1} sidewalls, though (Figure 4.32). They do not form rectangular etchings, but a parallelogram with angles of 109.47° and 70.53°, respectively. In addition, two more {1 1 1} planes are formed, with a 35.26° inclination in relation to the surface.

Example 4.6 Production of a self-supporting SiO_2 bridge on a (1 0 0)-Si wafer

The goal is to generate a self-supporting SiO_2 bridge over a channel in the (1 0 0)-Si wafer. Figure 4.34 shows the SiO_2 etch mask, a part of which is to form the self-supporting bridge. During the etching process, slow-etching {1 1 1} planes are generated and result in an actual etch stop. If the bridge is so small that the undercutting fronts meet (Figure 4.34a), before the two separate {1 1 1} fronts have completely formed, the goal cannot be reached. In Figure 4.34b, the bridge is too wide or not enough rotated and the two etchings are already completely enclosed by {1 1 1} areas before the fast-etching fronts had been able to meet.

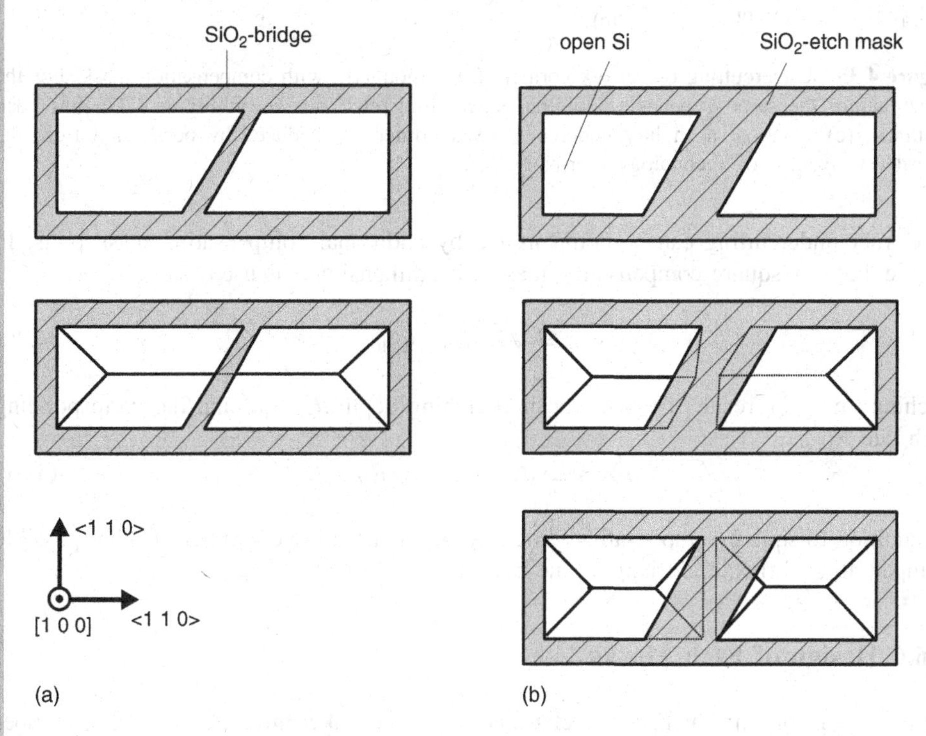

Figure 4.34 Undercutting of an SiO_2 bridge over an etched channel in (1 0 0)-silicon: (a) narrow bridge: undercutting etching fronts meet under the bridge und open it up completely; (b) too wide bridge: slow-etching {1 1 1}-planes have formed before the bridge is completely undercut

(c) Undercutting of convex edges

As described above, at convex corners planes are formed that have a high etch rate (see table 4.15). That leads to the rounding of such corners. For the construction of the resulting etch form, we will again use the WULFF-JACCODINE method (Figure 4.35). The corner point *A* in Figure 4.35a shifts under the mask due to fast-etching planes. The velocity corresponds to corner undercutting etch rate R_{cu}. Usually, $R_{Eu} > R_{\{1\,0\,0\}}$. For a 30 % KOH solution at 80 °C, the ratio $R_{cu}/R_{\{1\,0\,0\}} \approx 1.5$.

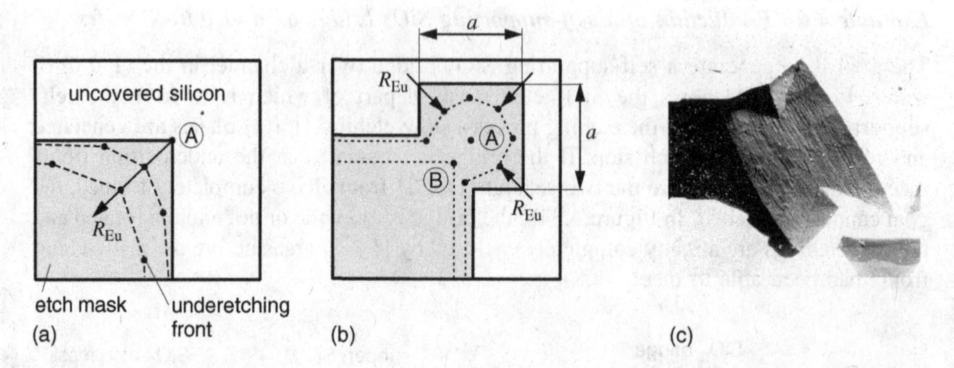

Figure 4.35 Undercutting of convex corners: (a) without; (b) with compensation mask. For the construction, the same dependence on the undercutting etch rate as in Figure 4.27a has been assumed; (c) image of a slightly undercut convex corner. Reproduced by permission from the Dresden University of Technology, Germany

Corner undercutting can be compensated by additional compensating mask parts. In Figure 4.35b, a square compensating mask with dimension a is used:

$$a = \sqrt{2} \cdot R_{\mathrm{cu}} \cdot t_{\mathrm{etch}}. \tag{4.30}$$

Etching time t_{etch} results from the desired etching depth $d_{<a\,b\,c>}$ and the corresponding etch rate $R_{\{a\,b\,c\}}$:

$$a = \sqrt{2} \cdot d_{<a\,b\,c>} \cdot R_{\mathrm{u}}/R_{\{a\,b\,c\}}. \tag{4.31}$$

In addition to square compensating structures, there are a large number of other possible compensating forms, e.g. triangles and bars [ELWENSPOEK98].

4.6.6 Design of Etch Masks

Form, dimension and thickness of etch masks have to take into account a large number of effects:

- orientation of the wafer surface and the mask edges;
- etching solution and etching temperature;
- etching time and etching depth;
- existence of convex and concave mask corners;
- etch stop layers.

Example 4.7 Dimensioning of an etch mask opening

The goal is to anisotropically remove a square membrane ($w = 500$ µm, thickness $d_{\mathrm{p}} = 25$ µm) in a (1 0 0) silicon wafer (diameter 100 mm; thickness $d = 525$ µm) (Figure 4.36). For etching, a 24 % KOH solution is used at a temperature of 85 °C.

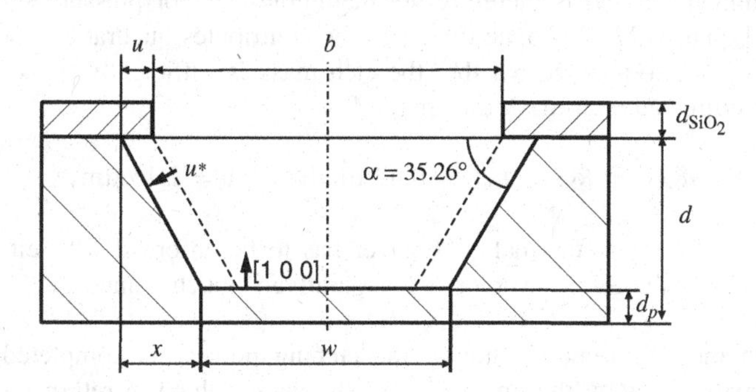

Figure 4.36 Anisotropically etched cantilever plate in (1 0 0)-Si

As etch resist, we use a 1-μm-thick thermal SiO_2 layer. The etch edges are <1 1 0>-oriented. When dimensioning etch mask opening b, we have to take into account etching depth d_{etch} and subsequently etching time t_{etch}, the geometric form of the etch groove with {1 1 1} sidewalls inclined at an angle of $\alpha = 54.74°$ as well as the undercutting of the sidewalls:

$$b = w + 2x - 2u. \tag{4.32}$$

The inclined sidewalls result in a widening of

$$x = d_{etch} / \tan\alpha. \tag{4.33}$$

Undercutting u is the projection of etching distance $u^* = R_{\{1\,1\,1\}} \cdot t_{etch}$ according to

$$\sin\alpha = u^*/u. \tag{4.34}$$

Thus, it results that

$$u = \frac{R_{\{1\,1\,1\}} t_{etch}}{\sin\alpha}. \tag{4.35}$$

Etching time t_{etch} results from the required etching depth d_{etch}:

$$t_{etch} = d_{etch} / R_{\{1\,0\,0\}}. \tag{4.36}$$

Equations (4.31) to (4.35) result in

$$b = w + 2d_{etch}\left(\frac{1}{\tan\alpha} - \frac{1}{\sin\alpha} \cdot \frac{R_{\{1\,1\,1\}}}{R_{\{1\,0\,0\}}}\right). \tag{4.37}$$

Table 4.13 provides the values for etch rates of 24 % KOH solution at 85 °C: $R_{\{1\,0\,0\}} = 100$ μm/h, $R_{\{1\,1\,1\}}/R_{\{1\,0\,0\}} = 1/40$, $R_{SiO_2} = 0.43$ μm/h. Thus, etching time is $t_{etch} = 5.0$ h. Therefore, the etch mask opening has to be $b = 1176$ μm ($x = 352.5$ μm, $u = 15.3$ μm). Due to the etching inclination, the silicon area required is larger than the membrane area. The undercutting $2\,u = 30.6$ μm amounts to approximately 6 %

of the plate dimension and is therefore not negligible, e.g. for pressure sensors, when according to Equation (7.125) plate dimension w contributes quadratic to sensitivity.

Finally, we have to analyze whether the etch mask is sufficiently thick. An etching time of 5 h requires an etch mask thickness of

$$d_{SiO_2} = R_{SiO_2} \cdot t_{etch} = 0.43\ \mu m/h \cdot 5.0\ h = 2.15\ \mu m \tag{4.38}$$

Therefore, the 1-μm-thick thermal SiO_2 layer has to be enforced with either a correspondingly thick CVD-SiO_2 layer or a comparatively much thinner Si_3N_4 layer, for instance.

If the etch mask is removed before the etching process is completed, the result would be convex edges on the upper face of the wafer which in return would lead to the formation of fast-etching planes.

4.7 DOPING

In semiconductors, adding small amounts of selected foreign atoms can change physical, and especially electrical characteristics. When doping with boron, the result will be p-conducting silicon, for instance; using arsenic and phosphorus will result in n-conducting silicon. Doping concentrations between 10^{14} and 10^{21} cm^{-3} lead to specific resistances between 10^2 and 10^{-4} $\Omega \cdot$ cm [SCHADE91]. This way, the temperature coefficient of the resistance can be set in the range of between $9 \cdot 10^{-3}$ and $0.3 \cdot 10^{-3}$K^{-1} (minimum for n-Si at 10^{-2} $\Omega \cdot$ cm).

In microsystem technology, doped layers are especially important as etch stop layers for anisotropic wet chemical etching (see Section 4.6.4). In addition to epitaxy, i.e. the growth of crystalline layers on a crystalline surface (see Section 4.4.6, normal-pressure CVD), particularly diffusion and ion implantation are widely used.

4.7.1 Diffusion

At high temperatures, foreign atoms can penetrate solid bodies and move there if there is a high concentration gradient of such particles in this room. Diffusion in solid bodies takes place through position and ring exchange with the neighboring lattice as well as vacancy or interstitial diffusion.

(a) Diffusion law

FICK's laws that are used to describe diffusion processes in liquids and gas can also be applied to solid bodies. In the following, we will look at a one-dimensional case as diffusion often takes place over large areas down to depth z of the silicon wafer. If there is a gradient of the particle density n (in m^{-3}) over distance z, there will be a particle flow density J_T (particles per area and time, in m^{-2}s^{-1}) towards the lower particle density

Table 4.16 Diffusion coefficients in silicon and SiO_2 [SCHADE91]

Atom		D_0 in cm²/s	E_a in eV	D in cm²/s	
				1000 °C	1100 °C
Single crystalline Si	B	0.7	3.46	$1.8 \cdot 10^{-14}$	$1.8 \cdot 10^{-13}$
	P	3.85	3.66	$1.3 \cdot 10^{-14}$	$1.5 \cdot 10^{-13}$
Poly-Si	B	$6.0 \cdot 10^{-3}$	2.39	$2.1 \cdot 10^{-12}$	$1.0 \cdot 10^{-11}$
	P			$10^{-12} \ldots 10^{-11}$	$10^{-12} \ldots 10^{-11}$
SiO_2	B	$5.2 \cdot 10^{-2}$	4.06	$4.3 \cdot 10^{-18}$	$6.4 \cdot 10^{-17}$
	P	$3.9 \cdot 10^{-9}$	1.63	$1.4 \cdot 10^{-15}$	$4.0 \cdot 10^{-15}$

(1st FICK's law):

$$J_T = -D\frac{\partial n}{\partial z}. \tag{4.39}$$

Here, D is the diffusion coefficient. As the probability follows BOLTZMANN distribution, D is exponentially related to the temperature:

$$D = D_0 \exp\{-E_a/kT\}, \tag{4.40}$$

where D_0 is a temperature-independent constant, E_a the activation energy, k the BOLTZMANN constant, and T the absolute temperature. Table 4.16 presents the diffusion coefficients of boron and phosphorus atoms in silicon and SiO_2. With the continuity equation of the time-related change of particle density in depth d_z

$$\frac{\partial J_T}{\partial z} = -\frac{\partial n}{\partial t} \tag{4.41}$$

the 2nd FICK's law, it results that

$$\frac{\partial n}{\partial t} = D\frac{\partial^2 n}{\partial z^2}. \tag{4.42}$$

(b) Dopant sources

Due to their low vapor pressure at ambient temperature, pure elements are hard to vaporize and are not suitable as doping sources. Therefore, it is common to use liquid sources (e.g. boroethane B_2H_6) which conduct a carrier gas N_2. This way, a vapor stream can reach the diffusion tube. It is mixed with oxygen. The result is the corresponding dopant oxide (B_2O_3) which is then reduced on the Si surface and incorporated – as an elementary dopant (B) – into the silicate (SiO_2). The B/SiO_2 layer then serves as predoping for the actual diffusion. Both of the following reaction equations illustrate the transition of liquid-source boroethane to atomic dopant:

$$B_2H_6(\text{gas}) + 3\ O_2(\text{gas}) \rightarrow B_2O_3(\text{solid}) + 3\ H_2O(\text{liquid})$$
$$2\ B_2O_3(\text{solid}) + 3\ Si(\text{solid}) \rightarrow 4\ B(\text{solid}) + 3\ SiO_2(\text{solid}).$$

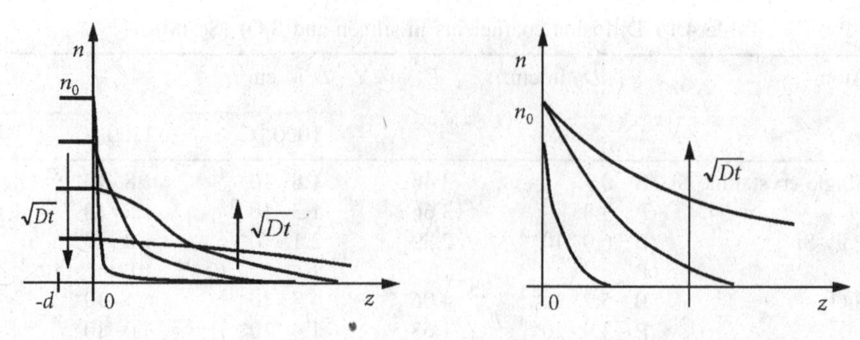

Figure 4.37 Diffusion profiles for diffusion from an (a) exhaustible and (b) inexhaustible source

For boron, also BF_3 and BBr_3 are used as dopant sources, for phosphorus PH_3 or $POCl_3$, and for arsenic AsH_3.

(c) Diffusion profile

The diffusion profile after the completion of the diffusion process depends, according to Equation (4.40), from diffusion temperature and boundary conditions.

When diffusing from an exhaustible source (Figure 4.37a), a predoping layer with thickness d has a dopant concentration n_0 on the silicon. The dopant is limited and increasingly transits from the predoping layer to the underlying Si. After an infinite time, there would be a constant concentration $n = n_0 \cdot d/(d + d_{Si})$ over the entire silicon thickness d_{Si}. Solving diffusion Equation (4.42) with the corresponding boundary and starting conditions ($n(z; t = 0) = n_0$ in the predoping layer; $n(z, t = 0) = 0$ in Si; $\partial n(z = 0, t)/\partial z = 0$), there results a GAUSSian function (Figure 4.37)

$$n(z, t) = \frac{n_0 d}{\sqrt{\pi Dt}} \exp\left\{-\frac{z^2}{4Dt}\right\}. \tag{4.43}$$

When doping with an inexhaustible source (Figure 4.37b), the surface concentration remains constant: $n(z = 0, t) = n_0$. Solving Equation (4.42) with starting condition $n(z > 0, t = 0) = 0$ results in the complementary error function:

$$n(z, t) = n_0 \cdot \text{erfc}\left(\frac{z}{2\sqrt{Dt}}\right). \tag{4.44}$$

Term $2\sqrt{Dt}$ in Equations (4.43) and (4.44) can be interpreted as diffusion length L:

$$L = 2\sqrt{Dt}. \tag{4.45}$$

In practice, diffusion is often carried out as two-step diffusion. During a predoping process with high predoping thickness, there will be diffusion from an inexhaustible source. After that, a deep diffusion process will take place where no dopant is provided (e.g., thermal treatment).

(d) Thermal budget

In microsystem manufacturing, diffusion is followed by a large number of process steps that change the doping profile. If an initial step with D_0t_0 results in a GAUSSian profile according to Equation (4.43) and during the subsequent diffusion steps there is no in- and outdiffusion at the boundary layer, the GAUSS characteristics of the dopant distribution will remain. After m more diffusion steps $D_it_i\,(i = 1, \ldots, m)$, the resulting profile will be

$$n(z,t) = \frac{2}{\pi} \cdot n_0(t=0) \cdot \arctan\sqrt{\frac{D^*t^*}{Dt}} \cdot \exp\left\{-\frac{z^2}{4(D_0t_0 + D^*t^*}\right\} \tag{4.46}$$

with

$$D^*t^* = \sum_{i=1}^{m} D_it_i. \tag{4.47}$$

The denominator term in the exponential term which determines the course of the profile behavior results from the sum total of all thermal diffusion loads:

$$D_0t_0 + D^*t^* = \sum_{i=1}^{m} D_it_i = (D \cdot t)_{tot} = \frac{L_{tot}^2}{4}. \tag{4.48}$$

Total diffusion length L_{tot} corresponds to diffusion length in Equation (4.45). With increasing temperature T, which leads to larger D_i, and progressing exposure time t_i, the concentration density in the semiconductor decreases. Due to this, $(Dt)_{tot}$ has to be restricted in order to not fall below a certain dopant concentration. In practice, often thermal budget TB is used. It comprises all thermal loads for the various temperatures T_i and must not exceed maximum value TB_{max}:

$$TB = \sum_{i=1}^{m} (t_iT_i) < TB_{max}. \tag{4.49}$$

4.7.2 Ion Implantation

During ion implantation, dopant ions are electrically accelerated and bombarded into the wafer surface. Deflecting the ion flow of the dopant in a magnetic field will lead to a magnetic mass separation, which results in extremely high purity levels. As the implantation process is a nonequilibrium process, it is possible to set doping concentrations n that exceed the thermodynamical solubility limit. Process temperatures can be kept very low in order to prevent indiffusion of damaging impurities.

(a) Doping profile

In the solid state body, the entering ions of the ion beam are decelerated as they collide with the body's nuclei and electrons and are subsequently scattered (Figure 4.38). The

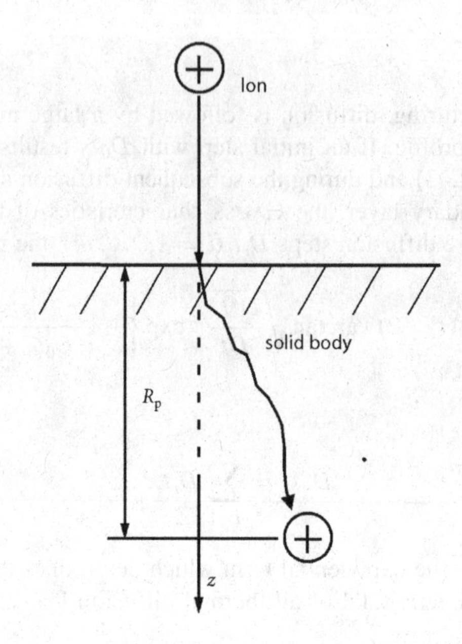

Figure 4.38 Projected range R_p of ions implanted in a solid body surface

average depth of ions is the projected range R_p, the scattering around this depth is ΔR_p. Dose n^* (number of implanted ions per area unit, in ions/m^2) is

$$n^* = \frac{I \cdot t}{e \cdot m \cdot A}, \tag{4.50}$$

where I is the ion beam density, t the implantation time, e the elementary charge, m the charging number per atom, and A the implanted area.

Dose n^* as well as acceleration voltage of the ions and thus their energy E determine the doping profile. In the depth, doping atoms follow an approximate GAUSSian distribution, which results in the following dopant concentration:

$$n(z) = \frac{n^*}{\sqrt{2\pi}\,\Delta R_p} \exp\left(-\frac{(z - R_p)^2}{2\Delta R_p^2}\right). \tag{4.51}$$

Table 4.17 presents values for projected range R_p and its scattering ΔR_p.

The depth of doped areas is determined by the ions' kinetic energy E. The goal is to maximize the probability of a collision of the incident ions with the crystal lattice and avoid the channeling effect where ions move along only little occupied crystal orientations. Therefore during the ion bombardment, the wafer is set in an inclined position where the angle for silicon is 7°.

A distinction is made between low- and middle-current implanters ($E < 180$ keV, ion current $I < 10$ mA), high-current implanters ($E < 120$ keV, $I = 10 \ldots 25$ mA) as well as high-energy implanters (E > 200 keV up till several MeV). The latter are suitable for generating buried layers in silicon or for doping silicon underneath thick oxide layers.

Table 4.17 Projected range R_p and its scattering ΔR_p for ion implantation in silicon

Ion	Relative	10 keV		100 keV		1000 keV	
	atomic mass	R_p in μm	ΔR_p in μm	R_p in μm	ΔR_p in μm	R_p in μm	ΔR_p in μm
B^+	11.009	0.0392	0.0167	0.3367	0.0782	1.70	0.1490
P^+	30.974	0.0169	0.0076	0.1358	0.0478	1.21	0.2136
O^+	16.995	0.0262	0.0117	0.2231	0.0627	1.30	0.1429
N^+	14.003	0.0291	0.0127	0.2470	0.0649	1.39	0.1438

Source: Ziegler, J. F. (Hg.) *et al.* 'The Stopping and Range of Ions in Solids', 321 pages, Pergamon Press (1985).

(b) Radiation damage and annealing

Semiconductor technology often uses doses n^* in the range of 10^{12} to $10^{15} \cdot \text{cm}^{-2}$ in order to achieve that there is a maximum volume concentration n below 0.1 atom-%. For larger doses and energies, surface sputtering and significant deviations from the GAUSSian profile can be observed. In addition, the already small doses of incorporated ions cause radiation damage in the crystal. At very high doses, the surface of the semiconductor may become amorphous.

In order to repair crystal damages and to allow the incorporated ions to diffuse to lattice positions, the semiconductor is annealed, i.e. it is subjected to temperature treatment. In order to cause as little diffusion effects as possible and thus keep the thermal budget low, rapid thermal annealing (RTA) is used. With tungsten halogenic lamp arrays, a wafer is heated on both sides to about 1000 °C with a heating rate of 150 °C/s and kept at this temperature during approximately 10 s.

(c) Application

There are several applications for ion implantation (Table 4.18).

4.7.3 Comparison of Doping Techniques

Diffusion is characterized by the following features:

- flat concentration change at pn-junctions;

Table 4.18 Applications for ion implantation

Application	Substrate	Ion	Doted area
Doping of semiconductors	Si	B^+	p-Si
Generation of buried insulating layers	Si	O^+	SiO_2
	Si	N^+	Si_3N_4
Production of compound semiconductors	Si	C^+	SiC
Production of conducting silicides	Si	Co^+	Co_2Si
Generation of insulating layers in metals	Al	N^+	AlN

- doping depth is determined by diffusion time and temperature; it can reach up to 100 μm;
- due to exponential temperature-dependence, the reproducibility of dopant concentration is poor;
- simple, economic procedure.

As opposed to this, ion implantation shows the following characteristics:

- steep pn-junctions, especially for small implantation energies E and thus low ΔR_p values;
- low doping depth of less than 1 μm, due to the R_p values that can be achieved;
- excellent reproducibility of dopant concentration using implantation energy and dose;
- high technical demand;
- process beam-borne and thus requires long process times.

4.8 BONDING TECHNIQUES

The production of three-dimensional structures is in many cases facilitated if the form is constructed using several, separately structured wafers. The structuring is achieved through chip bonding (die bonding) or on the wafer level through wafer bonding. The latter is more economic as several elements can be processed simultaneously.

For bonding two wafers, a large number of prerequisites have to be fulfilled. They affect the choice of a suitable bonding technique:

- rigidity of the bond (bonding energy of up to 1 J/m^2);
- hermetic tightness of the bond;
- electrically conducting or insulating bond;
- bond with good or poor thermal conductivity;
- low mechanical stresses through different coefficients of linear expansion of the bonding partners;
- maximum bonding temperature of 577 °C (eutectic temperature of AlSi), when already aluminum interconnections have been deposited on the Si wafer.

Gluing and glass soldering have been used for a long time for bonding silicon chips and wafers. They use, however, corresponding intermediate layers with a minimum thickness in the range of 10 μm. In microsystem technology, the goal is to achieve minimum thermally induced stresses and a high reproducibility. Therefore it preferably uses bonding techniques with extremely thin (in the range of micrometers) intermediate layers or no intermediate layers at all.

4.8.1 Eutectic Bonding

Eutectic soldering uses the fact that silicon can alloy with several metals. And for specific compositions, it will form a eutectic system. With gold, for instance, that will happen at 2.85 weight% Si. Here, the eutectic point is 370 °C and lies below that of silicon with aluminum (577 °C) (Figure 4.39). Eutectic bonding with Au intermediate layers can thus also be used for silicon wafers with Al interconnects.

Eutectic bonding is often used for bonding silicon wafers with Si- or glass wafers that have previously been coated with an AuSi3 film (Figure 4.40). For adhesion reasons, this AuSi layer preferably is sputtered. Ultrasound is used to remove the existing natural oxide layers during assembly resulting in a solid, hermetically tight connection.

4.8.2 Anodic Bonding

Anodic bonding uses electric fields for the bonding process between silicon and glass wafers. The thermal coefficient of expansion of glass has to be adapted to that of silicon,

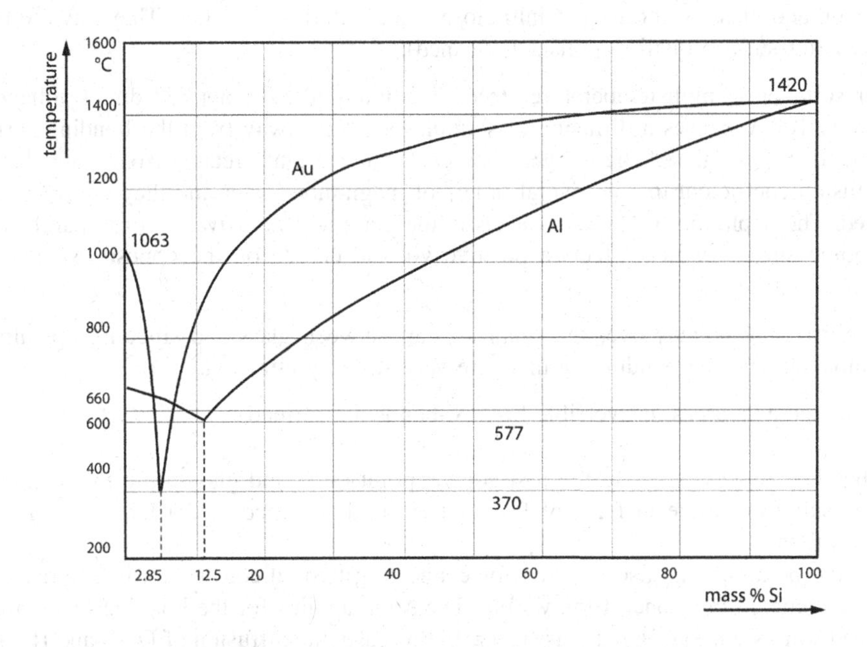

Figure 4.39 Phase diagram for Si-Au and Si-Al

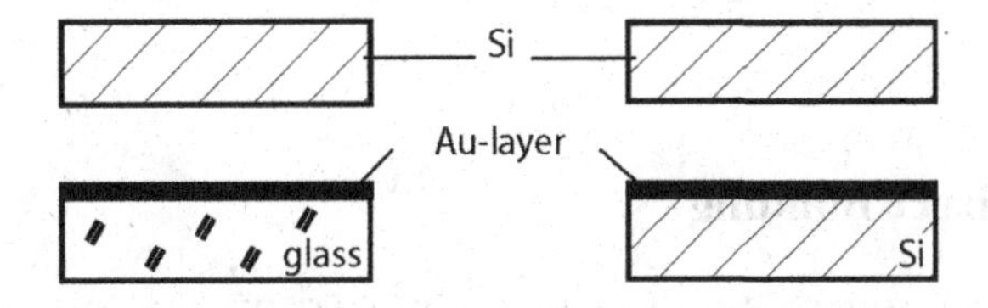

Figure 4.40 Bonding Si wafers to (a) Si or (b) glass using eutectic bonding

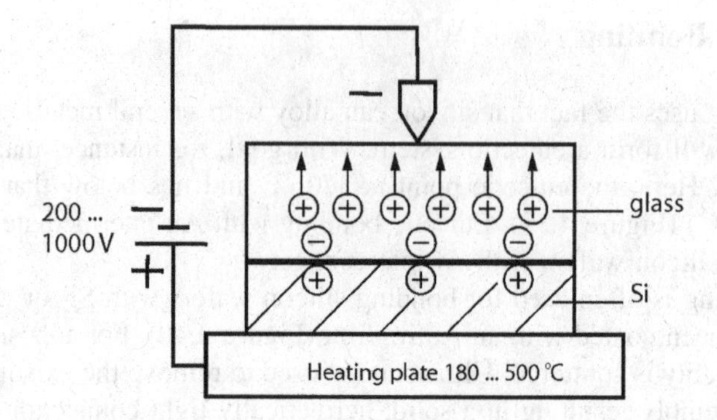

Figure 4.41 Anodic bonding of silicon and glass

and the glass has to have a sufficiently high concentration of alkali ions for the bonding mechanism. Such glasses are Pyrex, Rasotherm, Borofloat etc. (see Section 3.4).

Figure 4.41 shows a schematic of the bonding mechanism:

- Silicon and glass wafer are put into close contact with each other. They have to have a superior surface quality (polished, cleaned).
- For sufficiently high temperatures (see Equation 4.40), the applied direct voltage of several 100 V causes a diffusion of sodium ions Na^+ away from the bonding surface between Si and glass. The oxygen ions O^{2-} that the glass retains from NaO_2 have a diffusion coefficient that is several orders of magnitude lower and they are practically fixed. The depletion of Na^+ ions leads in the glass to a negative volume charge at the bonding surface which affects a positive volume charge on the opposite side in the silicon.
- This way, the voltage over the small air gap between glass and silicon drops almost completely and the bonding partners are very strongly attracted.
- On the contact areas, irreversible Si-O-Si-bonds are formed.

The bonding process starts at the first contact point of Si and glass close to the cathode and spreads from there in form of fronts. The bonding process of a larger wafer takes several minutes.

Anodic bonding can also be used for connecting two silicon wafers if a Pyrex glass layer is sputtered onto one of the wafers. The same applies for the bonding of thermally oxidized wafers without Pyrex glass layer. In this case, the diffusion of OH^--and H^+-ions instead of Na^+ leads to the bonding of the wafer. When using thin layers as intermediate bonding layer, the applied voltage must not exceed 20 . . . 50 V in order to avoid electrical breakdown.

4.8.3 Silicon Direct Bonding

Silicon direct bonding (SDB) describes the bonding of silicon wafers without any intermediate layer. If the wafers have a thin layer consisting of thermal or natural SiO_2, this process is called silicon fusion bonding (SFB). In general, the terms SDB and SFB are used interchangeably. The SDB technique requires extremely smooth and even wafer surfaces. This is achieved through chemical-mechanical polishing (CMP) and wet chemical cleaning processes. The bonding process takes place as follows [ELWENSPOEK98]:[8]

- The surfaces of both wafers are hydrophilized using a wet chemical process with a $NH_4OH/H_2O_2/H_2O$ solution or a plasma chemical procedure. On the surfaces, OH groups will form that bond with the external Si atoms.
- At temperatures above 120 °C, molecular water will be formed at the interface between the wafers

$$\equiv \mathrm{Si{-}OH} + \mathrm{HO{-}Si} \equiv \rightarrow \equiv \mathrm{Si{-}O{-}Si} \equiv + \mathrm{H_2O},$$

  which causes strong covalent bonds.
- The water vapor oxidizes the silicon near the interface. The remaining oxygen diffuses in the Si bulk.

A possible longer annealing between several 100 °C and 1000 °C leads to the formation of -Si-Si-bonds.

4.8.4 Comparison of Bonding Techniques

Table 4.19 compares the main characteristics of the different wafer bonding techniques.

Table 4.19 Comparison of characteristics of wafer bonding techniques (acc. [FUKUDA98])

Technique	Bonding partners	Intermediate layer	Temperature in °C	Surface preparation
Gluing	Si-Si SiO_2-SiO_2 Si-Glass	Glue	RT ... 200	Cleaning; if applicable deposit glue by spinning
Glass soldering	Si-Si SiO_2-SiO_2 Si-Glass	Glass solder (Na_2O-SiO_2)	200 ... 450	Screen-printing/applying the glass solder layer by spincoating
Eutectic bonding	Si-Si	Au Al	379 580	Vapor deposition or sputtering
Anodic bonding	Si-glass Si-Si	none Pyrex	>250 >300	
Silicon direct bonding	Si-Si SiO_2-SiO_2	none	200 ... 400	Plasma treatment, surface activation

[8] Gösele, U. *et al.* (1999) Fundamental issues in wafer bonding. *Journal of Vacuum Science and Technology*, A17 4, pp. 1145–52.

4.9 INSULATION TECHNIQUES

Insulating layers between two layers of mono-crystalline material have a large number of applications:

- Dielectric insulation instead of insulation via pn-junctions in silicon bulks makes it possible to use operating temperatures of several hundred °C instead of only 120...150 °C.
- Insulating layers of SiO_2 and Si_3N_4 can be used as thermal barriers.
- The selective etching characteristics of SiO_2 and Si_3N_4 as opposed to Si, make it possible to use etch stop layers in mono-crystalline silicon (cf. Figure 4.20) or as sacrificial layer in surface micromachining (see Section 4.10).

For mono-crystalline silicon layers on SiO_2, the abbreviation SOI (Silicon on Insulator) is used. In the following, we want to present techniques that can produce insulation layers (especially SiO_2 layers between two single crystalline Si layers).[9] It is an issue that silicon thin films can only be deposited in poly- or mono-crystalline form on SiO_2 layers. The recrystallization of such poly-crystalline layers puts high technical demand on devices and achieves only a limited homogeneity and quality. Therefore, such techniques (e.g. ZMR Zone-Melting Recrystallization) are not widely used.

4.9.1 SIMOX Technique

Due to its application in microelectronics, the SIMOX- (**S**eperation by **IM**planted **OX**ygen) technique is one of the most commonly used insulating techniques. It uses the implantation of oxygen ions O^+ in the silicon bulk and its transformation into SiO_2 (Figure 4.42).

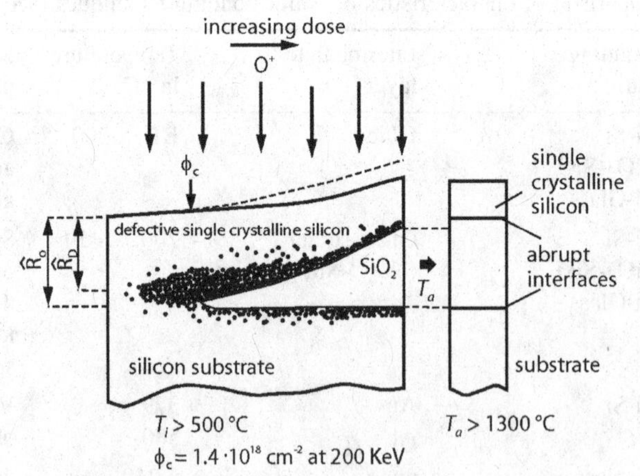

Figure 4.42 Generation of the SIMOX structure during oxygen implantation
Source: Celler, G. K., Cristoloveanu, S. (1993) Frontiers of silicon-on-insulator. *Journ. of Applied Physics* 93(9), pp. 4955–78.

[9]Celler, G. K., Cristoloveanu, S. (1993) Frontiers of silicon-on-insulator. *Journal of Applied Physics* 93(9), pp. 4955–78.

An O^+ ion implantation dose of $n^* > 10^{18}$ cm^{-2} is required to form a stoichiometric oxide in Si. This dose exceeds the values used in microelectronics by a factor of about 100. In order to prevent an amorphization of Si at such high doses, the implantation will be accompanied by the thermal annealing of radiation damages at 500...600 °C. After the implantation, a high-temperature annealing at approx. 1350 °C serves to bond the oxygen with silicon to SiO_2. At the same time, OSTWALD ripening takes place that grows SiO_2 precipitations above a critical size and dissolves those that are smaller. Above 1300 °C this critical radius approaches an infinite value, resulting in the formation of a planar SiO_2 layer with atomically sharp boundaries in the Si single crystal. Such layers are called buried oxide layers (BOX, Buried Oxide). Commercial SIMOX wafers have a silicon thickness of about 60...250 nm and the buried SiO_2 layer has a thickness of about 100 nm.

4.9.2 BESOI Technique

The Bond-and-Etchback technique for producing SOI (BESOI) uses two oxidized wafers that are directly bonded using silicon fusion bonding (see Section 4.8.3). Then, most commonly one of the two wafers is thinned to reach the thickness desired for the respective function elements (Figure 4.43). To achieve this, a highly doped or electrochemical etch stop layer is generated at this Si wafer, e.g. using high-dose implantation in silicon or growing an epitaxy layer on the silicon surface. Initially, the thinning is carried out mechanically, then wet chemical etching is used to reach the final thickness.

The bond-and-etchback technique can be directly applied for producing three-dimensional forms if one or both wafers are structured prior to bonding. In principle, also anodic bonding can be used in addition to SDB. Figure 4.44 shows that the modified BESOI technique (without oxide layers on the wafers) can be used to produce pressure sensor chips that – due to sidewalls inclined in opposite directions – require much less space than those produced using anisotropic etching.

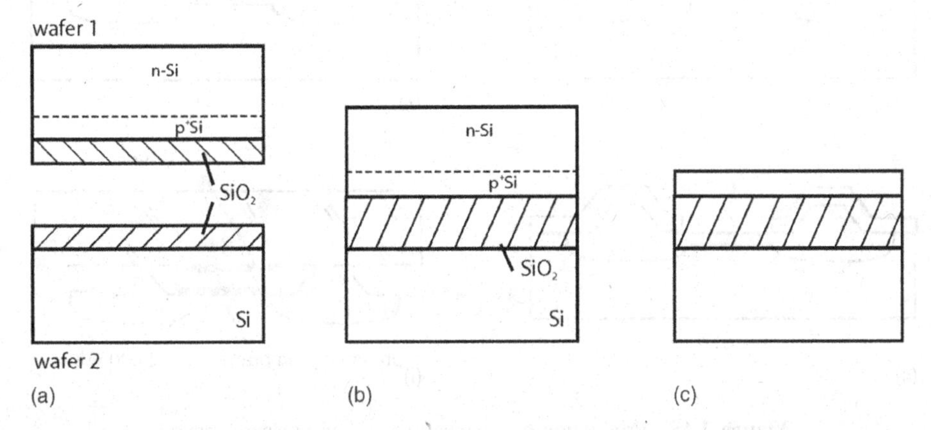

Figure 4.43 BESOI technique: (a) two oxidized original wafers, one with an etch stop layer (wafer 1: p^+n junction); (b) bonding of these two wafers; (c) etchback to the etch stop layer

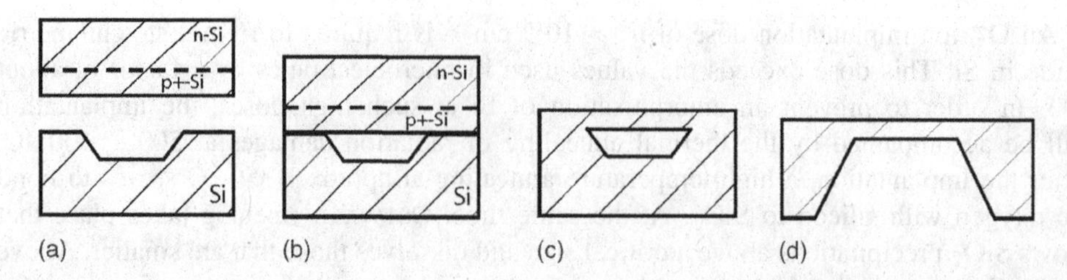

Figure 4.44 Production of a pressure sensor using bond-and-etchback: (a) preparation of the two wafers; (b) bonding; (c) back-etching; (d) size comparison with an anisotropically etched sensor chip

Example 4.8 Production of a capacitive silicon pressure sensor using Bond-and-etch-back[10]

The goal is to produce a capacitive silicon pressure sensor. The membrane of the silicon pressure sensor has the following dimensions: $200 \times 200\ \mu m^2$ and a thickness of 2 μm. The air gap between the electrodes is 5 μm. Figure 4.45 shows the

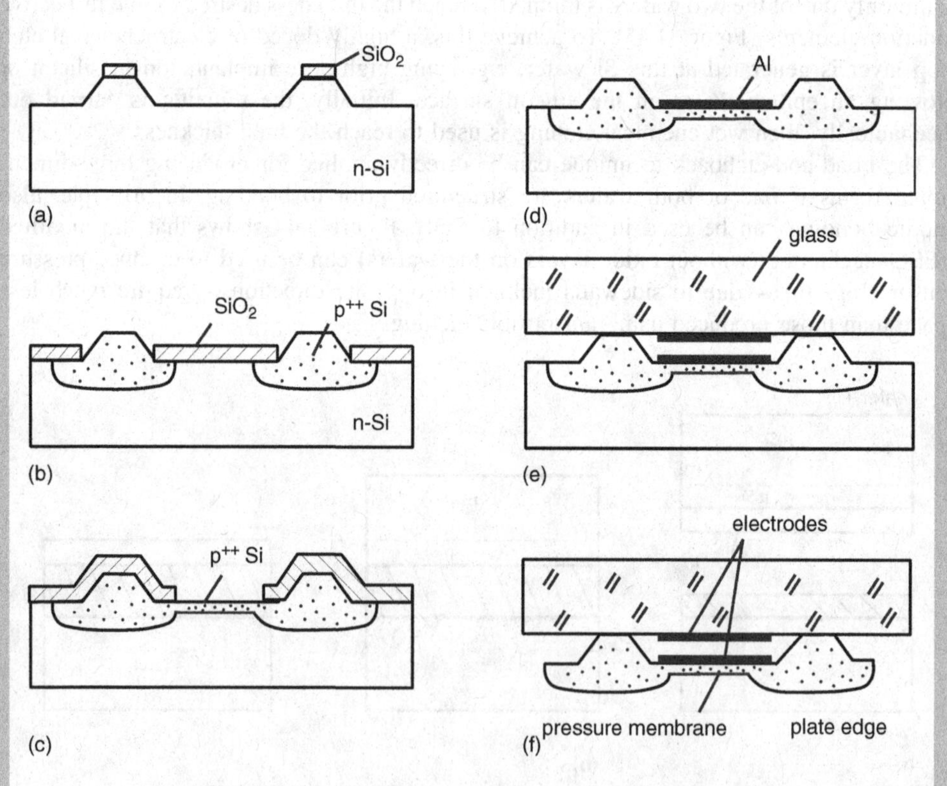

Figure 4.45 Production of a capacitive silicon pressure sensor

[10]Chao, H.-L., Wise, K.D. (1998) An ultraminiature solid-state pressure sensor for a cardiovascular catheter. *IEEE Transactions on Electron Devices* 35(2), pp. 2355–62.

manufacturing process of the sensor that requires five lithography masks: (a) etch mask, (b) and (c) implantation masks, (d), (e) metallization of movable and counter-electrode.

The following technological steps are necessary to reach the dimensions given above:

- The air gap distance results from the anisotropic etching step in Figure 4.45a and is reduced by the thickness of the two electrode layers (Figure 4.45d,e).
- The membrane dimensions are determined by the first ion implantation step of the solid edges in Figure 4.45b.
- The ion implantation depth in Figure 4.45c determines the membrane's thickness.

4.9.3 Smart-cut Technique

The Smart-cut technique[11] uses the implantation of gas ions that generate microcavities in the silicon wafer. If the wafer is heated to 400 ... 500 °C, pressure is built up and will lead to that the Si wafer breaks in the area of the highest dopant concentration (Figure 4.46). The process comprises the following steps:

- a 'donor wafer', used to generate the single crystalline silicon layer, is oxidized; this SiO_2 layer forms after the bonding the buried SiO_2 layer;
- ion implantation by the SiO_2 layer (Figure 4.46a: dose $> 5 \cdot 10^{16}$ cm^{-2});
- cleaning of the wafer and an oxidized carrier wafer; silicon fusion bonding,
- heating of the entire structure to 400 ... 600 °C, resulting in the separation of the 'donor wafer';
- polishing of the separated wafer area.

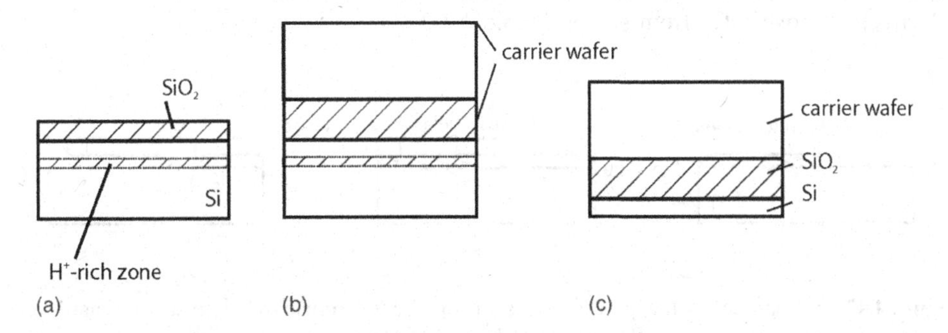

Figure 4.46 Smart-cut process: (a) ion implantation of H^+ ions by a SiO_2 layer into the silicon; (b) SDB with a carrier wafer; (c) separation of H^+-implanted area after heating

[11] Bruel, M. (1994) *Process for the production of thin semiconductor material films*. Patents FR 19910011491, US 5374564.

4.10 SURFACE MICROMACHINING

4.10.1 Principle

Surface micromachining is a technology for producing movable structures on the surface of silicon wafers. Figure 4.47 shows the general principle.

- At first, a sacrificial layer is deposited and patterned on the substrate surface.
- The function layer is deposited on the sacrificial layer and patterned according to its function.
- The sacrificial layer is removed. Those parts of the function layer that were deposited on the sacrificial layer will become movable, whereas the areas that are linked to the substrate surface will form the fixture of the mobile function elements.

The first time such a sacrificial layer technique was used was in 1965 for the production of a vibration sensor.[12] A cantilever made of gold (2.5 mm long, 3.4 μm thick) was used as a movable gate over a field effect transistor. Photoresist was used as sacrificial layer material. After this technique was rediscovered,[13] preferably the following material combinations are used:

- sacrificial layer: SiO_2, PSG (phosphosilicate glass);
- function layer: Poly-Si.

Poly-silicon naturally has material characteristics that resemble those of single crystalline silicon to a degree that undesired mechanical and thermal effects are small. A large number of procedures can be used to produce SiO_2 of a sufficient quality (Table 4.20). With etching solutions based on hydrofluoric acid (HF) or buffered hydrofluoric acid (BHF; aqueous solution of HF and ammonium fluoride NH_4F), it is possible to very selectively remove SiO_2 from silicon (Table 4.21).

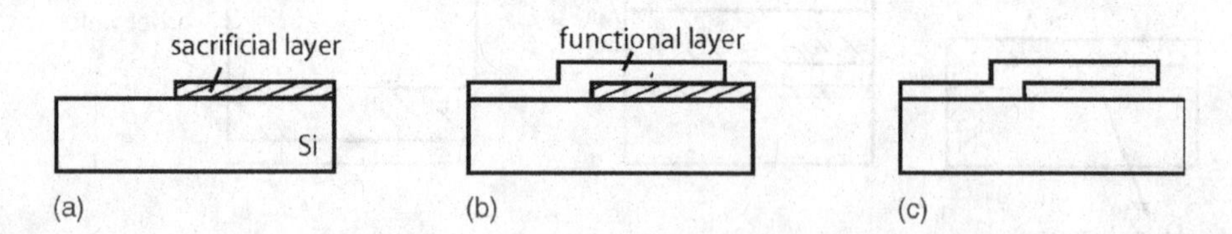

Figure 4.47 Schematic of the process steps of surface micromachining (example cantilever): (a) deposition and patterning of the sacrificial layer; (b) deposition and patterning of the function layer; (c) after etching the sacrificial layer

[12]Nathansen, H.C., Wickstrom, R.A. (1965) A resonant-gate silicon surface transistor with high-Q band pass properties. *Applied Physics Letters* 7, pp. 84–6.
[13]Howe, R.T., Muller, R.S. (1983) Polycrystalline and amorphous silicon micromechanical beams: annealing and mechanical properties. *Sensors and Actuators* 4, pp. 447–54.

Table 4.20 Characteristics of SiO_2 layers as sacrificial layer material

Production	Deposition temperature in °C	Deposition rate	Edge coating	Layer stress	Etch rate in 1 % HF solution in nm/s	Remarks
Thermal	800...1000	low	conform	−	0.1	very good layer quality, good thickness control
LPCVD	600...900	low	conform	−	low	good layer quality
LPCVD-LTO	400...500	low	poor	−...+	0.7 (25)[1]	high H-content
LPCVD-TEOS	650...750	medium	conform	−	medium	
APCVD	300...500	large	good	+	1 (15)[1]	layer impurities
PECVD	200...400	medium	conform	−	1.2 (40)[1]	incorporation of H, N into layer
PECVD-TEOS	350...400	medium	conform	−...+	0.3 (1)[1]	incorporation of H, N into layer
SOG	<200	large	planarizing	−...+	1.2	instable layers

− compressive stress; + tensile stress; LTO Low Temperature Oxide; TEOS Tetraethyl orthosilicate $Si(OC_2H_5)_4$; SOG Spin-on-Glass;
[1]PSG (SiO_2 deposited with 7 weight% P)
Source: Bühler, J., Steiner, F.-P., Baltes, H. (1997) Silicon dioxide sacrificial layer etching in surface micromachining. *Journal of Micromechanics and Microengineering* 7, R1–R13.

Table 4.21 Etch rates of buffered hydrofluoric acid (BHF; HF (50 %) : NH_4F (40 %) = 1: 5...1: 100)

Etched material	Etch rate in nm/min	Etched material	Etch rate in nm/min
(1 0 0)-Si	<0.1	SiO_2 (PECVD)	45...160
(1 1 1)-Si	<0.1	SiO_2 (PSG)	60
Poly-Si	0.2	Si_3N_4 (PECVD)	4
SiO_2 (thermal)[1]	22		

[1] Wet oxide, 1100 °C

Surface micromachining can even use several sacrificial and function layers, which means that it is possible to produce complex three-dimensional forms. A large variety of different forms can be carried out. The thickness of the sacrificial layer determines motion space or backlash, respectively, of the movable mechanical elements.

Example 4.9 Production of a gearwheel using surface micromachining

Figure 4.48 shows the steps for manufacturing a poly-Si gearwheel that is attached to a Si substrate rotating shaft. The two function layers #1 and #2 are used to form the gearwheel and the shaft for the gearwheel. In order for the gear wheel to rotate freely it has to be separated from substrate and shaft. This is achieved through two sacrificial layers that completely enclose the gearwheel. In order to keep friction between gearwheel and shaft low, the interfaces between function and sacrificial layers have to be very smooth.

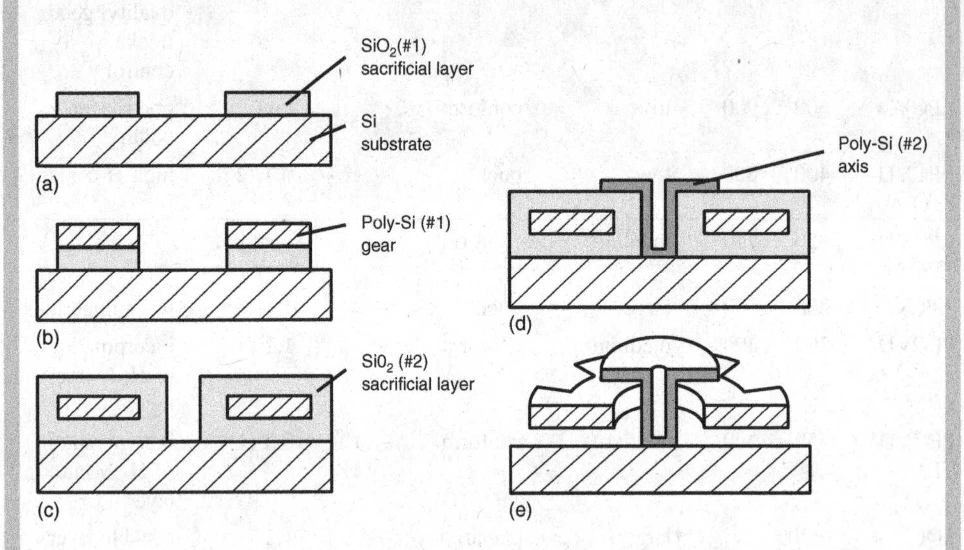

Figure 4.48 Production of a gearwheel: (a), (b) deposition and patterning of the first sacrificial and function layer, respectively; (c), (d) deposition and patterning of the second sacrificial and function layer, respectively; (e) baring the gearwheel through sacrificial layer etching

4.10.2 Production of Hollow Spaces

Many applications require closed, often even hermetically tight hollow spaces. Absolute pressure sensors, for instance, require vacuum on one side of the deformable pressure membrane. For separating sacrificial layers it is necessary, though, to have access underneath the function layers in order to etch out the sacrificial layers. Usually, there are specific sacrificial layer channels through which the corresponding etching solution can remove the sacrificial layer. These comparatively thin channels are later closed (Figure 4.49):

- Thermal oxidation: If sacrificial layer channels are incorporated into (single crystalline or poly-) silicon, they are closed by growing during thermal oxidation (see Figures 4.49d1 and 4.11).
- Layer deposition (Figure 4.49d2).

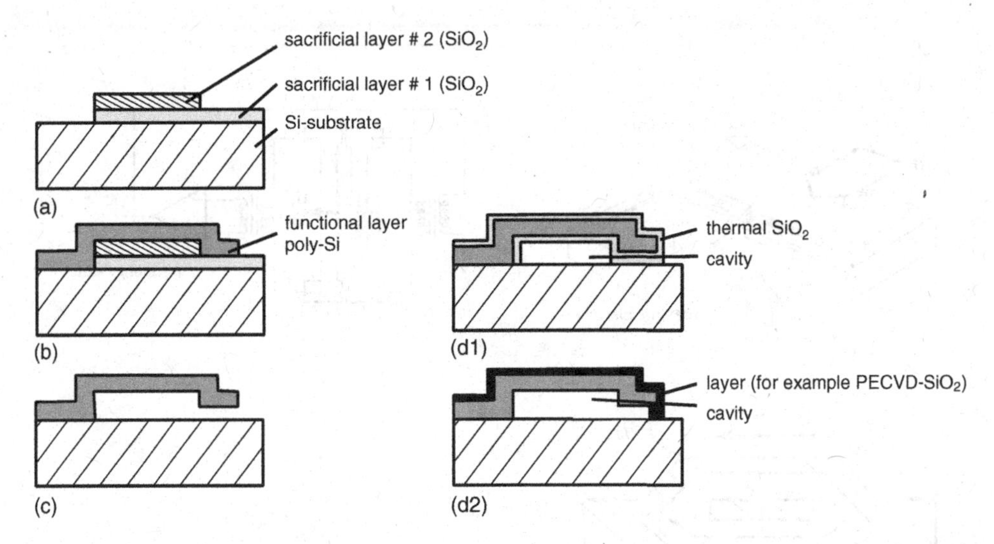

Figure 4.49 Production of closed hollow spaces applying surface micromachining: (a) deposition of two sacrificial layers in order to create sacrificial layer channels (#1) and hollow space (#2), (b) deposition of the function layer; (c) sacrificial layer etching; (d1) closure of hollow space through thermal oxidation; (d2) through layer deposition

Example 4.10 Production of a resonant pressure sensor

Figure 4.50 shows a resonant pressure sensor, where an H-shaped structure is bonded to the surface of a pressure plate at the four ends of the H and swings in resonance. The oscillation movement vertical to the pressure membrane is excited by alternating current flowing through the arms of the H-structure and is generated by LORENTZ force F_Z, when magnetic field B acts parallel to the membrane surface (cf. Section 1.4):

$$F_z(t) = I_y(t) \cdot l_y \cdot B_x. \tag{4.52}$$

A bending deformation caused by measuring pressure p changes the resonance frequency of the vibrating H-structure. A high-quality oscillation is achieved in vacuum. It is therefore useful to encase the H-structure according to Figure 4.50c. For the production, the following boundary conditions have to be taken into account:

- The oscillating H-structure (function layer #1) will consist of single crystalline silicon.
- Therefore, the first sacrificial layer has to be single crystalline silicon that is epitaxially applied – similar to the H-oscillator.
- In order to achieve a sufficient etching selectivity between function and sacrificial layers, the former has to be highly-doped with boron and the etch stop has to be applied to these highly-doped p^{++} layers (even Section 4.6.4), where p-Si is normally and p^{++}-Si is anisotropically and wet chemically removed with an etch rate that is several orders of magnitude lower.
- A second function layer, also p^{++}-Si, is used for the casing of the oscillating structure.

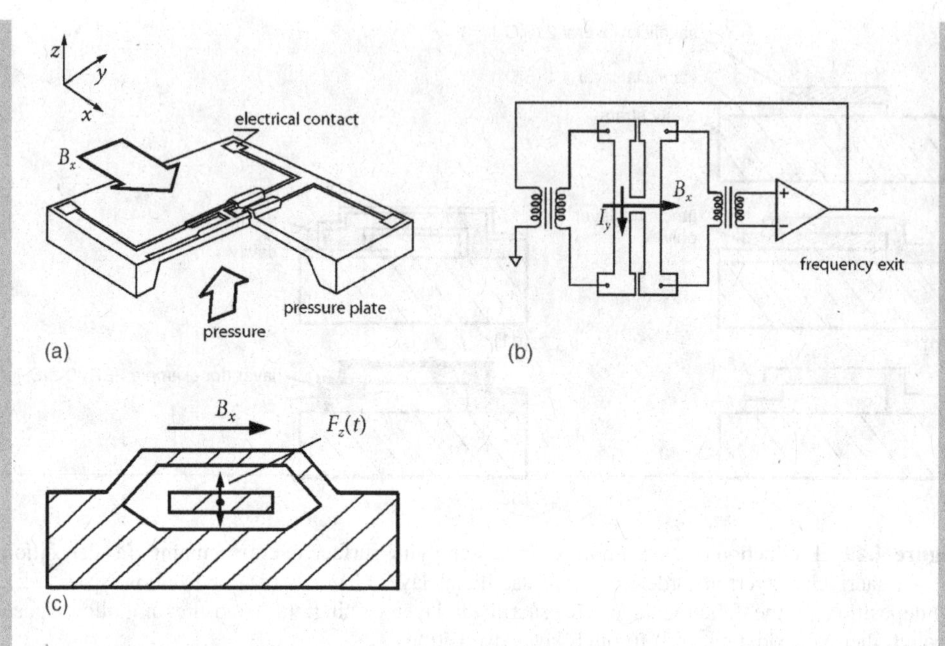

Figure 4.50 Resonant pressure sensor: (a) structure; (b) schematic of the function principles; (c) cross-section of oscillating structure
Source: Ikeda, K., Kuwayama, H., Kobayashi, T., Watanabe, T., Nikshikawa, T. *et al.* (1990) Three dimensional micromachining of silicon pressure sensor integrating resonant strain gauge on diaphragm. *Sensors and Actuators* A 21–23, pp. 328–31.

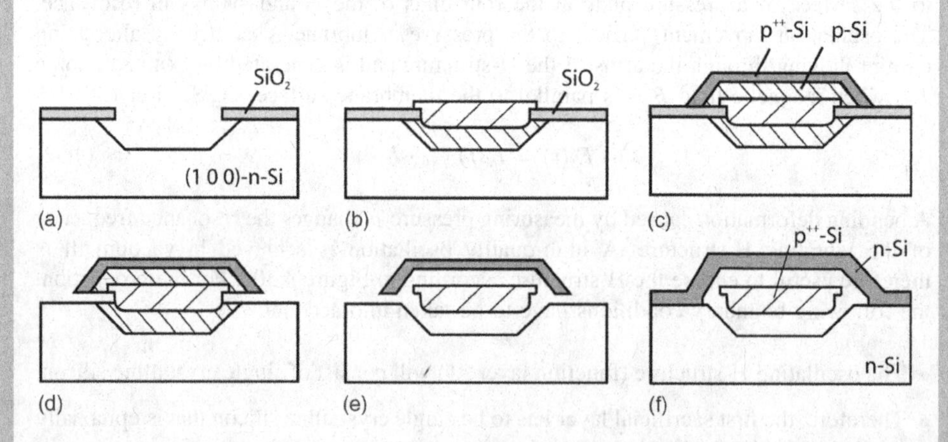

Figure 4.51 Production of the resonant pressure sensor in Figure 4.50c (explanation in text)

Figure 4.51 shows the production steps. A thin SiO_2 layer is first used as an etch mask for creating an etching channel to etch the sacrificial layer (Figure 4.51a) and later as the etching channel to etch the sacrificial layer (Figure 4.51d). Then a sacrificial layer of p-Si is epitaxially deposited. It has to be thicker than the oscillation amplitude of the p^{++}-layer (oscillating H-structure) that has been grown epitaxially (Figure 4.51b). The following p-Si sacrificial layer later allows H-structure to vibrate upwards, whereas the p^{++} layer applied to it will later form the lock of the casing (Figure 4.51c). Now, the sacrificial layers will be removed from SiO_2 (Figure 4.51d) and p-Si (Figure 4.51e). Hydrofluoric acid, for instance, is used for that as it has a high etching selectivity between SiO_2 and p- or p^{++}-Si, respectively, as well as KOH solution which etches p-Si but retains the p^{++}-Si function layers. The structure is closed via epitaxial deposition of n-doped silicon (Figure 4.51f). Annealing in nitrogen atmosphere causes the charge carriers to diffuse from the p^{++}-Si-cover layer to the above situated n-Si layer where they recombine. After the annealing, the resulting cover layer thus becomes n-conducting which means that only the oscillation structure remains p^{++}-conducting. This way, the current can be directly applied to the oscillating H-structure which secures the electrical insulation during pn-junctions to the surrounding n-Si.

4.10.3 Adhesion of Movable Structures

(a) Adhesion through liquid films

In surface micromachining, the movable structures are laid bare at the end of the production process during sacrificial layer etching. If SiO_2 is the sacrificial layer, aqueous HF solution is used for this (Table 4.21). The structures are then rinsed in water and dried. In the small gaps, the liquid deploys capillary force that cause the mobile structures to adhere to the directly opposite, fixed areas (Figure 4.52, Appendix E). This process is often referred to as sticking.

Adhesion will occur if the adhesion force is larger than the force of the elastic displacement of the surface-micromechanical structure. Critical length l_{crit}, where adhesion will occur, is[14]

$$l_{\text{crit}} = \sqrt[4]{\frac{3}{2} \cdot \frac{Eh^3d^2}{\gamma_S}}. \tag{4.53}$$

Here, E is YOUNG's modulus, h the cantilever thickness, d the gap distance, and γ_S the area-related adhesion energy. If adhesion is caused by a liquid film, adhesion energy is determined by surface tension γ_{FA} between liquid and air as well as contact angle φ between liquid and structural element:

$$\gamma_S = 2\gamma_{\text{FA}} \cos\varphi. \tag{4.54}$$

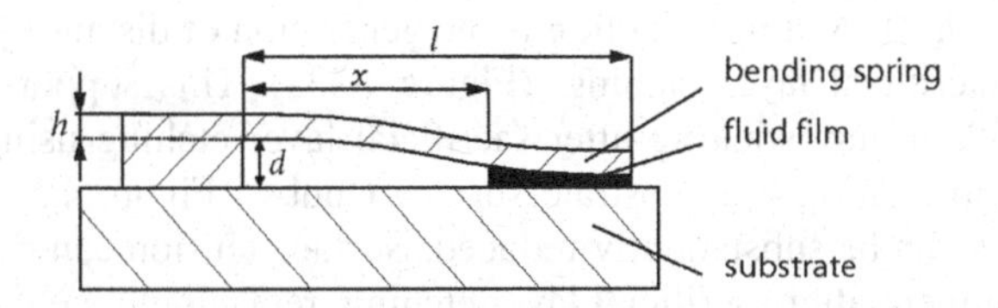

Figure 4.52 Adhesion of a cantilever spring through a liquid film on the substrate

[14] Tas, N., Sonnerberg, T., Jansen, H., Legtenberg, R., Elwenspoek, M. (1996) Stiction in surface micromachining. *Journal of Micomechanics and Microengineering* 6, pp. 385–97.

Factor 2 takes into account that two liquid menisci form in the gap. This way, Equation (4.53) becomes

$$l_{crit} = \sqrt[4]{\frac{3}{4} \cdot \frac{Eh^3d^2}{\gamma_{FA} \cos\varphi}}. \tag{4.55}$$

Neglecting the intrinsic stress and the membrane stress caused by the elongation of the neutral axis, for beams clamped on two sides, l_{crit} exceeds the value in Equation (4.55) by factor 2.5 and the critical radius of circular plates by factor 1.7.

Example 4.11 Adhesion of a cantilever spring

A cantilever spring or beam made of poly-Si and clamped on one side has a thickness of $h = 2$ µm and a YOUNG's modulus of $E = 160$ GPa. The gap d caused by the removed sacrificial layer is 2 µm. Surface tension is $\gamma_{FA} = 146$ mJ/m^2 and the contact angle $\varphi = 0$ (water, 20 °C). Thus, critical length according to Equation (4.54) is

$$l_{crit} = \sqrt[4]{\frac{3}{4} \cdot \frac{160 \cdot 10^9\ \mathrm{Nm}^{-2} \cdot (2 \cdot 10^{-6})^3 \mathrm{m}^3 \cdot (2 \cdot 10^{-6})^2 \mathrm{m}^2}{146 \cdot 10^{-3}\ \mathrm{Nm}^{-1} \cdot \cos 0}} = 72\ \mu\mathrm{m}.$$

In most applications, the structures have larger lateral dimensions resulting in sticking.

(b) Avoiding adhesive structures

Based on Figure 4.52 and Equation (4.55), the following measures can be adopted to avoid sticking of surface micromechanical structures on silicon substrate or on neighboring structures:

- more rigid structure,
- smaller adhesion areas,
- lower surface tension,
- avoiding liquids during sacrificial layer etching and cleaning.

Achieving a higher intrinsic stiffness of the structure is difficult as critical length l_{crit} is only proportional to $h^{3/4}$. A simpler option is the generation of distance pillars of polymer material prior to sacrificial layer etching. (Figure 4.53a). This support structure can be easily removed without any sticking after sacrificial layer etching using plasma etching in oxygen. By prestructuring the substrate surface ('nubs', Figure 4.53b), adhesion area and adhesion forces can be substantially reduced. Surface tension can be reduced through special surface treatment after sacrificial layer etching, replacing the etching solution with a suitable liquid. This results in hydrophobic, low-adhesion, self-assembling mono-layers, for instance.

The generation of liquid films can be avoided using gaseous HF vapor instead of aqueous HF solution. The same applies to rinsing and cleaning processes, where – instead

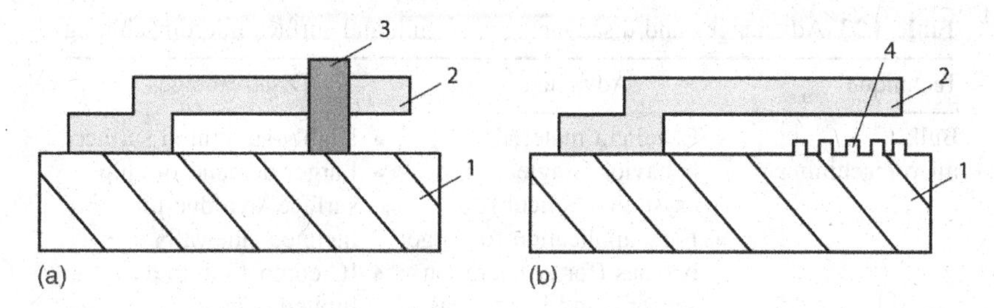

Figure 4.53 Avoiding of sticking of surface micromechanical structures by (a) higher rigidity using distance columns; (b) reducing the contact area through nubs; 1: substrate, 2: cantilever spring, 3: distance column, 4: nubs

Table 4.22 Technique for avoiding sticking when opening up surface micromechanical structures

Sublimation	Super-critical drying with CO_2 applications
1. Sacrificial layer etching with HF- or BHF-etching solution	
2. Rinsing with de-ionized water	
3. Rinsing the water with IPA (Isopropylalcohol)	
4. Rinsing the IPA with cyclohexane	4. Rinsing the IPA with liquid CO_2
5. Freezing of cyclohexane at $-10\ ^{\circ}C$	5. Increasing temperature (>35 °C) and pressure (>75.8 bar)
6. Sublimation of cyclohexane	6. Relieving super-critical CO_2 at 35 °C and normal pressure
7. Return to ambient temperature	

of cleaning solutions – sublimation or super-critical drying through CO_2 applications is used (Table 4.22).

4.10.4 Comparison of Bulk and Surface Micromachining

Table 4.23 compares advantages and disadvantages of surface micromachining with those of bulk micromachining. Which of the two techniques will be chosen depends on the specific application, structural forms and the available processing equipment.[15]

Therefore, the majority of pressure sensors is currently produced using bulk technology (with the piezoresistive measuring principle dominating), whereas acceleration sensors are predominantly manufactured using surface micromachining (and operate mainly capacitively).

[15]French, P.J., Sarro, P.M. (1998) Surface versus bulk micromachining: the contest for suitable applications. *Journal of Micromechanics and Microengineering* 8, pp. 45–53.

Table 4.23 Advantages and disadvantages of bulk and surface micromachining

Technique	Advantages	Disadvantages
Bulk micromachining	• Excellent material behavior (single crystalline silicon) • Easy application for large batches (for acceleration sensors) and large areas (for capacitive and electrostatic transformers)	• Highly-structured surface • Larger demand of chip surface area due to inclined sidewalls • IC-compatible to a limited degree • Limited structural variability
Surface micromachining	• Smaller chip areas are possible • IC-compatible • Higher structural variety • High material variety	• Small areas and masses • No single crystalline materials can be used

4.11 NEAR-SURFACE MICROMACHINING

Surface micromachining has the big advantage that the entire wafer processing is carried out on the surface of the silicon wafer. Double exposure is avoided and, due to the application of thin layers, etching times are low. Its disadvantage, however, is that function elements consist of poly-Si and not of single crystalline silicon. Near-surface micromachining intends to generate movable structures of single crystalline silicon in the near-surface area of Si wafers.

4.11.1 Principle

In order to generate completely movable structures on the Si wafer surface, it is necessary to open up the area around them. Figure 4.54 shows the two principal process steps for achieving that:

1. Starting from the etch mask on the wafer surface, deep trenches are anisotropically etched (Figure 4.54a). The trench walls will later delimit the target structure.
2. Deep down in the wafer, the target structure has to be undercut. To achieve that the silicon is isotropically etched starting from the bottom areas of the trenches (Figure 4.54b). Lateral etch removal opens up the structure. In order to protect the lateral boundary areas from further etching they are passivated – similar to the wafer surface – by an etch resist.

Both etching steps can use the same etch mask. Therefore, the techniques of near-surface micromachining are often referred to as one-mask technique.

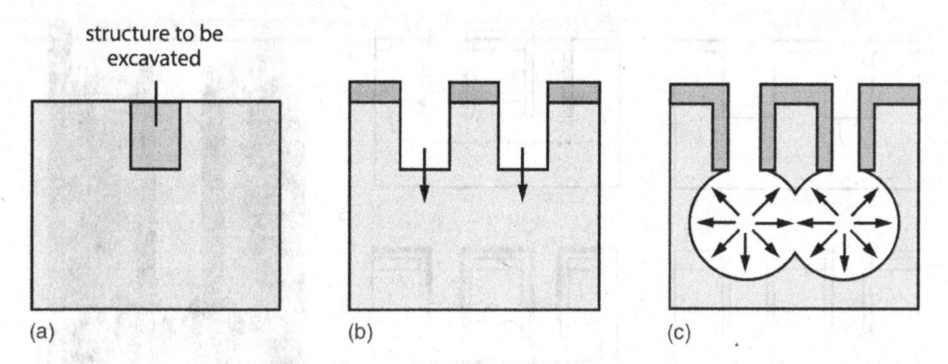

Figure 4.54 Schematic representation of near-surface micromachining: (a) initial structure; (b) anisotropic etching process for lateral boundary areas; (c) isotropic etching process for undercutting of the target structure

4.11.2 Techniques

The principle presented in Figure 4.54 will be illustrated with two of the most commonly known techniques.

(a) SCREAM process

The acronym SCREAM stands for **S**ingle **C**rystal **R**eactive **E**tching **A**nd **M**etallization. This process was introduced by Z.L. Zhang and N.C. MacDonald in 1992. It was the first technique of near-surface micromachining.[16,17]
Figure 4.55 shows a schematic presentation of the essential manufacturing steps:

- production of the SiO_2 etch mask (thermal oxidation);
- deep reactive ion etching (DRIE) for generating etch trenches ($Cl_2/BCl_3/H_2$ RIE) (Figure 4.55a);
- passivation of sidewalls in the etch trenches using thermal oxidation (150 nm, 1000 °C, wet) (Figure 4.55b);
- anisotropic reactive ion etching in order to remove the oxide layer from the bottom areas of the structure and to deepen the structure (CF_4/O_2) (Figure 4.55c);
- isotropic etching of Si-bulks starting from the opened-up bottom of the etch trenches (SF_6) (Figure 4.55d);
- magnetron sputtering of Al (Figure 4.55e).

[16]Zhang, Z.L., MacDonald, N.C. (1992) A RIE process for submicron, silicon electromechanical structures. *Journal of Micromechanics and Microengineering* 2, pp. 31–8.
[17]MacDonald, N.C. (1996) SCREAM MicroElectroMechanical Systems. *Microelectronic Engineering* 32, pp. 49–73.

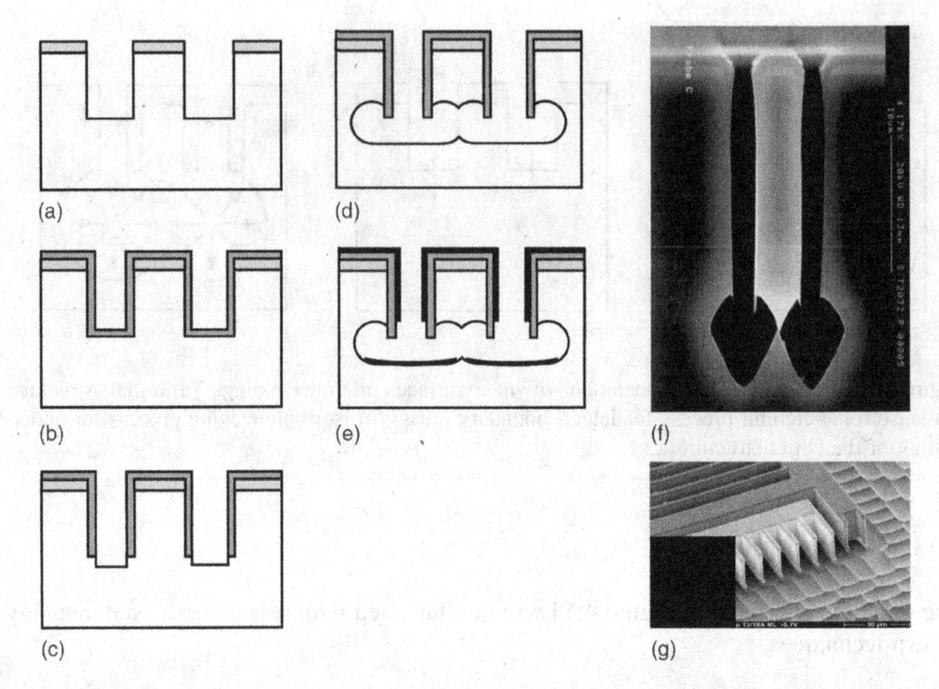

Figure 4.55 SCREAM technique: (a) etching of lateral boundary trenches using DRIE; (b) oxidation of the surface; (c) continuation of DRIE; (d) isotropic etching of the etch trench bottom; (e) metallization and contacting of the structure; (f) and (g) cross-section and lateral view of the bared structure. Images 4.55f and 4.55g are reproduced by permission from the Center for Microtechnologies, Chemnitz University of Technology, Germany

If the trench walls are metallized, the structure is very suitable for electrostatic drives. The electrostatic forces increase in proportion to the electrode area. Through a high aspect ratio (etch depth to width of structure) it is possible to use large electrode areas, which is particularly advantageous for interdigital structures.

(b) SIMPLE process

The SIMPLE technique (**Si**licon **M**icromachining by **P**lasma **E**tching) applies doped and epitaxial layers on the wafer surface. These layers are used as etch stop layer for the function elements to be opened up and as fast-etching layer for the isotropic lateral etching used for structure baring. Figure 4.56 shows a schematic presentation of the individual process steps:[18]

- The original wafer has a buried highly-doped n^+-Si layer (As or Sb, implantation energy 100 ... 200 keV, dose $3 \cdot 10^{15} \ldots 3 \cdot 10^{16}$ cm^{-2}) for the lateral undercutting etching process and, on this, an epitaxially deposited silicon layer. Its thickness determines the thickness of the mechanical function elements to be structured (Figure 4.56a).

[18]Li, Y.X., French, P.J., Sarro, P.M., Wolffenbuttel, R F. (1996) SIMPLE – A technique of silicon micromachining using plasma etching. *Sensors and Actuators* A57, pp. 223–32.

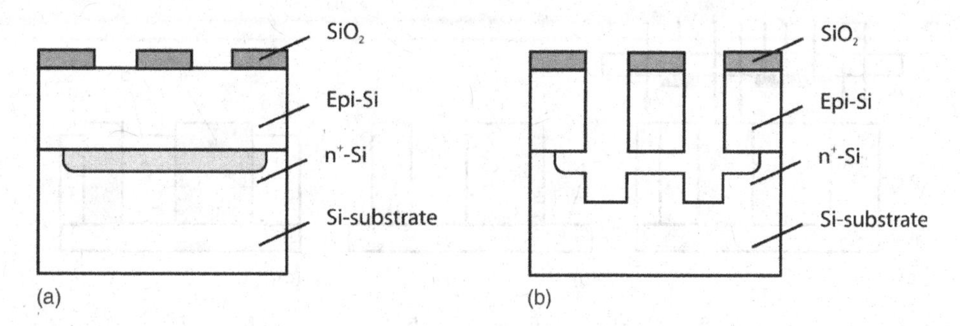

Figure 4.56 SIMPLE process: (a) original wafer; (b) after etching in the Cl_2/BCl_2 plasma

- The wafer is etched chemically with gas in a Cl_2/BCl_3 plasma. Cl_2/BCl_3 has the characteristic to isotropically remove p-conducting or slightly n-conducting silicon (Figure 4.56b). This produces deep etchings in the epitaxial Si and the Si substrate whereas the buried n^+-Si layer is laterally etched off opening up the mechanical structure.

In order to produce – similar to the SCREAM process – electrostatic transducers, a metallization step can be added, analogous to Figure 4.55e. In comparison to the SCREAM process, the SIMPLE technique has several advantages:

- joint etching process for the generation of separating trenches and lateral undercutting;
- less process steps;
- the thickness of the mechanical function elements can be set, and reproduced, through the thickness of the Epi-Si-layer.

4.12 HARMST

4.12.1 Definition

The acronym HARMST comprises techniques of **H**igh **A**spect **R**atio **M**icro **S**ystem **T**echnologies. These are techniques that can produce form and function elements with high ratios of structural height and width. In principle, it subsumizes even those techniques that use anisotropic etching processes in order to produce deep etch trenches (HART High Aspect Ratio Trenches [ELWENSPOEK98], see Section 4.5.3) as well as the processes of near-surface micromachining (see Section 4.11).

4.12.2 LIGA Technique

LIGA describes a process where mechanical elements are produced using (X-ray deep) **LI**thography, (micro-), '**G**alvanik' (German for electroplating) and '**A**bformung' (German for molding). This technique was developed in the late 1970s by BECKER, EHRFELD and

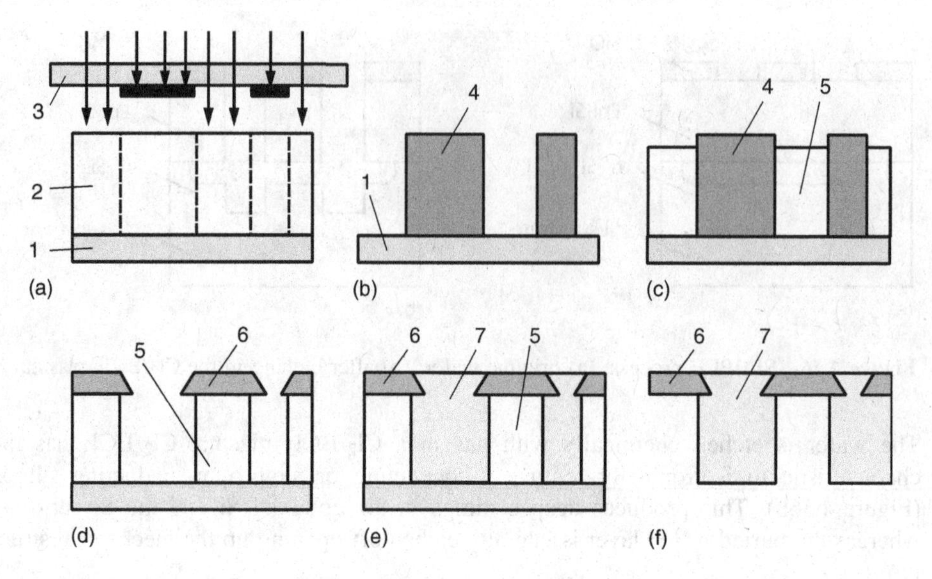

Figure 4.57 LIGA basic process: (a) X-ray deep lithography; (b) resist development; (c) galvanic deposition; (d) form production for shaping; (e) molding; (f) removal from form; 1: substrate, 2: photoresist, 3: mask, 4: resist structure, 5: metal structure, 6: carrier plate, 7: moulding mass

MÜNCHMEYER in the former Kernforschungszentrum (Nuclear Research Centre) Karlsruhe, Germany, in order to produce ultra-precision nozzle arrays for nuclear isotropic separation.[19]

Originally, the development focused on metal and plastic structures using microelectronic techniques (e.g. X-ray lithography), though. This way, LIGA constitutes a rather separate structuring technique.

Figure 4.57 shows the basic procedure of LIGA processes [BRÜCK01][EHRFELD02][MENZ95]:

- A substrate coated with an X-ray sensitive photo resist (PMMA polymethyl methacrylate) is exposed to synchrotron X-ray radiation (Figure 4.57a). The high-energy X-rays make it possible to expose a resist thickness of up to 1 mm. The highly-parallel synchrotron radiation generates extremely sharp transitions between exposed and unexposed areas in the range of micrometers. Exposure inclination is less than 0.1°.
- The photoresist structure is developed (Figure 4.57b) and serves as a form mask for the subsequent electroplating and molding processes. Mainly, electrodeposition is used to produce metallic submaster masks that are used during the actual shaping process.
- For this, a micro-electroplating process is used to fill the openings of the resist mask with metallic material (pure metals or alloys, Figure 4.57c). The submaster form is completed through removal of the photoresist material (Figure 4.57d).

[19] Becker, E., Ehrfeld, W., Münchmeyer, D. (1982) Production of separation-nozzle systems for uranium enrichment by a combination of X-ray lithography and galvanoplastics. *Naturwissenschaften* 69, pp. 520–3.

- By filling the metal form and removing the elements from the metal form, a large number of identical form or function elements can be reproduced (Figure 4.57e, f). Commonly, molding of polymers (e.g. PMMA, polycarbonate, polyamide) is used, but it is also possible to apply sintering processes with ceramics or processes with glass. As X-ray exposure produces mold inclinations of 0.1°, it is possible to easily remove the shaped structures from the metal form.

Using synchrotron X-ray for PMMA-deep exposure it is possible to achieve structural heights of up to 1 mm and aspect ratios of up to 100. During the galvanic and molding step, these extreme characteristics and precision of the lithography are transferred to the target structure. Due to the principle of LIGA, the pattern of the lithography mask continues deep down into the photoresist, limiting the range of producible forms. It is, therefore, not a genuine three-dimensional manufacturing technique, but is rather called $2\,^1\!/_2$-dimensional. Using multiple-inclined exposure and integrating sacrificial layers, it is possible to produce genuine three-dimensional and completely movable function structures.

X-ray deep lithography requires synchrotron radiation, which is expensive. LIGA is therefore only attractive for the production of large batches where one exposed PMMA structure in Figure 4.57b can be used for the production of a large number of submaster masks. For applications with lower demands, it is also possible to use UV radiation instead of X-ray. Due to the lower photon energy, the resist thickness that can be exposed is much smaller, though. For such application, photoresist SU-8 is commonly used (see Section 3.4.2a).

4.13 MINIATURIZED CLASSICAL TECHNIQUES

There are two principle options to achieve miniaturization of mechanical, optical and fluidic components:

- applying technological procedures of microelectronics to produce three-dimensional function and form elements, as has been described in the previous sections;
- further development of classical techniques of precision mechanics. Watchmakers were pioneers of this second approach. They already designed, manufactured and assembled microproducts when micromanufacturing and microproduction was not yet part of the common technical language.[20]

The size of microcomponents is not clearly defined. Typical dimensions range from micrometers to millimeters. The specific characteristics of such components consist of small volumes, small weight and low tolerances. As micromanufacturing techniques, we refer to those that are scaled to smaller dimensions and qualify to be used for the production of microcomponents or microstructures, respectively. That does not mean that something existing has been only shrinked, but something new has been developed. Miniaturization includes certain physical or structure-related scaling effects (see Chapter 2), that cause components, manufacturing techniques and tools to change their characteristics. An

[20]Kocher, H. (1963) *Die Stellung der Mikrotechnik in der metallverarbeitenden Industrie* (The situation of microtechnology in the metal-working industry). *Technische Rundschau* 19, pp. 17–8.

example for a physically caused effect is the rapidly increasing ratio of component surface and component volume. Therefore thermal processes, such as heating or cooling, behave differently compared to the macro-range resulting in faster component cooling-down during micromolding or microforming, for instance. An example of a structure-related effect is that the grain size of a material composite does not decrease with component dimension which affects the cutting edge of micro cutting tools or friction processes in the micro-range.

Modern commercial micromanufacturing techniques either complete the lithographically based manufacturing processes derived from microelectronics or constitute an alternative to those. They are flexible, they can produce a large variety of forms with a wide range of materials and they can be cost-efficient already for small to medium-sized batches. Micromanufacturing techniques are used in many different applications. Units produced range from mass production in the automotive industry to small and middle-sized batches in mechanical engineering or medical technology. The decision to apply micromanufacturing techniques depends, in addition to the desired component function, the material and the later integration of the component into a microproduct, above all on the required number of units and the related costs.

Table 4.24 presents the most important micromanufacturing procedures of the first three main groups of manufacturing. The techniques of the main groups 'Separating' can not only be used for producing individual microcomponents made of metal, plastics, ceramics or silicon, but also microstructured tools, referred to as molding forms. These are integrated into machines used for micromanufacturing procedures of the main groups 'Molding' and 'Forming' and allow the production of shaped microcomponents in small to medium unit numbers according to the 'negative-positive principle'. Therefore, the procedures of the main groups Molding and Forming are also called secondary manufacturing techniques or replication techniques, whereas those of the main group Separating are referred to as primary manufacturing techniques.

In the following, we will describe three techniques each representing one of the main groups: micro injection molding, micro hot embossing and micro cutting. In addition, we will present examples of microcomponents that were produced using different micromanufacturing techniques.

4.13.1 Micro Injection Molding

In micro injection molding (Figure 4.58), granule is melted in the plastifying unit of a injection molding machine and injected with high pressure into the hollow of a completely

Table 4.24 Micromanufacturing techniques from the first three main groups

Primary transforming	Transforming	Separating
Micro injection molding	Micro hot embossing	Micro cutting (milling, turning, drilling, grinding, thread cutting)
Micro powder injection molding	Micro imprinting	Micro eroding
Micro casting	Precision cutting	Micro laser ablation

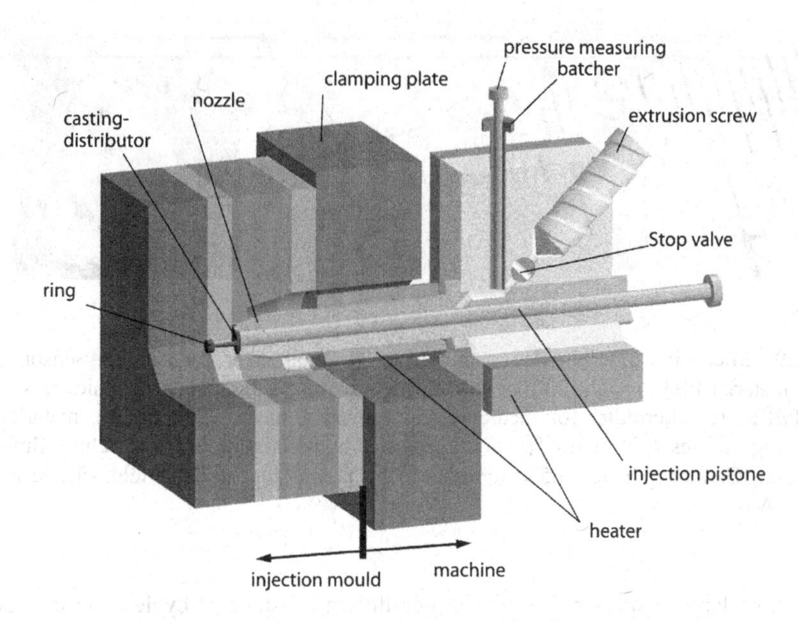

Figure 4.58 Micro injection molding machine. Reproduced by permission from Battenfeld, plastic machines, Ges.m.b.H., Austria

closed injection molding tool. The tool is tempered and contains molding forms that constitute a negative of the injection molded component. The plastified plastic mass completely fills the tool's hollow space. Has the plastic mass sufficiently solidified, the tool is opened and the injection molded piece is removed from the mold.

There are different forms and sizes of micro injection molding pieces. Typical dimensions for microcomponents are exterior dimensions in the millimeter range, wall thickness of some ten micrometers and component weights of less than 100 mg. In addition, there are larger, often also injection molded components with dimensions in the centimeter range including structures in the micrometer range, for instance in fluidic or spectrometer applications. They are referred to as microstructure components; their weight lies in the gram range. Another group comprises even larger microstructure components: dimensions in the range of decimeters and microstructured surfaces with a comparatively low aspect ratio, e.g. CDs (weight of a CD: approx. 15 g), DVDs, holograms. A fourth group are micro precision parts: components with precision mechanical dimensions and tolerances in the micrometer range, e.g. glass fiber connectors with a component weight of 6 g.

As the volume of microcomponents is very small (Figure 4.59), injection molding machines have to be adapted accordingly, in order to ensure that the exact dosing and injecting of minimal melting amounts is reproducible. Injection molding of components with very precise surface structures uses microstructured molding forms in combination with evacuated tools and dynamic tool tempering (variothermal process control). In this case, the tool wall is heated to the melting temperature of the plastic material and, after the injection process, cooled down again below the setting temperature of the plastic. The heating supports the complete filling of the form and prevents that the outer layer of the injection molding piece congeals too fast. The possible cycle time of micro injection molding is mainly determined by the aspect ratio of the injection workpiece and the dynamics of the tool tempering. For simple structures, minimum cycle time is

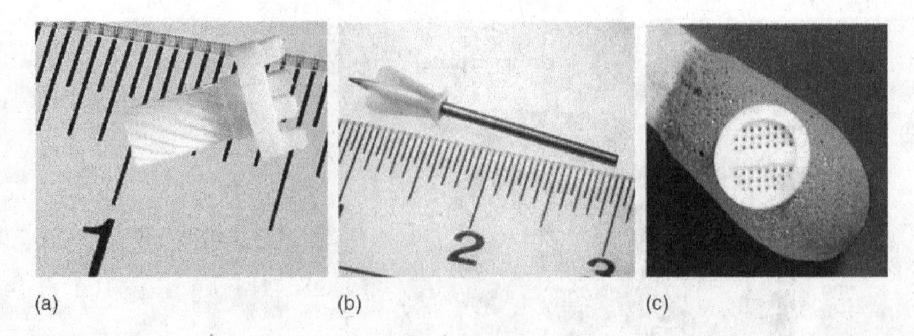

Figure 4.59 Micro injection molded components: (a) tooth gear for rotation sensor, module 0.11 mm, material PBT, weight 20 mg; (b) impeller for blood pumping, wall thickness 0.2 mm, material PEEK. (c) Microfilter for medicine and acoustics, mesh width 80 μm, material POM, weight 0.9 mg. Images 4.59a and 4.59b are reproduced by permission of Horst Scholz GmbH+Co. KG, Kronach, Germany. Image 4.59c reproduced by permission of Battenfeld, plastic machines Ges.m.b.H, Austria

5...15 s. For larger aspect ratios and larger flowing distances, cycle time can even lie in the range of minutes. Injection molded pieces can be made of all thermoplastic synthetic materials. Typical materials are polyethylene (PE), polypropylene (PP), polystyrene (PS), polymethyl methacrylate (PMMA), polycarbonate (PC), polysulfone (PSU), polybutylene terephthalate (PBT), polyetheretherketone (PEEK), polyoxymethylene (POM), liquid crystal polymere (LCP), polyamide (PA).

In the injection molding tool, assembling processes can also be carried out. Here, the additional components will be positioned in the injection molding tool and combined to a perfect junction with the injection molded piece. As opposed to this, it is also possible to intentionally combine incompatible materials using multi-component injection molding und thus producing components that are movable in relation to each other.

In addition to the actual injection molding process, modern manufacturing cells for injection molding of microcomponents enable additional functions such as optical quality control, handling, packaging and stacking of components. They can produce under clean-room conditions.

Injection imprinting has an intermediate position between injection molding and hot embossing. The plastic molten mass is injected into the not completely closed tool. Directly after filling the tool's hollow space, the movable tool component with the stamping matrix is pressed into the still plastic molding mass resulting in an embossed area contour with increased precision. Injection stamped parts usually have less tension than normally produced injection molded parts because the injection pressure during mold filling is lower. Injection imprinting is especially used for optical function components and memory media made of PC (CD, DVD).

In addition to synthetic materials, even metallic and ceramic materials can be used for injection molding if they are pulverized and mixed with a thermoplastic binder. This technique is called micro powder injection molding (MicroPIM). A suitable pressure and temperature is applied to inject the powder mixture into the tool. After removing the injection molded component from the mold, there are two thermal processing steps, de-binding and sintering. At first, the binder is removed from the injection molded piece. The subsequent sintering shrinks the part by about 20 % to its final size.

4.13.2 Micro Hot Embossing

During hot embossing (Figure 4.60), a thermoplastic substrate is introduced into the opened pre-tempered shaping tool and then heated to its softening temperature. After this, a large force and vacuum is used to press the shaping tool into the substrate. This way, the structure of the shaping tool is represented with high detail precision in the synthetic material. Upper and lower part of the tool are not completely closed; there remains a residual layer between the two embossing contours. After the tool has cooled down, the solidified workpiece is removed from the tool. As opposed to injection molding, hot embossing uses pre-manufactured substrates. They can have a surface of up to 150×150 mm^2. Hot embossing is particularly suitable for structuring plane components used in microfluidics (Figure 4.61) and microoptics. As during the embossing process, much less material is flowing than during injection molding, the components have a low internal stress, which is of particular importance for their optical characteristics. Important criteria for good embossing results are a temperature control that is adapted to material and component as well as a very precise driving movement of the embossing tool with velocities in the range of μm/s. Embossing temperatures for typical thermoplastic substrates may reach 250 °C. Hot embossing can be cost-efficient, even for small series production with up to 1000 pieces.

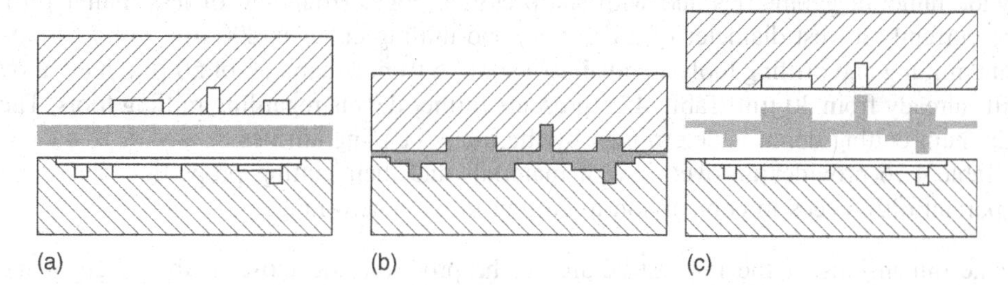

Figure 4.60 Hot embossing : (a) opened tool, introducing the substrate; (b) closed tool, hot embossing in vacuum; (c) opened tool, removing the workpiece

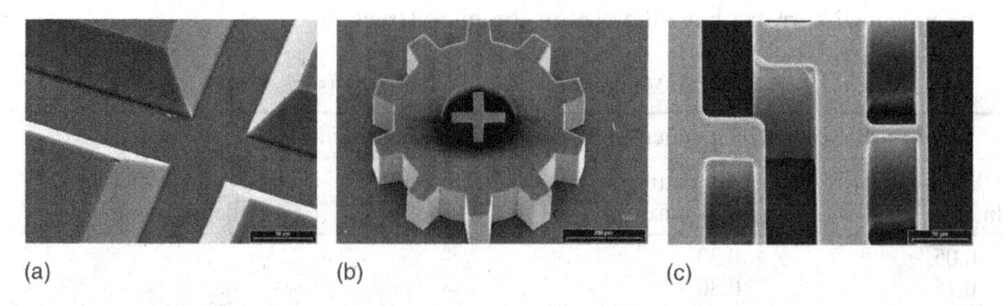

Figure 4.61 Hot embossing: (a) fluidics structure, tool anisotropically etched silicon wafer, material PMMA, structural width (50...100) μm, height 50 μm; (b) gearwheel, tool LIGA molding form, material PMMA, structural width 30 μm (cross), height 150 μm; (c) part of an acceleration sensor, tool LIGA molding form, material PMMA, structural width 8 μm (cross), height 150 μm. Reproduced by permission of JENOPTIK Microtechnology GmbH, Germany

4.13.3 Micro Cutting

Micro cutting refers to the cutting of material with miniaturized tools that have been further developed from classical procedures such as milling, turning, drilling, grinding. It is a cost-efficient technology for processing plastics, non-iron metals and also other tool steel and as well as for producing complex three-dimensional geometries in large varieties: channels with different cross-sections, crossing channels, trenches, concentric grooves on different surface profiles, bent areas, through holes and pocket holes. Typical tasks comprise the production of microcomponents such as optical frames, assembly plates, grippers as well as shaping tools for injection molding or hot embossing. For deep structures such as channels, the lateral dimensions depend on the diameter of the tool, for raised structures such as bridges above all on the material's rigidity. Typical lateral dimensions are 30...500 μm, typical tolerances are a few micrometers. The precision of the structure depends on the machine's precision and the deformation of the tools.

Micro cutting uses tools of high-alloy steel, solid carbide metal or diamond. Diamond tools produce a high surface quality, which is particularly important for optical components. They are suitable for processing plastics and non-iron metals. However, they are not suitable for processing steel due to the high wear. For steel and other materials difficult to cut, hard alloy and solid carbide metal tools are used. They only achieve a limited surface quality, though. Diamond tools for turning and milling are available in a wide range of geometries and with sharp cutting edges (rounding of less than 1 μm). The currently lowest diameter of a diamond end-milling cutter is 200 μm. Solid carbide metal milling and drilling tools are available starting from a diameter of 50 μm, hard alloy drills already from 30 μm. Table 4.25 presents further details regarding milling tools. The maximum cutting depth is determined by the tool's bending stiffness.

Table 4.26 provides an overview of materials and their cutting properties. There are important differences in comparison to cutting in the macro-range:

- The dimensions of the microstructures to be produced are close to the range of the crystalline grain structure of some materials.
- In the micro-range, material characteristics change.
- The ratio of cutting edge rounding and cutting depth is substantially changed.
- The produced burr can have the size of the microstructure.

Table 4.25 Milling tools for micro cutting (selection)

Hard-metal end-milling cutter		Diamond end-milling cutter	
Diameter in mm	Maximum cutting depth in mm	Diameter in mm	Maximum cutting depth in mm
0.05	0.30	–	–
0.15	0.30	–	–
0.20	0.50	0.20	0.50
0.30	1.00	0.30	0.80
0.50	1.50	0.50	1.50
0.80	3.00	0.80	1.80
1.00	3.00	1.00	2.20

Table 4.26 Materials, tools and cutting properties

Material	Microcomponent		
	Tool	Cutting property	Burr formation
Polymethyl metacrylate (PMMA)	Diamond; Solid carbide metal; Hard metal	Good	Low
Polycarbonate (PC)			Strong
Polyetheretherketon (PEEK)			Medium
Aluminium, high-strength		Very good	Low
Brass (CuZn)			Medium
Copper (Cu)		Medium	Strong

Table 4.27 Typical parameters of miniaturized classical shaping procedures

Characteristics	Micromanufacturing techniques		
	Micro injection molding	Micro hot embossing	Micro cutting
Minimum structural width	1 μm	0.05 μm	Bridge 30 μm Channels 60 μm
Maximum structural depth	a few mm	a few mm	0.1...6 mm
Maximum aspect ratio	smaller than 10	up to 100	1.5...30
Component tolerance	larger than 2 μm	1...2 μm	1...2 μm
Material	thermoplastic materials	all thermoplastics	plastics, metals
Economic batch size	10000	10	1...100

Table 4.27 provides typical parameter of the described micromanufacturing techniques. Figure 4.62 shows tools and product examples for micro cutting.

Example 4.12 Cutting speed and rotational speed at micro cutting

For micro cutting, the comparatively high rotational speed puts special demands on the machine tools. The smaller the tool diameter the larger the rotational speed required in order to reach the optimum cutting speed. It depends on the material and tool. If a diamond end mill with a diameter of $d = 200$ μm is used and the required cutting speed is $v = 100$ m/min, $n = v/\pi d$ results for the work spindle of a micro machine tool in an rpm of $n \cong 160000\ \text{min}^{-1}$. Such high rpm in combination with other criteria such as component dimensions and component tolerances require new machine concepts in comparison to traditional cutting, especially regarding the high stiffness of machine and tool, revolving and positioning precision, low vibration levels, high-resolution controls and distance-measuring systems as well as work fixture.

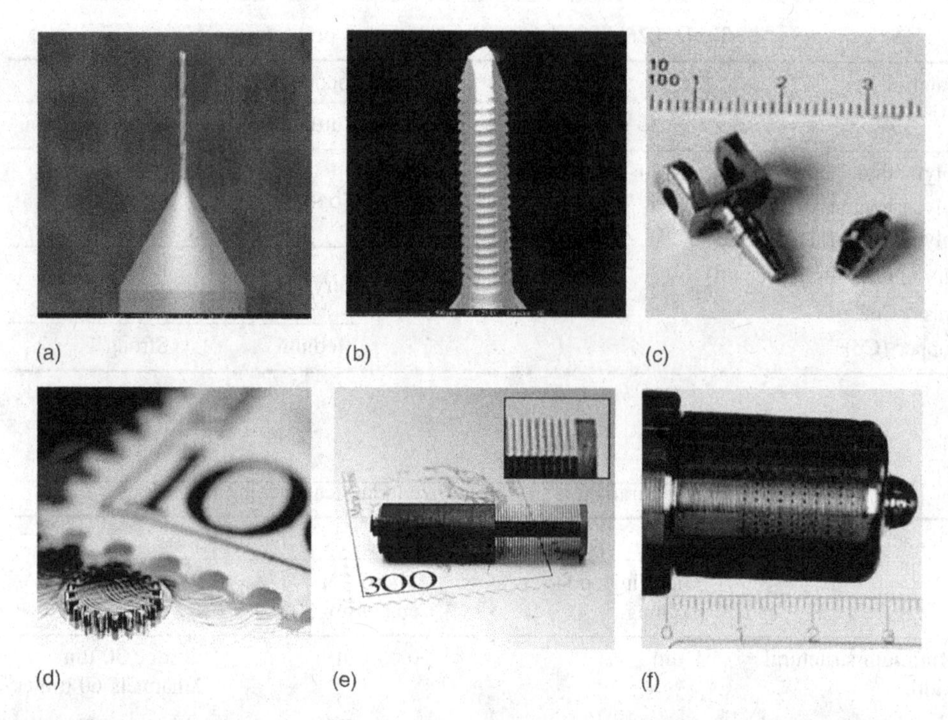

Figure 4.62 Micro cutting: (a) hard-metal drill ø 50 μm; (b) hard-metal tap S 0.306 × 0.08. Reproduced by permission from DIXI polytool, Le Locle, Switzerland; (c) finger joint and tooth root, material titanium; (d) electrode for eroding micro gearwheels, ø 2.5 mm, material copper, milled with end-milling cutter ø 0.1 mm; (e) graphite electrode for producing cutter heads for dry-shaving devices, sawed, lamellar thickness 0.1 mm; (f) diesel engine injection nozzle, borehole ø 0.36 mm, material steel 53 HRC. Reproduced by permission from Kern Mikro- und Feinwerktechnik GmbH & Co. KG, Murnau, Germany

4.14 SELECTION OF MICROTECHNICAL MANUFACTURING TECHNIQUES

In the previous sections, we have presented a large number of different techniques for the production of microtechnical function and form elements. Often, several process steps have to be combined in a process sequence in order to generate complicated three-dimensional forms. Table 4.28 summarizes the decisive issues that have to be taken into account when selecting a suitable manufacturing technique for the production of function and form elements.

Example 4.13 Production of a nozzle structure

Nozzles for ink-jet print heads constitute an important branch for microsystem technology. In the example, we will use the target structure in Figure 4.63 and develop a manufacturing technology for a nozzle made of single crystalline silicon. For a long life cycle and high reliability, the nozzle structure will be coated with a SiO_2 layer.

Table 4.28 Decisive issues for the selection of manufacturing techniques for the production of microsystem

1. Material of function and form elements, e.g.
   - Classical structure materials (Table 4.26)
   - Silicon (bulk micromachining, surface micromachining with epitaxial layers)
   - Poly-silicon (surface micromachining)
   - SiO_2 (thin film technology)
   - Glass (wafer bonding to silicon)
2. Analysis of the basic form: Decision on shaping (Table 4.27, Table 4.29)
   - Holes, trenches, pits
   - Vias
   - Thin plates, membranes
   - Cantilever, torsion springs
   - Bridges
   - Mesas
3. Synthesis of total form
   - One- and two-sided processing
   - Multi-step shaping
   - Two- or multi-wafer structure (including wafer bonding, see Section 4.8)
4. Form limitation by structuring
   - laterally by: etch mask, etch stop layers, anisotropy of etching technique
   - vertically by: etch stop mechanisms
5. Selection of process steps in Table 4.1, taking into consideration:
   - Maximum process temperature
   - Process time
   - Conformity (Section 4.4.2)
   - Thermal budget (Section 4.7.1d)
   - and many more.

(a) Basic forms and synthesis of total structure

Figure 4.63a presents the desired form. It consists of a groove and a hole (Table 4.28, point 2) and, according to Table 4.28, point 3, two approaches can be used:

- through double-sided processing in a wafer (local thinning from one side, hole from the opposite side); or
- in a two-wafer process (e.g. one etching per wafer, followed by bond-and-etchback).

In the following, we will apply the first approach.

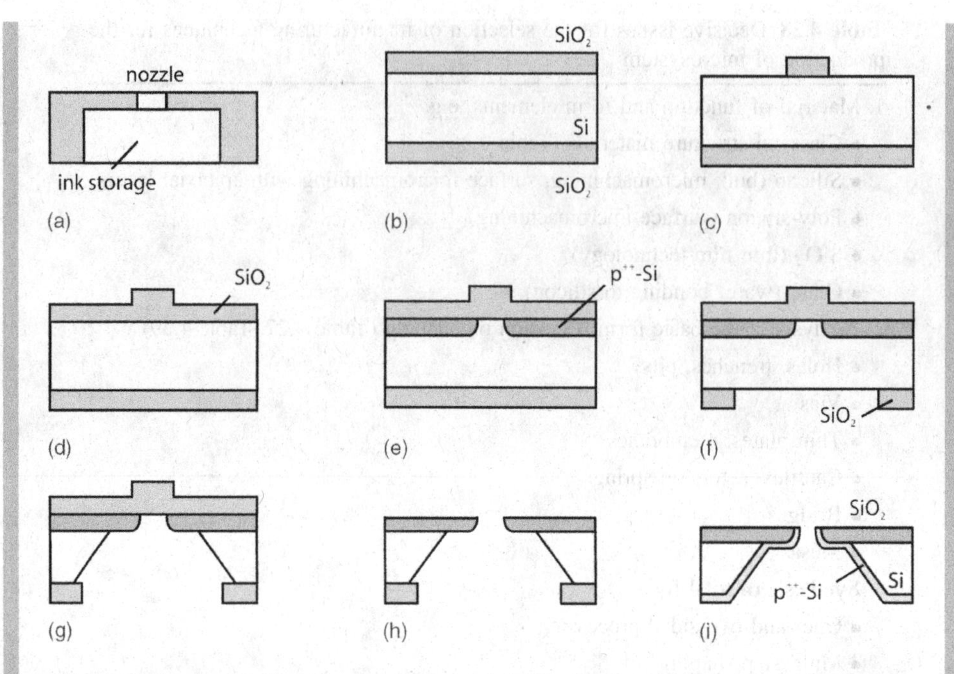

Figure 4.63 Production of a nozzle for an ink-jet print-head (explanations in the text)

(b) Shape limitation

We will use anisotropic deep etching to shape the nozzle structure. The nozzle opening will be laterally limited by etch stop layers (p^{++}-Si highly doped with boron), whereas the reservoir can be set geometrically through etch mask limitation.

(c) Selection of process steps

For the geometric delimitation of nozzle opening and reservoir, the corresponding p^{++}-Si areas on the front side will be doped and an etch mask will be generated on the backside. The following steps are required for this:

- Generation of a thermal oxide layer with low thickness where through-implantation is not possible. For this, we use two oxidation steps with intermediate structuring (Figure 4.63b-d). Implantating through the thin SiO_2 layer makes it possible to approach the concentration maximum of the dopant closer to the surface.
- High-dose implantation with subsequent annealing (Figure 4.63e). This defines the later nozzle opening.
- Structuring of the SiO_2 etch mask on the back of the wafer using wet chemical etching in BHF solution (Figure 4.63f).
- Anisotropic silicon etching, e.g. with KOH etching solution (Figure 4.63g). Now, the shapes of the reservoir and the nozzle opening appear.
- Removal of the SiO_2 layers on front and back using wet chemical etching with BHF solution; thermal oxidation of the entire structure (Figure 4.63i).

EXERCISES

4.1 The anisotropic etching of (1 1 0)-silicon results in cavities that are enclosed by six {1 1 1}-oriented sidewalls (Figure 4.33). Calculate the angles of these areas in relation to each other and to the wafer surface.

4.2 On a (1 0 0)-Si-wafer, there is an etch mask with <1 1 0>-oriented mask edges (Figure 4.64). In order to protect the convex edge, a long rectangular compensation etch mask is added. Carry out a quantitative construction of the undercutting process. Use the etch rate diagram in Figure 4.21.

4.3 A pressure sensor is supposed to have a square membrane with a dimension of 500 μm and a thickness of 25 μm. For stability and assembly reasons, the thick rim around the membrane must amount to 100 μm. Compare the space requirements for the production of a standard wafer with a diameter of 150 mm, if it is manufactured either with anisotropic wet chemical deep etching or as a two-wafer process through bond-and-etchback.

4.4 In the SIMOX process, $2 \cdot 10^{18}$ oxygen ions per cm^2 are implanted in Si with an energy of 200 keV. After high-temperature annealing (6 h, 1300 °C), a 400-nm-thick stoichiometric SiO_2 layer that has a sharp boundary to the Si is formed. Above the oxide layer, there is a 200-nm-thick Si layer. Using these data, estimate projected range R_P and scattering ΔR_P for oxygen implantation in silicon (for 200 keV).

4.5 Sacrificial layer etching used in surface micromachining for opening up movable structures may cause these structures to stick to the substrate. Explain why nubs according to Figure 4.53b can be used to achieve critical cantilever lengths that by multiples exceed those resulting from Equation (4.53).

4.6 Techniques of near-surface micromachining allow the generation of structures that are only supported by SiO_2 bridges with Al conducting paths (AIM Air-gap Insulated Microstructures, Figure 4.65). Develop a simple production technology.

4.7 Develop a manufacturing technology for a hemisphere with a radius of 200 μm and a wall thickness of 5 μm, made of single crystalline silicon!

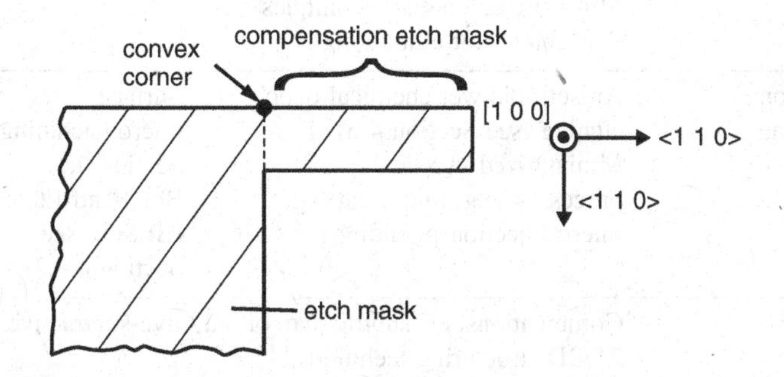

Figure 4.64 Convex etch mask corner with compensation mask

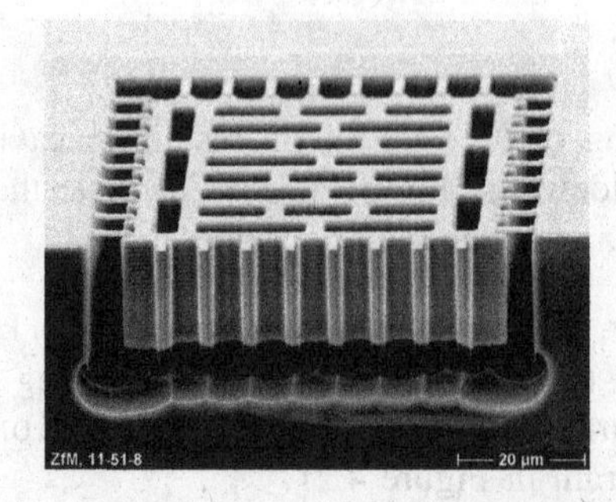

Figure 4.65 Air-gap insulated microstructure produced with near-surface micromachining. Structural height 20 µm, lateral dimensions 60 × 60 μm^2, bridge width 2 µm. Reproduced by permission from the Center for Microtechnologies, Chemnitz University of Technology, Germany

4.8 There are a large variety of ways to produce pressure sensors (Table 4.30). Develop manufacturing technologies for all shown variants.

4.9 Develop a manufacturing technology for the piezoresistive acceleration sensor with an endstop in Figure 4.66.

4.10 Provide a graphic representation of the relative proportion of tolerance to rated dimension, according to ISO 286, for components with rated dimensions of 6 . . . 1 mm and IT qualities of IT 5 . . . IT7! What trend can be discerned, what development

Table 4.29 Categorization of shaping techniques

Technique	Subtractive	Additive-subtractive
Two-dimensional (2D-) structuring	–	Thin film technology in micro-electronics
Quasi-three-dimensional (21/2D-) structuring[a]	Isotropic wet chemical etching (see Section 4.5) Anisotropic dry etching (see Section 4.5.3) Near-surface micromachining (see Section 4.11; e.g. SCREAM-, SIMPLE-process) Miniaturized classic techniques (e.g. micro hot embossing)	HARMST techniques (see Section 4.12, e.g. LIGA process)
Three-dimensional (3D-) structuring	Anisotropic wet chemical deep etching (see Section 4.6) Miniaturized classical processes (e.g. micro cutting, micro injection moulding)	Surface micromachining (see Section 4.10) Bond-and-Etchback (BESOI, see Section 4.9.2)
	Combinations of subtractive or additive-subtractive 3D- and 21/2D-structuring techniques	

[a] simple continuation of even, lithographically represented patterns in depth

Table 4.30 Structuring of thin membranes for pressure sensors

Processing	Si membranes	Poly-Si membranes	SiO_2 or Si_3N_4 membranes
Double-sided processing	SiO_2, Si_3N_4 Si SiO_2, Si_3N_4	poly-Si Si SiO_2, Si_3N_4	SiO_2, Si_3N_4 Si SiO_2, Si_3N_4
One-sided processing (surface micromachining)	Si [a] porous Si	poly-Si SiO_2, Si_3N_4 Si	SiO_2, Si_3N_4 [b] Si
Two-wafer processing (bond-and-etchback)	p^{++}-Si glass, Si		Si SiO_2, Si_3N_4 Si

[a] Prior to the removal of the porous silicon;
[b] Combination of anisotropic wet chemical etching and sacrificial-layer techniques.

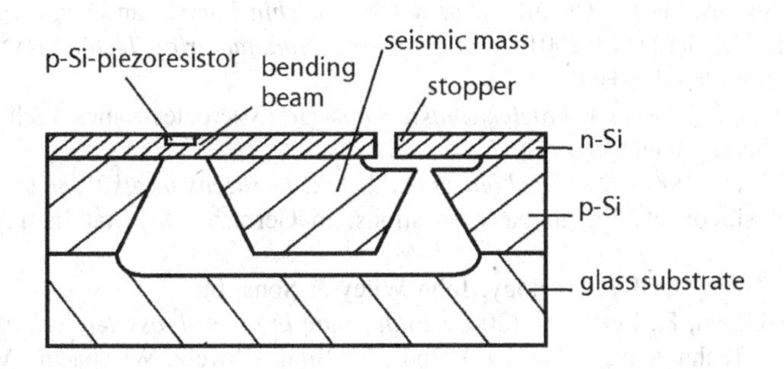

Figure 4.66 Piezoresistive acceleration sensor
Source: Barth, P.W., Pourahmadi, F., Mayer, R., Poydock, J., Petersen, K. (1988) *Technical Digest. IEEE Solid State Sensor and Actuator Workshop*; Hilton Head Island, June 6–9, 1988, pp. 35–6

requirements result regarding tolerating as well as measuring and testing technology for microcomponents with dimensions below 1 mm?

REFERENCES

[ALBERS05] Albers, J. (2005) *Kontaminationen in der Mikrostrukturierung* (Contaminations in Microstructuring; in German). München, Wien: Carl Hanser Verlag.

[BRÜCK01] Brück, R., Rizvi, N., Schmidt, A. (2001) *Applied Microtechnology*. München, Wien: Carl Hanser Verlag.

[CAMPELL01] Campell, S. A. (2001). *The Science and Engineering of Microelectronic Fabrication*. 2. Aufl. New York, Oxford: Oxford University Press.

[EHRFELD02] Ehrfeld, W. (ed.) (2002) *Handbuch Mikrotechnik* (Handbook Mocrotechnology; in German). München, Wien: Carl Hanser Verlag.

[ELWENSPOEK98] Elwenspoek, M., Jansen, H. V. (1998) *Silicon Micromachining*. Cambridge University Press.

[FRÜHAUF05] Frühauf, J. (2005) *Shape and Functional Elements of the Bulk Silicon Microtechnique*. Berlin: Springer.

[FUKUDA98] Fukuda, T., Menz, W. (eds) *Micro Mechanical Systems. Handbook of Sensors and Actuators*, Bd. 6. Amsterdam: Elsevier.

[GAIL02] Gail, L., Hartig, H.-P. (2002) *Reinraumtechnik* (Clean Room Technology; in German). Berlin: Springer.

[GARDNER01] Gardner, J. W., Varadan, V. K., Awadelkarim, O. O. (2001) *Microsensors, MEMS and Smart Devices*. Chichester: John Wiley & Sons, Ltd.

[GERLACH05] Gerlach, G; Werthschützky, R. (2005) 50 years of piezoresistive sensors – History and state of the art in piezoresistive sensor technology. In: Sensor 2005, *12th International Conference, Nürnberg, 10-12 May 2005. Proceedings*, vol. I. AMA, pp. 11–16.

[KERN93] Kern, W. (ed.) (1993) *Handbook of Semiconductor Cleaning Technology*. Park Ridge: Noyes Publishing.

[MADOU97] Madou, M. (1997) *Fundamentals of Microfabrication*. Boca Raton: CRC Press.

[MENZ05] Menz, W., Bley, P. (1995) *Mikrosystemtechnik für Ingenieure* (Microsystem Technology for Engineers; in German). 2nd edn, Weinheim. VCH Wiley.

[NISHI00] Nishi, Y., Doering, R. (2000) *Handbook of Semiconductor Manufacturing Technology*. New York, Basel: Marcel Dekker.

[OHRING92] Ohring, M. (1992) *The Materials Science of Thin Films*. San Diego: Academic Press.

[QUIRK01] Quirk, M., Serda, J. (2001) *Semiconductor Manufacturing Technology*. Upper Saddle River, Columbus: Prentice Hall.

[SCHADE91] Schade, K. (1991) *Mikroelektroniktechnologie* (Microelectronics Technology; in German). Berlin: Verlag Technik.

[SEIDEL86] Seidel, H. (1986) *Der Mechanismus des Siliziumätzens in alkalischen Lösungen* (The mechanism of silicon etching in basic solutions; in German. Dissertation. Freie Universität Berlin.

[SZE00] Sze, S. (2000) ULSI Technology. John Wiley & Sons, Inc.

[VÖLKLEIN05] Völklein, F., Zetter, T. (2005) *Einführung in die Mikrosystemtechnik* (Introduction to Microsystem Technology; in German). 2nd edn, Braunschweig, Wiesbaden: Vieweg.

5 Packaging

Packaging integrates the function components into technical systems (microsystems), where system function have to be maintained and ensured independent of ambient and operating conditions.

It is a system technology (see Figure 1.4) and it has a wide variety of tasks regarding system functions. Estimates state that only one third of total costs refers to the production of the silicon chip, whereas one third is assigned to packaging and testing, each [RAI00].

5.1 TASKS AND REQUIREMENTS

5.1.1 Tasks

Packaging has to carry out a wide variation of tasks which result from the required system functions (Table 5.1, Figure 5.1) as well as from product requirements for the different areas of application (Table 5.2).

In order to carry out such a variety of taks, microsystems are often – similarly to electronic devices – built hierarchically with each level being assigned different functions. In packaging, we generally talk about four levels:

1. packaging on the chip (e.g. passivating layers on the chip surface, on the individual chip or still on the wafer array);
2. chip assembly (e.g. bonding to substrate; packaging of an integrated circuit);
3. module assembly (e.g. printed circuit board with sensors, ICs and discrete components);
4. system assembly (e.g. device assembly, motherboard, backplane).

Each level applies different techniques. The deposition of passivation layers uses thin film techniques (see Section 4.4), the bonding of microsystem chip and carrier substrate uses the corresponding bonding techniques (see Section 4.8). Module assembly often uses soldering and gluing, based on thin- or thick-film technique, respectively.

Table 5.1 Functions of packaging

Level	Functions
Mechanical	Geometric arrangement of function and form elements in the system; mechanical actuations
Electrical	Energy/power supply, signal distribution
Sensoric	Communication of sensor signals
Thermal	Eliminating heat loss
Protective	Protection of all sub-components and the total system against disturbances affecting the function (mechanical, chemical, electromagnetic)
Compatibility	With the environment; bio-compatibility

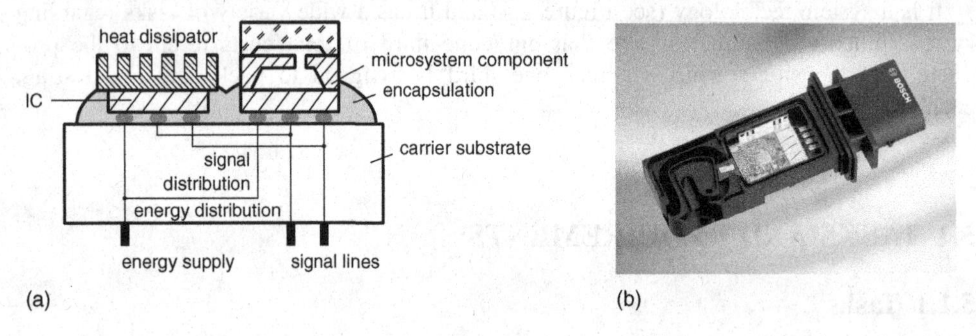

Figure 5.1 Microsystem packaging with mechanical, electrical, sensoric, thermal and protection functions: (a) principal structure (according to [TUMMALA01]; (b) example air mass flow meter (Robert Bosch GmbH). Reproduced by permission from the Robert Bosch GmbH, Germany

Table 5.2 Requirements regarding products in different market segments

Area	Temperature range in °C	Mechanical impact load	Relative humidity in %
User	0... +60	Drop test (1 m on concrete)	normal
Industry	−20... +80	≤5 g	85 (at 85 °C)
Automotive (engine-related)	−40...+150	≤3 g	85 (at > 100 °C)
Aviation and aeronautics	−55...+125	≤1500 g	85 (at 85 °C)
Information and communication	−40... +85	Drop test (1 m on concrete)	85 (at 85 °C)
Medical	−20... +80	≤5 g	normal

Example 5.1 Packaging for piezoresistive pressure sensors

Figure 5.2 shows the cross-section of an assembled piezoresistive pressure sensor, the exterior of which resembles DIL (Dual In Line) package. Leading the measured pressure to the absolute-pressure sensor requires a specific pressure inlet. The core of the sensor is the piezoresistive silicon pressure sensor chip (see Section 7.2.5). The measured signal should only be affected by measuring signals and not by thermal or mechanical deformations. Therefore the sensor chip is bonded to a counter-body with a respective temperature coefficient of linear expansion (mostly silicon or glass) and soft-glued into the housing.

The electrical contacting of the piezoresistors in the silicon chip to the exterior electrical connections is carried out via wire bridges. As the pressure medium should not affect the semiconductor surface (e.g. by corrosion), the sensor chip is embedded into a protection gel which – due to its elasticity – ensures that the reduction of the measuring pressure on its way to the silicon bending plate is negligible.

As there is only little heat loss generated in piezoresistive sensors, we do not need a special heat dissipation measures. In addition, the good thermal conductivity of silicon will conduct the heat loss generated in the piezoresistors to the thick chip rim.

Table 5.3 provides an overview of the requirements due to system functions and application specification.

Example 5.1 illustrates clearly that the demands in packaging exceed those in electronics and microelectronics by far for the following reasons:

- In addition to electrical power supply and other electrical signals, there are different non-electrical (mechanical, optical, fluidic) signals and variables.

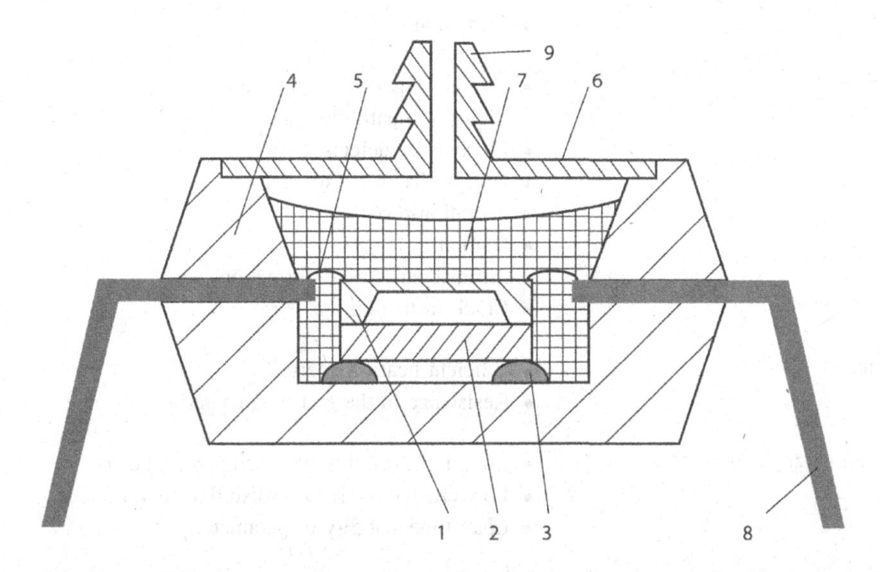

Figure 5.2 Assembled pressure sensor chip in the package: 1 sensor chip, 2 counter-body, 3 glueing points, 4 IC housing, 5 electrical wire bonding, 6 cover, 7 protection gel, 8 electrical connections (lead frame), 9 pressure inlet

Table 5.3 Requirements regarding packaging solutions

Level		Requirements
General		• Low costs • Small size • High reliability and quality • Low impact of disturbances
Electrical	Chip metallization	• Good adhesion, delamination-resistant • Assignment precision • Scratches, interruptions • Short circuits • Thickness • Contact resistance • Edge coating • Laser trim
	Bonding wires	• Strength • Adhesion • Positioning accuracy • Bi-metal contamination (KIRKENDALL effect) • Dimensions (height, bending)
	Wires	• Strength • Adhesion • Solderability • Contaminations • Corrosion
Mechanical		• Height, dimensions, precise orientation • Media compatibility • Mechanical defects • Uniformity of passivations • Vibration-resistance • Damping • Intrinsic stresses, deformations • Crack formation
Thermal		• Efficient heat dissipation • Resistance to thermal load changing
Function parameters, impact		• Long-term stability of function parameters • Low cross-sensitivity to disturbance variables • Long-time stability of parameters

Table 5.4 Failure mechanisms in microsystems

Overload		Wear		
Mechanical	Electrical	Mechanical	Electrical	Chemical
Breakage	Electromagnetic compatibility	Crack formation	Short circuits due to hillock and whisker formation	Corrosion
Plastic deformation	Electrostatic breakdown	Abrasion	Electromigration	Diffusion of media (e.g. humidity)
Delamination	Electromigration			

- Microsystems need to be protected against specific variables (e.g. electro-magnetic radiation, temperature and humidity), whereas other variables are expected to affect the system, at least locally (e.g. measuring parameters in sensors, chemical species in lab-on-chip systems). This requires selective protection measures.
- Progressive miniaturization even of non-electrical components.

5.1.2 Packaging for Reliability

Error or failure mechanisms occur at the lowest level of packaging. Due to operating conditions, failures are caused by overload or wear (Table 5.4).

In order to comply with the main task of packaging, microsystems have to have a structure that ensures that function variables and disturbance variables do not have a damaging effect on the function parameters, at least over a specific operating period. This requires a specific design for the packaging – a Design for Reliability. This puts specific demands on the selection of materials for function and form elements and on the technologies used for manufacturing microsystem components as well as on packaging (Table 5.5). Therefore, microsystem design has always to be combined with packaging design.

5.2 FUNCTIONS OF PACKAGING

Table 5.1 has provided an overview of the functions that microsystem packaging has to fulfill. In the following, we will take a closer look at selected functions without entering into all problems in a detailed way. We refer to [TUMMALA01] and [RAY00] (especially regarding microsystems) as well as to [TUMMALA97] and [SCHEEL97] (for electronic components and units).

5.2.1 Mechanical Links

One task of packaging is to geometrically arrange and fix the function components of microsystems. For this, usually a carrier or a substrate is used, which can even be part of the housing.

Table 5.5 Microsystem design for reliability

Level	Error mechanism	Design for reliability
1. + 2. Chip packaging	Corrosion	Sealing and encapsulation
	Brittle fracture	Reduction of mechanical stresses, avoiding defects
	Electromigration	Application of other materials; reduction of current density
3. Function group packaging	Crack formation	Load decrease; application of high-temperature materials
	Fatigue fracture and spreading	Reduction of mechanical stress load; limitation of temperature range; application of other materials, geometries and dimensions.
	Delamination	Improved adhesion; reduction of film stresses
	Interdiffusion	Reduction of temperature
	Radiation damages	Screening
4. System packaging	Corrosion	Avoiding and reducing defects; reduced humidity range; reduced temperature
	Abrasion, wear	Decreased tension; reduced friction

(a) Substrate materials

For carriers or substrates, it is possible to use the silicon chip itself (for monolithic integration), circuit board material, ceramics, plastics or metals (Table 5.6).

(b) Monolithic versus hybrid integration

In principle, it would be possible for many applications to monolithically integrate all function components of a microsystem, i.e. within a single silicon chip (Figure 5.3). For microelectronic components, there are standard technologies available (e.g. CMOS, Bi-CMOS); micromachining techniques (see Chapter 4) offer a wide range of technologies for creating three-dimensional forms. Monolithic integration has the following advantages [WOLFFENBUTTEL96]:

- minimum size and minimum weight through highest-level integration;
- high reliability due to reduced number of components;
- optimum temperature adjustment due to uniform carrier material silicon;
- easy realization of sensor and actuator arrays;
- low parasitic inductivities, capacities and resistors;

Table 5.6 Substrate materials for microsystems

Substrate	Characteristics
Silicon	• Hard and brittle • Bonding of Si chips with Si or glass applying gluing, glass soldering, eutectic bonding, anodic bonding or silicon direct bonding (see Section 4.8) • Hermetically sealed contacts are possible • Metallization on Si for wire bonding • Mainly used for chip-size packaging (CSP) of subsystems in microsystems
Ceramics	• Hard and brittle • Ceramic wiring carrier • Ceramic packaging, mainly two-piece (carrier, cap) • Electrical interconnects can be generated using thin- and thick-film techniques
Circuit board material	• Consists of carrier (glass reinforced laminate) and binder (epoxy resin) • Flexible circuit boards based on plastic foil (PE, PI, PTFE) available • Interconnects through laminating and photolithographic structuring
Plastics	• Molding to coat function components that are electrically contacted to a lead frame • Function components are subjected to the harsh conditions of injection moulding • Not hermetically sealed
Metal	• Robust, easy to process • Can be hermetically sealed and used in harsh environments (stainless steel) • Suitable for smaller batch sizes

- high electromagnetic immunity due to small dimensions;
- lower costs for large batch sizes.

There are also several disadvantages to this method:

- Monolithic integration of micromechanical with electronic function elements requires a complex manufacturing technology consisting of basic microelectronic techniques and micromachining. This means that fixed costs FC_m, which are independent of batch size, are high. In addition, there are also variable costs per unit VC_m for material, labour etc. Total cost per unit amounts for number of produced units n to

$$C_m = \frac{1}{n} FC_m + VC_m. \tag{5.1}$$

Figure 5.3 Examples of hybrid integrated acceleration sensors. Reproduced by permission from the Robert Bosch GmbH, Germany

Accordingly, per unit costs for hybrid integrated microsystems is

$$C_{\mathrm{h}} = \frac{1}{n} FC_{\mathrm{h}} + VC_{\mathrm{h}}. \tag{5.2}$$

Figure 5.4 presents the correlation between cost C and number n of units produced. As fixed costs FC_{m} are much higher for monolithic integration than for hybrid integration; monolithic integration becomes only economically feasible when the number of units produced exceeds critical number n_{crit}:

$$n_{\mathrm{crit}} = \frac{FC_{\mathrm{m}} - FC_{\mathrm{h}}}{VC_{\mathrm{h}} - VC_{\mathrm{m}}}. \tag{5.3}$$

- The integration of micromechanics and microelectronics requires additional mask levels for semiconductor processing. This reduces the combined yield according to Equation (4.2) substantially (see Example 4.2).
- Monolithically integrated microsystems in general require the integration of analogous and digital circuit design (non-recurring engineering). This increases complexity and thus the costs of the design.
- Especially in microsensors, the packaging has to allow the impact of measuring parameters but not that of any other, disturbing parameter. To achieve such separation is often rather costly. In hybrid integrated microsystems, electronic circuit component can be protected much more easily from disturbing impacts due to the modular structure.

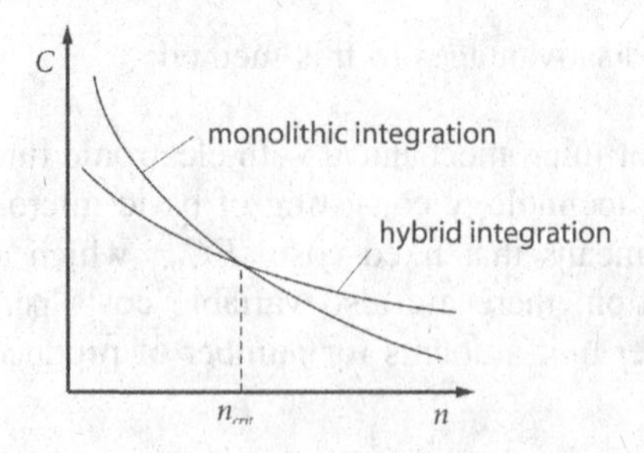

Figure 5.4 Relation of per unit cost C to number n of units produced

- The hybrid structure of a microsystem makes it possible to flexibly create system families for different operating conditions, electrical output signals forms (analogous, digital) and measuring ranges (for sensors).

As opposed to the semiconductor component market, the microsystem market is much more diversified and often requires only small batch sizes. Therefore, the major part of applications uses hybrid integration. Currently, completely monolithic integrated solutions are only used in the automotive industry for standard pressure, acceleration and rpm sensors with production units of more than 10^7.

Example 5.2 Monolithic integration of capacitive acceleration sensors

The monolithic integration of capacitive acceleration sensors requires an investment in semiconductor equipment of 10 million €. Taking into consideration a reduced yield, production cost per sensor chip is 10 €. Packaging and encapsulation cost another 2 € per unit.

The hybrid variant requires an investment of 1 million € for packaging. The production of a sensor chip costs 4 €, assembly and packaging another 5 €. Each sensor will also be equipped with a signal processing circuit which has to be purchased and costs 6 €.

The task is to calculate what number n_{crit} of units produced is necessary for monolithic integration to become economically feasible. Based on the given prices we have:

- for the monolithic variant:

$$FC_{\text{m}} = 10^7 \text{€} \qquad VC_{\text{m}} = (10 + 2)\,\text{€} = 12\,\text{€}$$

- for the hybrid variant

$$FC_{\text{h}} = 10^6 \text{€} \qquad VC_{\text{h}} = (4 + 5 + 6)\,\text{€} = 15\,\text{€}$$

Applying Equation (5.3), it results that

$$n_{\text{crit}} = \frac{FC_{\text{m}} - FC_{\text{h}}}{VC_{\text{h}} - VC_{\text{m}}} = \frac{9 \cdot 10^6\,\text{€}}{3\,\text{€}} = 3 \cdot 10^6.$$

Under the given conditions, monolithic integration becomes only economically feasible for a minimum production number of 3 millions.

(c) Thermal adjustment

When assembling microsystems, we often combine materials with different thermal coefficients of linear expansion. (see Table 6.7). Similar to the deposition of thin layers on substrates (Section 3.5), there is a deformation of the compound system due to the thermically caused expansion difference (see Equation (3.32))

$$\Delta\varepsilon = (\alpha_1 - \alpha_2)\Delta T$$

Here, α_1 and α_2 denominate the expansion coefficients of the compound partners and ΔT the temperature difference to the bonding temperature. In practice, there are mainly to commonly used ways for preventing deformations or the introduction of mechanical stresses in function and form elements of the microsystem:

- thermal adjustment through a fixed bond of elements with adapted thermal coefficients of linear expansion (e.g. anodic bonding of silicon and glass; silicon direct bonding);
- mechanical stress decoupling through a highly elastic, very soft intermediate layer (e.g. soft glue above glass temperature T_G, cp. Figure 3.10).

5.2.2 Electrical Connections

Packaging has the task to supply microsystem components with power and to contact them to each other and with the system environment via electrical signal lines.

Table 5.7 shows important techniques for electrically contacting silicon chips on substrates.

Wire bonding (Chip & Wire) is currently still the most important technique for contacting semiconductor chips. It uses micro wires with a typical diameter of 20...25 μm, which are laid in form of wire bridges. The energy required for the bonding is provided by ultrasound (ultrasound or US bonding), by a combination of heat and pressure (thermocompression or TC bonding) or heat, pressure and ultrasound (thermosonic or TS bonding). Al or AlSi1 wire are suitable for US bonding, Au-wire for TC- and TS-bonding. A disadvantage is that the wire bridges can be carried out only one after the other and not simultaneously.

For TAB-bonding, soldering or TC-bonding is used to carry out contacting with a flexible switching carrier. Contacting can take place simultaneously during one manufacturing step. A disadvantage is the comparatively high space requirements.

Table 5.7 Contacting techniques of unpackaged chips on substrates

Technique	C&W Chip and Wire	TAB Tape Automated Bonding	FC Flip Chip
Schematic diagram	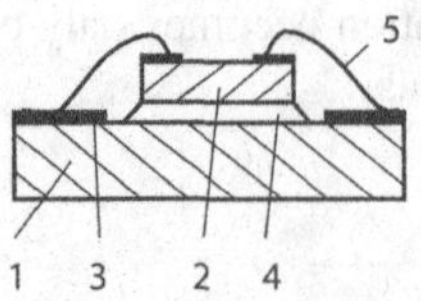 5 1 3 2 4	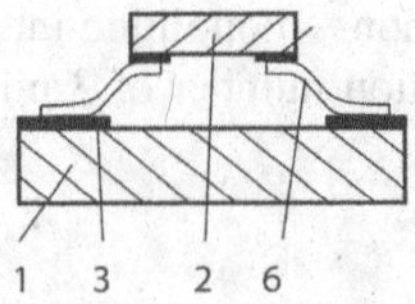 1 3 2 6	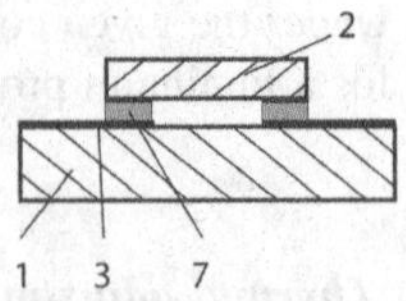 2 1 3 7
Connection	Serial	Parallel	Parallel
Electrical connection chip-substrate	Wire bridges	Flexible switching carrier	Solder bumps
Contacting	TC- or US-wire bonding	Soldering or TC bonding	Soldering, gluing or TC bonding

1 Substrate, 2 Si chip, 3 Bond pad, 4 Glue, 5 Wire bridge, 6 Flexible switching carrier, 7 Solder bump.

Flip-chip bonding also makes it possible to carry out all contacting during a single manufacturing step. It requires less substrate area, though, which allows high packaging densities. Soldering bumps are deposited on the chips of a wafer array and after turning the chips around, these can be contacted with the interconnects on the substrate. A variety of metal contact systems as well as conducting polymers (polymer-FC technique) can be used as bumps. Due to temperature differences and the resulting expansion differences between chip and substrate, there can occur large shear deformations in the bumps. These can be reduced when the shear forces not only act on the small cross-sectional area of the bumps, but on large cross-sectional areas. In order to achieve that, the entire gap between chip and substrate is filled with a so-called underfiller. In addition, it also constitutes a better protection of the contacts from humidity and other chemical species.

5.2.3 Heat Dissipation

Microsystem technology mainly uses heat sinking for dissipating heat (see Section 6.3.2), microfluidics even uses convection.[1] Heat conducting paths that are technically important are via (Figure 5.5):

- the silicon chip itself;
- special heat sinks (mainly made of copper);
- electrical contacts (bond wires, lead frames, bumps, soldering joints).

Heat dissipation via plastic housings and embedding, circuit boards or ceramic substrates is poor due to their limited thermal conductivity (Table 5.7).

For air gaps in the micrometer range, as e.g. for bolometers or electrostatic actuators, even the thermal conductivity of air may be of importance. Filling the gaps with gas that has a higher coefficient of thermal conductivity can increase heat dissipation; evacuating air (vacuum) can reduce it.

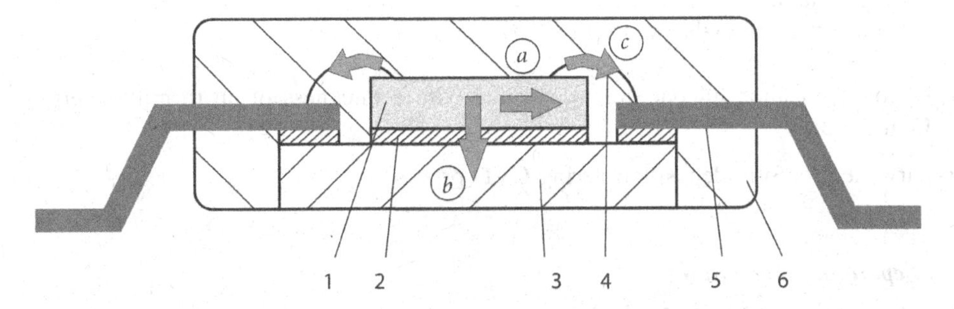

Figure 5.5 Heat conducting paths from the silicon chip with dissipation sources: (a) lateral in the chip; (b) to the heat sink; (c) via metal contacts; 1: Si chip, 2: isolating bonding layer, 3: heat sink (copper), 4: bonding wires, 5: contacts, 6: encasing/housing

[1]This is used, for instance, in air flow sensors that apply the anemometer principle.

5.2.4 Encapsulation and Packaging

Encapsulation and packaging provide protection for sub-components and the entire microsystem against disturbances that affect their functions. On the one hand, this refers to chemical influences like

- humidity;
- corrosive operating and measuring media (e.g. pressure measuring in the process control engineering);
- environmental contaminations (ionic contaminations such as sodium, potassium and chlorine ions caused by biogene (e.g. sweat) or non-biogene (e.g. salt water) sources);
- atmospheric components (e.g. NO_X and SO_2 in air and emissions)

and, on the other hand, to mechanical causes such as

- mechanical damages;
- mechanical loads due to stress concentration in specific system areas (e.g. elimination of damaging shear stresses in flip-chip contacts using underfiller in the bump areas between chip and substrate).

(a) Coating

The simplest form of protecting against media impact is the coating of microsystem components or the entire microsystem. The coats will form a barrier to the corresponding chemical species. Passivating layers can already be generated at the chip level and can be applied prior to wafer dicing. In silicon micromachining, double layers of silicon dioxide (thermal SiO_2 as quasi-perfect defect-free layer) and silicon nitride (chemically almost inert)[2] are particularly suitable. For silicon chips, we make a distinction between primary passivation (below the metallization level) and secondary passivation (protection including metallization).

Also organic materials are used for coating, such as [RAI00]:

- silicone and fluor silicone gels (deposited with a thickness of up to millimeters, see Figure 5.2);
- parylene (conform deposition during CVD process).

(b) Separating membranes

Pressure sensors often use stainless steel separating membranes for coupling measuring pressure (see Figure 3.1). The separating membrane has to be elastic to an extent that

[2] Nakladal, A., Sager, K., Gerlach, G. (1995) Influences of humidity and moisture on the long-term stability of piezoresistive pressure sensors. *Measurement* 16, pp. 21–9.

its flexibility remains negligible in relation to the silicon bending plate. Therefore, it has commonly the form of a corrugated membrane. Pressure transfer is carried out via an oil-filled hollow space, where the oil filling has to be free of air or gas bubbles. In comparison to coating, separating membranes are much more resistant to chemical impacts. Their disadvantages are packaging costs, the required structure size and the limited operating temperature range.

(c) Encapsulation

Microsystems can be encapsulated using [TUMMALA01]

- sealing;
- injection molding and pressing (see Figure 5.6); as well as
- application as liquids and hardening.

Typical materials are:

- Epoxy resins: These are the most common materials. They polymerize fast and clean and without formation of volatile components. Anhydrides, amines or phenols are used as hardeners.
- Cyanate ester: They have a higher glass temperature (190 °C ... 290 °C) and lower water absorption than epoxy resins.
- Urethanes: They show excellent adhesive characteristics and an outstanding film conformity. Acrylate groups can be crosslinked using UV.

(d) Hermetic sealing

According to [TUMMALA01], packaging can be considered hermetically sealed if the volume flow (leakage rate) for the diffusion of helium is lower than 10^{-8} cm^3s^{-1}. Hermetically sealed housings also prevent the diffusion of humidity and water und thus

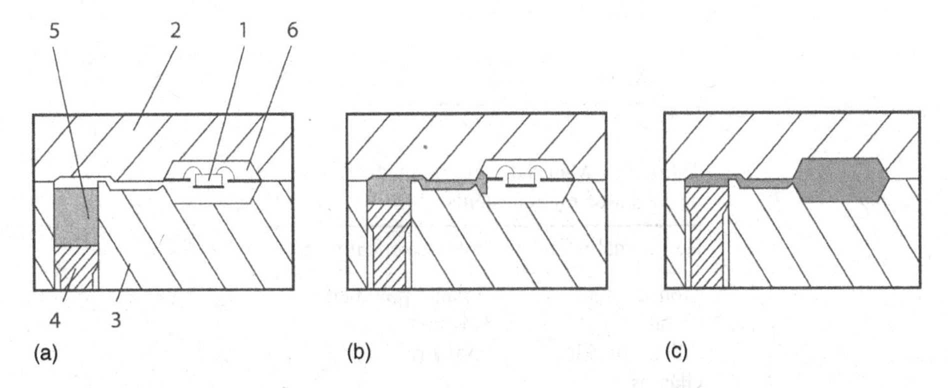

Figure 5.6 Sealing of microsystems using injection molding in three different phases: 1: microsystem; 2, 3: upper and lower casting mold part; 4: stamp; 5: pourable sealing mass; 6: hollow space

increase long-term stability of the parameters of electronic components and microsystems. Gettering materials in the housing absorb remaining or through-diffusing gas or water molecules und improve the characteristics even further.

Table 5.8 provides humidity diffusion rates through typical housing materials while Table 5.9 shows that inorganic materials are particularly suitable for hermetic sealing. Suitable techniques are:

- soldering,
- welding,
- glass soldering,
- anodic bonding,
- silicon direct bonding.

Often accelerated durability tests are carried out to test the tightness of packages; they consist of the following sub-tests:[3]

- thermal shock: cyclical tests between −65 °C and 150 °C (cycle time 10 s);
- salt fog tests: 0.5 ... 3% NaCl (pH = 6.0 ... 7.5) during 24 hrs at 35 °C;
- humidity test in the autoclave: 100 % r.H., 121 °C, 2 atm.

The latest developments in the area of plastics have produced materials that are able to survive the test conditions mentioned above.

Table 5.8 Humidity diffusion rates through housing materials for electronics and microsystem technology (according to [SCHEEL97])

Material	Diffusion rate in $g \cdot cm^{-1} \cdot s^{-1} \cdot Pa^{-1}$
Silicon	10^{-4}
Epoxy resin	10^{-8}
Glass	10^{-12}
Metal	10^{-14}

Table 5.9 Materials for hermetically sealed packages of microsystems

Hermetically sealed	Not hermetically sealed
Monocrystals	Organic polymers
Metals	Silicones
Silicon nitride	Epoxy resins
Glasses	

[3] US military standard MIL-STD-883.

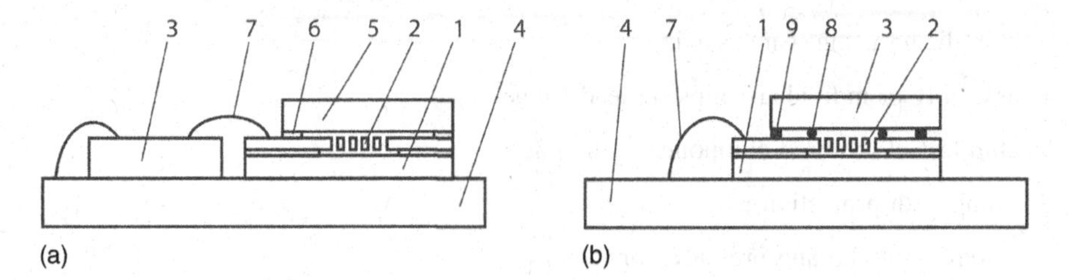

Figure 5.7 Packaging solutions for capacitive surface micromachined acceleration sensors. 1: sensor chip; 2: sensor structure; 3: signal evaluation IC; 4: substrate; 5: encapsulation; 6: polymer sealing; 7: wire bond bridge; 8: solder bump; 9: sealing

Example 5.3 Assembly of microsensors

A packaging will be used for acceleration sensors with surface micromachined interdigital capacitor structure (Figure 1.9d). It has to fulfill the following requirements:

- protection of movable interdigital capacitor structure using (if possible hermetic) encapsulation;
- integration of separately manufactured sensor chip and signal processing IC.

In principle, there are several solutions to this problem.[4] Figure 5.7a shows the simplest variant where sensor chip and IC are arranged next to each other on a substrate and are electrically contacted via wire bonding. Such an arrangement puts special requirements on chip size and design of the signal evaluation IC. A cap protects the moveable mechanical structure of the sensor chips. For a good thermal adjustment to the sensor chip, it is also made of silicon. The package is sealed through a polymer seal ring. It is very thin, but laterally expanded which results in a quasi-hermetic sealing.

Figure 5.7b shows a variant where the IC itself operates as mechanical protection for the sensor chip. As both chips consist of Si, the thermal adjustment of the chip compound is optimal. The signal processing IC is linked as a flip-chip bond 'face down' to the sensor chip. Simultaneously to the bumps, it is possible to generate a seal ring that seals the structure hermetically. The total area required for the chip structure mainly consists of the area for the sensor chip. The yield of the total arrangement corresponds to the product of the yields for sensor chip and IC.

Example 5.4 Manufacturing process for piezoresistive pressure sensors

We want to use the pressure sensor in Figure 5.2 to provide an example of a manufacturing process starting with the completed wafer production and ending in a ready-to-deliver sensor:

1. bonding of wafer to sensors and counter-bodies
2. wafer test

[4]Delapierre, G. (1999) MEMS and microsensors from laboratory to industry. In: Transducers '99, The 10th International Conference on Solid-State Sensors and Actuators. *Digest of Technical Papers*, Vol. 1. Sendai, Japan, June 7–10, pp. 6–11.

3. chip dicing using wafer sawing
4. assembly of individual chips on lead frame
5. chip test, if required component trimming
6. filling with protection gel
7. assembly of cap and pressure connector
8. final test

EXERCISES

5.1 Show the electrical, mechanical, sensoric, thermal and protective functions of the packaging using the following examples:
- yaw-rate sensor (Figure 1.7),
- ink-jet printer head,
- infrared bolometer focal plane array.

5.2 Which economic measures would have to be adopted to make the monolithic integration of capacitive acceleration sensors in Example 5.2 already economically feasible for 1 million units?

5.3 Show for the pressure sensor in Figure 5.2 how the two strategies of thermal adjustment in Section 5.2.1c have been applied.

5.4 The piezoresistors in piezoresistive sensors generate heat loss. Show the thermal conducting paths for heat dissipation for the sensor in Figure 3.1.

5.5 What causes fatigue and delamination (see Table 5.5)?

5.6 Why is the packaging of inertial sensors (e.g. acceleration or yaw-rate sensors) technically less demanding than that of pressure sensors?

REFERENCES

[HANKE94] Hanke, H. (ed.) (1994) *Baugruppentechnologie der Elektronik: Hybridträger* (Cicuit Boards in Electronics: Hybrid Carriers; in German). Berlin: Verlag Technik.

[RAI00] Rai-Choudhury, P. (ed.) (2000) *MEMS and MOEMS Technology and Applications.* Bellingham: SPIE Press.

[SCHEEL97] Scheel, W. (ed.) (1997) *Baugruppentechnologie der Elektronik: Montage* (Cicuit Boards in Electronics: Assembly; in German). Berlin: Verlag Technik and Saulgau: Eugen G. Lenze Verlag.

[TUMMALA97] Tummala, R. R., Rymaszewski, E. J., Klopfenstein, A. G. (eds) (1996) *Microelectronics Packaging Handbook* 3 vol., 2nd ed. New York: Chapman & Hall.

[TUMMALA01] Tummala, R. (ed.) (2001) *Fundamentals of Microsystems Packaging*. New York.

[WOLFFENBUTTEL96] Wolffenbuttel, R. F. (ed.) (1996) *Silicon Sensors and Circuits: On-Chip Compatibility. Sensor Physics and Technology*, vol. 3 London: Chapman & Hall.

6

Function and Form Elements in Microsystem Technology

Function and form elements are the basic components for structuring micromechanical, microoptical, microfluidic and other microcomponents. Microcomponents are complete function units that interact to carry out a desired system function. Microcomponents have structures in the micrometer range; their technical function is based on their micro-design. This definition [NEXUS02] characterizes that in microsystem technology function and form are very closely interconnected: structure, material and technology of an element have a stronger interrelation to each other and to the intended function than in other classical disciplines, such as precision engineering, electrical or mechanical engineering. The term 'function and form element' is an expression of this situation. An illustrative example is a resistor integrated into silicon (Figure 6.1). Resistance results from $R = \rho l/A = \rho l/hb$, where l, b and h are length, width and thickness, and ρ the specific resistance. Ratio ρ/h is determined by the technology (ion implantation: energy, kind of ions, substrate), ratio l/b by the design (mask): The value of R can only be exactly set through a precise interaction of design, material and technology.

6.1 MECHANICAL ELEMENTS

Micromechanical function and form elements can be categorized into two groups: static and dynamic elements (Table 6.1). In the following, we will consider oscillatory spring-mass-arrangements as representative examples of micromechanical function and form elements. They are often applied as core elements in sensor and actuator transducers (Chapter 7), e.g. as signal transformers of physical units such as acceleration, yaw rate, inclination, force, pressure as well as energy transformers for controllable mirrors or valves. There, a cantilever element links a movable mass to a fixed frame and generates a restoring force that is related to the displacement of the mass.

Typical cantilever elements in silicon micromechanics are cantilever beams, torsional beams and membranes. Different suspension designs of movable mass at frame and different dimensioning make it possible to influence important parameters, such as sensitivity, cross-sensitivity, eigenfrequency or maximum allowable stresses in the springs, through the construction of the elements.

Introduction to Microsystem Technology: A Guide for Students Gerald Gerlach and Wolfram Dötzel

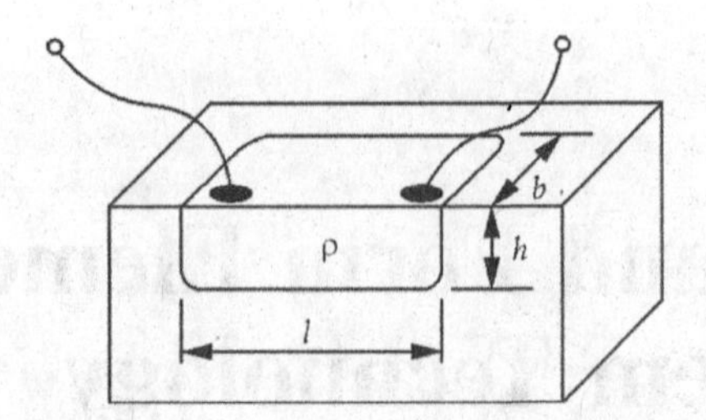

Figure 6.1 Resistor integrated in silicon

Table 6.1 Categorization of micromechanical function and form elements.

Kind	Static, fixed	Dynamic, movable
Example	Frames, bars, caps, hinges, stops, channels, cavities, nozzles	Beams, bridges, membranes, plates, gearwheels, levers
Functions	Mechanical stability, bearing, limitation of moving range, encapsulation, influence on damping, microfluidics	Transformers, resonators, straight-line mechanisms, transmissions

Especially during an early design phase, analytical descriptions can be helpful for selecting principally suitable function and form elements. Using simplified assumptions and taking into consideration the available technology, they can be dimensioned in a way that they can approximately comply with the specified function characteristics. Given specifications can also often be achieved by modifying already proven form elements that are stored in model libraries. A comprehensive behavior analysis for proving function characteristics and mechanical reliability requires numeric calculation programs, e.g. based on the finite element method (see Chapter 8).

Figure 6.2 represents basic forms of micromechanical function and form elements that are commonly used in bulk technology.[1] These are spring-mass-systems with a few degrees of freedom of motion which are determined by the position and form of the spring. Often it is sufficient to include the preferred direction of motion into the analysis. In the simplest case, the elements can be assumed to be spring-mass-damper-systems with one degree of freedom. In case of translation, the motion equation

$$m\ddot{x} + k\dot{x} + cx = F_{\mathrm{exc}}(t) \tag{6.1}$$

is valid, with displacement x, spring stiffness c, damping constant k and excitation force $F_{\mathrm{exc}}(t)$.

Figure 6.3 shows microscope images of manufactured elements for four basic forms of cantilever and torsional spring joints. For the calculation of mechanical characteristic values of such micromechanical elements, the same physical laws apply as in the macroworld.

[1]For characteristic values of function and form elements in Figure 6.2, see Mehner, J. (1994) *Mechanische Beanspruchungsanalyse von Siliziumsensoren und -aktoren unter dem Einfluss von elektrostatischen und Temperaturfeldern* (Mechanical load analysis of silicon sensors and actuators under the influence of electrostatic and temperature fields). PhD Thesis, Chemnitz University of Technology, Germany.

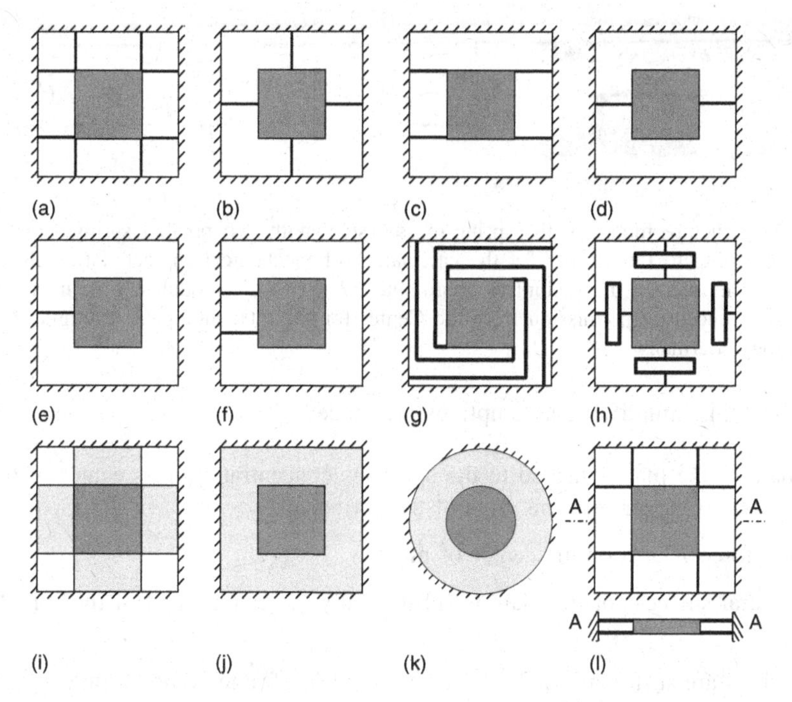

Figure 6.2 Basic figures of function and form elements in bulk technology

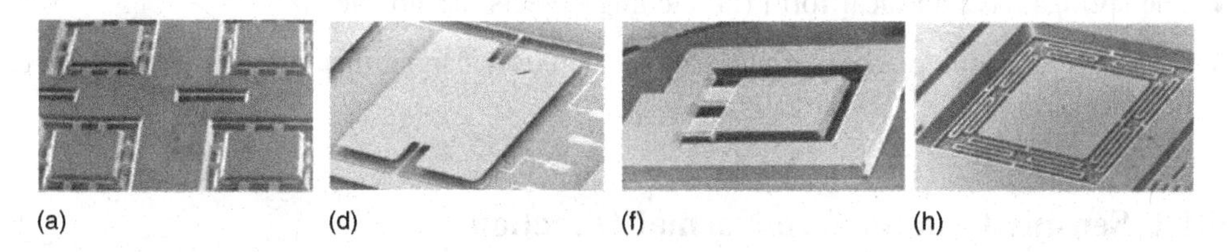

Figure 6.3 Microscope images of micromechanical elements (a), (d), (f) and (h) in Figure 6.2. (a) eight-point suspension of the plate with two parallel springs on each side allows vertical motion with little displacement (bending of the springs); (d) double-sided suspension of the plate allows a tilting motion around the longitudinal axis (torsion of the springs); (f) the one-sided suspension of the plate with two parallel springs allows a perpendicular displacement (bending of the springs); (h) the four-point suspension of the plate with meander springs allows tilting motions around two rotational axes as well as vertical motions (torsion and bending of the springs). Reproduced by permission from the Center for Microtechnologies, Chemnitz University of Technology, Germany

The laws of continuum mechanics can be applied down to a material thickness of a few hundred atom layers. For silicon with a lattice constant of 0.543 nm, the distance between two neighboring atoms is 0.235 nm (see Section 3.3). That means that continuum mechanics can be used for calculating dimensions as small as 50 nm (corresponds to ca. 200 atom layers).

The example of the elements in Figure 6.2f will be used to calculate important characteristic values. Figure 6.4 presents the corresponding structure model consisting of two parallel cantilever springs that are stiffly clamped on one side and an attached rigid plate.

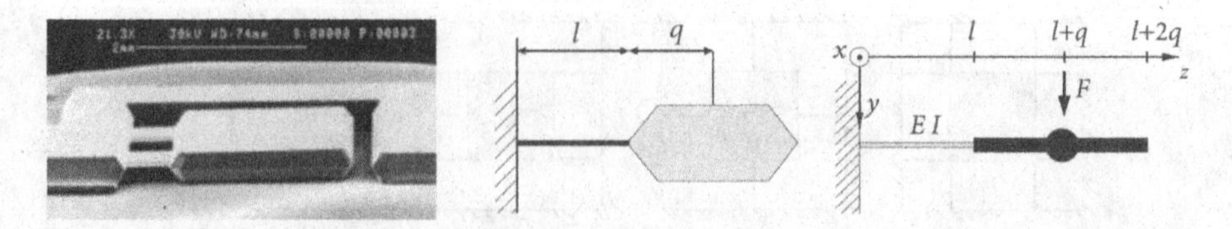

Figure 6.4 Structure model of the spring-mass-system with two parallel springs suspended on one side according to Figure 6.2f for the calculation of mechanical characteristic values: x, y, z constitute the normalized dimensionless coordinates. E YOUNG's modulus, I moment of inertia, F force. Reproduced by permission from the Center for Microtechnologies, Chemnitz University of Technology, Germany

The following simplifying assumptions are made:

- The mass of the plate attached to the spring is concentrated in its center of mass and is substantially larger than the mass of the springs.
- External force F acts on the center of mass.
- The flexural stiffness of the plate is substantially larger than that of the springs (rigid plate).
- The springs are stiffly attached: $y' = 0$ at $z = 0$ (no frame deformation).
- There is a linear spring behavior (only small displacements).
- The springs have an ideal form (no etching defects, no notches).
- Length and width of the springs are substantially larger than their thickness (pure bending).

6.1.1 Sensitivity in Surface Normal Direction

The sensitivity in surface normal direction, also just called sensitivity, is

$$S_y = \frac{\text{Output value}}{\text{Input value}} = \frac{\text{Displacement } y}{\text{Force } F_y}. \tag{6.2}$$

The ratio of spring displacement and acting force is

$$\frac{y''(z)}{(1 + y'(z)^2)^{3/2}} = -\frac{M_x(z)}{EI_{xx}}, \tag{6.3}$$

where M_x is the bending moment around the x-axis, E is YOUNG's modulus, and I_{xx} the area moment of inertia in relation to the x-axis. The denominator term to the left describes the effect of large displacements when leaving the linear range. Applying the simplifying assumption $y' \ll 1$, it results that

$$y''(z) = -\frac{M_x(z)}{EI_{xx}}. \tag{6.4}$$

Regarding the springs ($0 < z < 1$), it is valid that

$$y'' = -\frac{M_x(z)}{EI_{xx}} = -\frac{F_y(z - l - q)}{EI_{xx}}. \tag{6.5}$$

Double integration and taking into account the conditions for the stiff clamping at the left side of the spring $y'(z = 0) = 0$ und $y(z = 0) = 0$ results in the following equation for the displacement of the springs

$$y = \frac{F_y}{EI_{xx}}\left(-\frac{z^3}{6} + \frac{lz^2}{2} + \frac{qz^2}{2}\right). \tag{6.6}$$

Regarding the stiff plate ($1 < z < 1 + 2q$), it is valid that $y'' = 0$, $y' = a$ and

$$y = az + b. \tag{6.7}$$

Due to the transitional conditions spring–plate, for $z = l$, displacement y and slope y' of spring and plate have to be identical. This condition is provided by constants a and b in Equation (6.7) and thus the displacement of the plate's center of mass is:

$$y(l + q) = \frac{F_y l^3}{3EI_{xx}}(1 + 3r + 3r^2) \tag{6.8}$$

with $r = q/l$. This results in the sensitivity of

$$S_y = \frac{y(l + q)}{F_y} = \frac{l^3}{3EI_{xx}}(1 + 3r + 3r^2). \tag{6.9}$$

The sensitivity is identical to compliance n or to the reciprocal value of stiffness or spring constant c of the element, respectively: $S_y = n = c^{-1}$.

It applies to the area moment of inertia of a rectangular spring that

$$I_{xx} = \frac{bh^3}{12}. \tag{6.10}$$

Here, spring width is b and spring thickness is h. For number w of parallel springs, it is possible to use w as a factor in Equation (6.10) or as quotient in Equation (6.9), respectively. In the given example, the two-step anisotropic etching process produces a spring cross-section of a double trapezium as shown in Figure 6.5. The trapezium angle is approximately 25° (fast-etching {1 1 3}-plane, see Section 4.6). As opposed to Equation (6.10), here it applies to the area moment of inertia that

$$I_{xx} = \frac{bh^3}{12}(1 + 0.53\lambda) \tag{6.11}$$

with $\lambda = h/b$ [MEHNER00]. As an approximation, it is possible to use a rectangular cross-section of $\lambda < 0.2$. Thus, using Equation (6.10) it applies for the sensitivity in surface

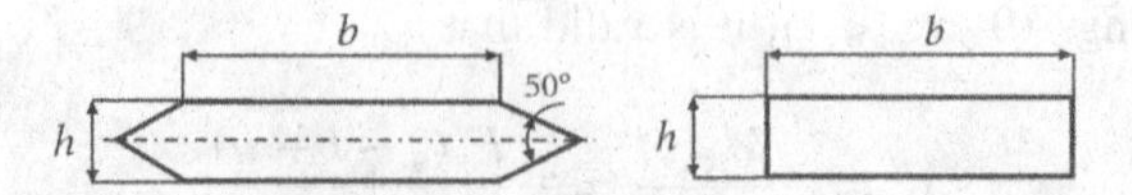

Figure 6.5 Technologically produced spring cross-section of the element in Figure 6.2f in comparison to a rectangular cross-section

normal direction that

$$S_y = \frac{2l^3}{Ebh^3}(1 + 3r + 3r^2). \tag{6.12}$$

Equation (6.12) shows that spring thickness h and spring length l as well as the respective tolerances have the largest effect on sensitivity.

For very wide and short springs or for membranes that are firmly attached on all sides, the fixture reduces the lateral deformation and results in a forced plane state of deformation. YOUNG's modulus will be used for the calculation

$$E' = E/(1 - \nu^2) \tag{6.13}$$

with ν being the orientation-related coefficient of lateral deformation or POISSON's ratio. For bulk technology and its typical < 1 1 0 >-oriented springs, ν becomes 0.063 and hardly results in any difference between E and E'.

Example 6.1 Sensitivity of special-design cantilever elements

The sensitivity in surface normal direction is, among others, related to spring length l and its tolerance. When using a design with very short springs and very long plates with $q \gg l$, Equation (6.12) results in

$$S_y = \frac{2}{Ebh^3}(l^3 + 3ql^2 + 3q^2 l) \approx \frac{6q^2}{Ebh^3} l. \tag{6.14}$$

Here, sensitivity is predominantly related to one term which – as opposed to Equation (6.12) – contains spring length, only as a linear variable.

6.1.2 Transverse Sensitivity

Transverse sensitivity S_T is the displacement of the element due to a force acting vertically to the surface normal direction (in Figure 6.4 in directions x and z). For a force acting in direction x, the springs are displaced in an s-shape, in accordance to a parallel spring arrangement with stiffly clamped spring ends. Transverse sensitivity in direction x is

$$S_{\mathrm{T}x} = \frac{x(l + q)}{F_x} = \frac{l^3}{24EI_{yy}} \text{ or } S_{\mathrm{T}x} = \frac{l^3}{2Ehb^3}. \tag{6.15}$$

A displacement in direction x occurs together with a transverse response in direction z:

$$\Delta z = \frac{F_x^2 l^5}{960 E^2 I_{yy}{}^2} = \frac{3x^2}{5l}. \tag{6.16}$$

In practice, this transverse response can be neglected for the bulk elements shown here due to the dimensions of the parallel springs.

Transverse sensitivity $S_{\mathrm{T}z}$ in direction z is uncritical if the spring's center of mass is located in the spring plane, which is the case here. If the center of mass is moved away from the spring plane by distance e_y due to an asymmetric design or an additional mass on the plate, an interfering force produces a moment $M_x = F_z e_y$ in direction z. This interfering force would lead to a displacement in direction y, which could not be distinguished through measurements from a displacement produced by an excitation in the surface normal direction.

Example 6.2 Sensitivity and transverse sensitivity of cantilever springs

Cantilever springs are supposed to have a minimum transverse sensitivity and a high sensitivity in surface normal direction. The ratio of transverse sensitivity and sensitivity in surface normal direction shows how such springs have to be dimensioned in order to comply with this requirement:

Ratio $\frac{S_{\mathrm{T}x}}{S_y} \sim \frac{h^2}{b^2}$ becomes small for thin and wide springs.

6.1.3 Eigenfrequency

Using a simplified calculation, Equation (6.9), it applies for the first eigenfrequency of the element in Figure 6.2f that

$$\omega_{\mathrm{e}} = \sqrt{\frac{c}{m}} = \sqrt{\frac{3EI_{xx}}{l^3(1+3r+3r^2)m}} = \sqrt{\frac{Ebh^3}{2l^3(1+3r+3r^2)m}}. \tag{6.17}$$

Here, we neglect the rotational degree of freedom intrinsically related to a translational motion of a mass that is stiffly suspended on one side.

6.1.4 Damping

Damping is neglected in Equation (6.17). The dynamic behavior is largely affected by damping, though. In principle, damping is caused by

- viscous damping through the surrounding media (gas);
- losses due to electric circuitry;
- energy losses via the suspension;

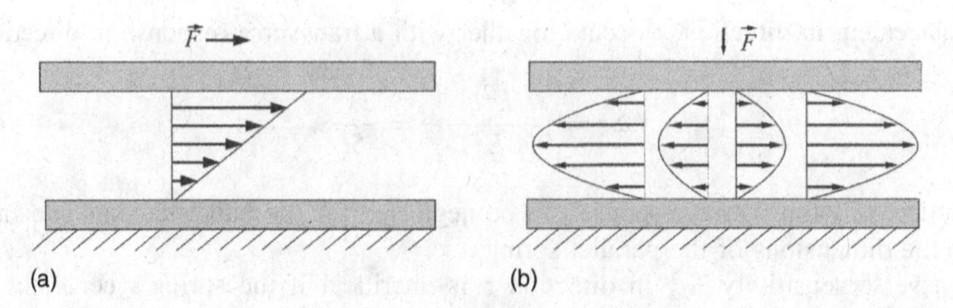

Figure 6.6 Damping effects for micromechanical oscillators

- thermoelastic losses (thermal compression);
- plastic deformation, internal friction (for silicon, not known yet).

Viscous damping has the largest effect. For micromechanical oscillators, where masses move in relation to fixed plates with small gap distances, two cases have to be distinguished. They are different regarding the direction of the plate motion (Figure 6.6).

- If the plate moves parallel to the fixed area (Figure 6.6a), the pressure in the air gap will not change. The damping results from the gravity forces in the fluid. This effect is called *slide-film damping*.
- If the plate moves vertically to the fixed area (Figure 6.6b), there will be pressure on the gas. The gas partially escapes from the gap between moving and fixed plate and produces frictional losses. This process is called *squeeze-film damping*. Especially for narrow gaps and high oscillation frequencies, part of the gas is not able to escape from the gap and is compressed.

These two physical phenomena generate a force onto the plate that can be described as a complex quantity. The real part acts as damping k_S, the imaginary part as an additional spring constant c_S, that changes the system's stiffness. Both effects depend on pressure and frequency. The frequency of the oscillating mass at which real and imaginary part of the reaction force are identical is called *cut-off frequency*. For frequencies below cut-off frequency, the damping part ($K_{\text{eff}} = k + K_s(\omega)$) is predominant; for working frequencies above cut-off frequency the stiffness portion ($c_{\text{eff}} = c + c_S(\omega)$) dominates. Typical plate dimensions in the millimeter range and typical gaps of 1 ... 6 μm result in cut-off frequencies below 1 kHz. Due to the described effects, the eigenfrequency of micromechanical oscillators can be decreased or even increased by damping:

$$\omega_e = \sqrt{\frac{c_{\text{eff}}}{m} - \left(\frac{k_{\text{eff}}}{2m}\right)^2}. \tag{6.18}$$

In addition to damping constant k, we use dimensionless damping ratio D:

$$D = \frac{k}{2\sqrt{mc}}. \tag{6.19}$$

The effect of gas in narrow gaps and vertically moving plates is called *molecular* or *viscous flow*, with a transitional range between them. For characterizing the three ranges, we use KNUDSEN number, which describes the ratio between free range of the air molecules and gap size (see Section 6.2).

6.1.5 Quality Factor

Mechanical quality factor Q is

$$Q = \frac{m\omega_e}{k} = \frac{c}{k\omega_e} = \frac{1}{2D} \tag{6.20}$$

Real quality factors of micromechanical oscillators amount to $10 \ldots 100$ for air and atmospheric pressure, $10^4 \ldots 10^6$ in vacuum.

6.1.6 Amplitude Response

The curve of the proportion of sensitivity $S_y = y/F_y$ in relation to frequency ω during a sinusoidal excitation $F_y(t) = F_{y\,\max}$ provides the amplitude response. Figure 6.7 shows a typical curve with resonance changing with damping.

The curve shows that static sensitivity ($\omega \to 0$) corresponds to the reciprocal value of spring stiffness and that, at low frequencies, the oscillator has a linear transfer function, i.e. it is equally sensitive to all frequencies. The behavior remains linear up to a frequency that is called critical frequency $\omega_c \leq \omega_e$. Depending on excitation frequency and damping, there can occur too large or too small displacements for $\omega > \omega_c$. According to the required precision, it can be assumed for the calculations that $\omega_c = (0.2 \ldots 0.7)\,\omega_c$. From this perspective, a maximum eigenfrequency is desirable. This means a small mass m and a high spring stiffness c. The latter will result in a low static sensitivity though. As can be seen from Equations (6.12) for sensitivity and (6.17) for eigenfrequency, a desired high sensitivity and, at the same time, a desired high eigenfrequency are contrary to each other.

The special characteristics of mono-crystalline silicon (fatigue-free, very low internal damping) make it possible to specifically dimension micromechanical oscillators for

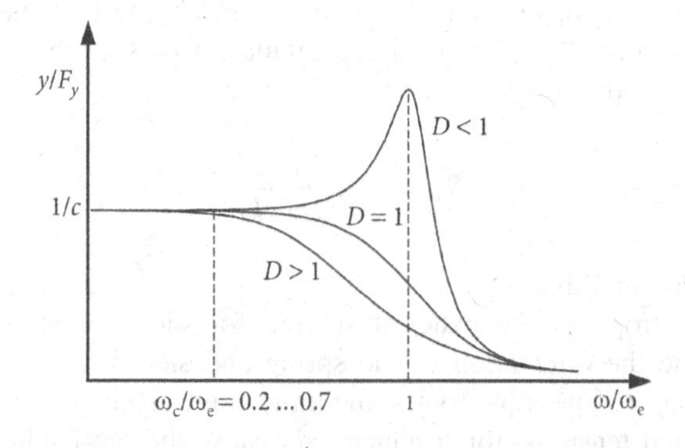

Figure 6.7 Amplitude response of micromechanical oscillators

sensor and actuator applications operating within resonance $\omega = \omega_e$. This means, that for actuators, it is possible to produce large displacements using minimum energy input. Sensors can use the fact that the oscillator's sensitivity in small areas $\Delta f = f_e/Q$ exceeds that for any other frequency by the approximate factor Q. This means that for each excitation the oscillator basically selects the frequency parts around its own eigenfrequency. Thus it is frequency-selective.

6.1.7 Stress at the Fixation Point

The stress in the spring cross-section (bending stress) produced by displacement reaches its maximum at the fixation point. It results from

$$\sigma_b = \frac{M_b}{W_b} = \frac{F(l+q)}{W_b}, \tag{6.21}$$

with bending moment M_b and section modulus W_b. For a rectangular cross-section the latter is

$$W_b = \frac{bh^2}{6}\, w, \tag{6.22}$$

where w is the number of the parallel springs.

The parameters presented in Sections 6.1.1 to 6.1.7 have been discussed using the example of the element in Figure 6.2f. They can be arrived at in a similar way if the force acting on the movable mass produces a tilting movement around the spring axis (z-axis). In this case, there is no bending stress, but torsional stress affecting the spring joint between movable mass and frame. This results in different equations for the parameters, such as sensitivity, eigenfrequency or stress at the fixation point. The sensitivity of the element in Figure 6.2d, for instance, results from

$$S_{\text{rot z}} = \frac{\text{Angle of twist } \phi}{\text{Torque } M} = n_{\text{rot z}} = \frac{1}{c_{\text{rot z}}} = \frac{l}{2G \cdot I_t}. \tag{6.23}$$

Here, l is the spring length, G the shear modulus and I_t the torsional resistance of the spring cross-section. For springs with a rectangular cross-section $b \times h$, it results that $I_t = bh^3k_1$. Thus, sensitivity becomes

$$S_{\text{rot z}} = \frac{l}{2Gbh^3k_1}. \tag{6.24}$$

Factor k_1 is provided in Table 6.2.

Due to the anisotropy of the material silicon, for shear modulus G it is therefore necessary to include the orientation of the spring and side ratio b/h. For springs with $< 110 >$-orientation, it is possible to assume an average shear modulus G_a (Table 6.3).

Technology-related tensions, for instance, can cause the pre-loading of the spring by a longitudinal force F, thus increasing spring stiffness $c_{\text{rot z}}$ by the term $F(b^2 + h^2)/6l$.

Table 6.2 Factors k_1 and k_2 in relation to side ratio b/h

b/h	∞	10	8	4	2	1
k_1	0.333	0.313	0.307	0.281	0.229	0.140
k_2	0.333	0.313	0.307	0.282	0.246	0.208

Table 6.3 Average shear modulus in relation to side ratio b/h

b/h	∞	4	2	1.5	1	0.1	0
G_a in GPa	50.9	52.5	55.0	57.3	62.3	77.5	79.6

This produces a lower sensitivity in comparison to Equation (6.24):

$$S_{\text{rot z}} = \frac{6l}{12Gbh^3k_1 + F(b^2 + h^2)}. \tag{6.25}$$

The eigenfrequency of the element in Figure 6.2d is

$$\omega_e = \sqrt{\frac{c_{\text{rot z}}}{J}} = \sqrt{\frac{24Gbh^3k_1}{lm(d^2 + a^2)}}. \tag{6.26}$$

Here, J is the mass moment of inertia of the plate suspended on two sides, m its mass, d and a its thickness and width. The tension at the fixation point[2] results from

$$\tau_t = \frac{M}{W_t} = 2G\frac{h}{l}\frac{k_1}{K_2}\varphi. \tag{6.27}$$

Factor k_2 is provided in Table 6.2. For a side ratio of $b/h \geq 4$, we can use $\tau_t = 2G(h/l)\varphi$ for the calculation.

Also in surface technology, it is possible to consider oscillatory spring-mass-arrangements as typical representatives of micromechanical function and form elements. As opposed to bulk technology, they move within the wafer plane. Similar to Figure 6.2, there are also model libraries for surface technology with basic elements and construction details such as stoppers or damping structures, including their analytic and numeric description. Often existing models and technologically proven constructional forms can be used and adapted as variants in order to fulfill specific given requirements. Such requirements frequently include parallel moved masses with large displacements for maximum linear motions and constant spring stiffness. Movable masses that are used for electrostatic drives or for capacitive detection, often show a comb structure (Figure 6.8).

Table 6.4 shows structural forms for parallel motions. They can be characterized by parameters according to Sections 6.1.1 to 6.1.7 [MEHNER00]. For a two-spring guided motion, the displacement of the mass results in an s-shaped bending of the spring suspended on one side (*s-spring*) and therefore no linear motion. Equation (6.16) can be used to calculate the transverse response. Such a transverse response as well as a possible

[2] In Chapter 1, torsional stress τ is called σ_4.

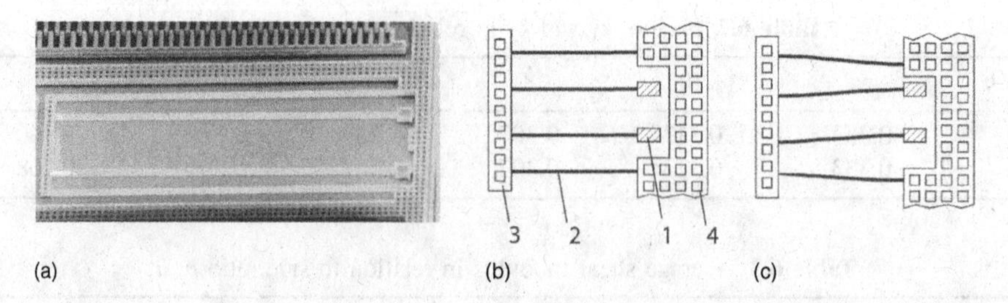

Figure 6.8 Parallel moved mass in surface technology: (a) microscope image; (b) not displaced; (c) displaced. 1: fixed anchor; 2: spring; 3: crossbar; 4: moved mass. Reproduced by permission from the Center for Microtechnologies, Chemnitz University of Technology, Germany

Table 6.4 Structural forms and characteristics of parallel motions in surface micromachining.

Two-spring guidance	Four-spring guidance	Bow-spring guidance	crossbar guidance
$\hat{X}\sin(\omega t)$			
• Linear spring stiffness • Transverse response • Insensitive to material tensions • Large sagging	• Progressive spring stiffness • No transverse response • Material tensions change spring stiffness • Small sagging	• Linear spring stiffness • No transverse response • Insensitive to material tensions • Large sagging	• Linear spring stiffness • No transverse response • Insensitive to material tensions • Small sagging

rotation of the mass are particularly critical for comb structures. There, the narrow gaps between the intertwining combs can easily get jammed. For four-spring guided motions, buckling effects have to be taken into consideration. The total spring stiffness of a bow spring guidance results from the parallel connection of the four bow-type springs. It consists of two S-springs connected in series. This structure has a low lateral stability and is receptive to rotation around the plate normal. Due to the additional crossbars, the crossbar guide – known from precision mechanics – shows a substantially higher stiffness towards transverse forces and rotation around the plate normal and a lower static sagging. It is insensitive to material tensions. Due to its advantages, it is frequently applied.

Example 6.3 Eigenfrequency of micromechanical oscillators

Figure 6.9 shows a micromechanical oscillator with linear stepwise structure consisting of silicon. It is used as the basic element of a frequency-selective vibration sensor. The oscillator has a mass of $m = 20\ \mu g$, the length of one cantilever band is $l = 400\ \mu m$.

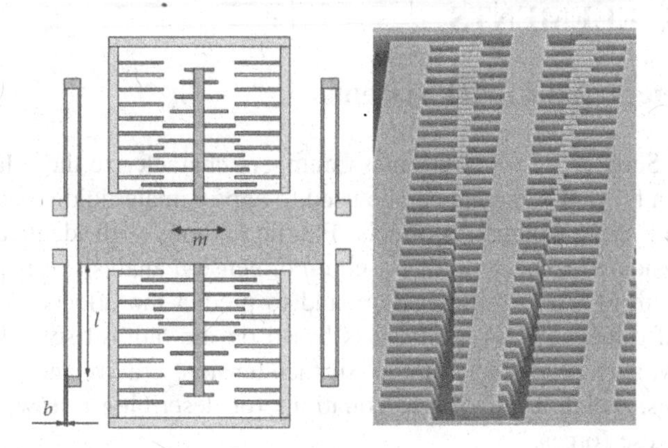

Figure 6.9 Micromechanical oscillator with linear stepwise comb elements. Reproduced by permission from the Centre for Microtechnologies, Chemnitz University of Technology, Germany

Further data known: spring height $h = 50\ \mu\text{m}$ (not shown in the figure), spring width $b = 3\ \mu\text{m}$, YOUNG's modulus $E = 169$ GPa.

The eigenfrequency of the shown oscillator results from

$$f_e = \frac{1}{2\pi}\sqrt{\frac{c}{m}}. \tag{6.28}$$

The displacement of the mass results for each cantilever with a length l in an s-shaped deformation (S-spring) with

$$c_s = \frac{12EI}{l^3} = \frac{Ehb^3}{l^3}. \tag{6.29}$$

For a bow-type spring:

$$\frac{1}{c_{bow}} = \frac{1}{c_s} + \frac{1}{c_s} \text{ or, respectively, } c_{bow} = \frac{c_s}{2}. \tag{6.30}$$

The mass is suspended to the frame on four bow-type springs in a parallel arrangement, which results in

$$c_{tot} = 4c_{bow} = 2c_s = \frac{2Ehb^3}{l^3}. \tag{6.31}$$

Total spring stiffness is $c_{tot} = 7.13\ \text{Nm}^{-1}$. This results in an eigenfrequency of the resonator of $f_e = 3$ kHz.

A reduction of the eigenfrequency by 5 %, for instance, from 3 kHz to $f_e^* = 0.95 f_e = 2.85$ kHz requires a spring stiffness of

$$f_e^* = \frac{1}{2\pi}\sqrt{\frac{c_{tot}^*}{m}} = 0.95\frac{1}{2\pi}\sqrt{\frac{c_{tot}}{m}} \tag{6.32}$$

i.e. $c^*_{tot} = 0.95^2 c_{tot} = 6.43\ \text{Nm}^{-1}$ (compare also $\Delta c/c = 2\Delta f_e/f_e$).

6.2 FLUIDIC ELEMENTS

6.2.1 Parameters and Model Systems

As described in Section 2.2, dimensionless numbers characterize the relationship of different effects of a flow. Parameters can be used to approximate fluidic behavior and judge the importance of specific physical effects. Fluidic systems with identical dimensionless numbers are considered to be similar. Based on dimension analysis, it is possible to compare fluidic system of varying dimensions and to predict the effects of miniaturization. A large group of dimensionless numbers is based on the ratio of several forces that are present in a flow, such as inertia, friction, surface tension, gravity and pressure. NAVIER-STOKES equations, which are the basic equations for describing a flow, characterize the equilibrium of these forces.

The ratio of inertia and viscous friction determines whether a flow is laminar or turbulent. If friction prevails, turbulence in the flow is dampened by frictional forces and the flow is laminar. For higher velocities, disturbances due to inertial forces are larger than frictional forces. The flow becomes instable or turbulent. The ratio between inertial forces and viscous frictional forces is called REYNOLDS number:

$$Re = \frac{\text{Inertial force}}{\text{Viscous frictional force}} = \frac{\rho u D_\mathrm{h}}{\mu} = \frac{u D_\mathrm{h}}{\nu}. \tag{6.33}$$

Here, u is the mean fluid velocity and D_h the hydraulic diameter or another characteristic dimension. Density ρ, dynamic viscosity μ and kinematic viscosity $\nu = \mu/\rho$ are constant fluid characteristics. The hydraulic diameter is determined by channel cross-section A and circumference U:

$$D_\mathrm{h} = 4A/U. \tag{6.34}$$

In the macro-range, we assume a critical REYNOLDS number of 2300 as the transition point between laminar and turbulent flow. In microfluidic components, the REYNOLDS number is much smaller than this critical number. The fluid behavior in most microfluidic components and systems can therefore be assumed to be laminar. In practical applications, this often causes problems when mixing flows.

The ratio of pressure and inertia is expressed in EULER number or cavitation number:

$$\mathrm{Eu} = \frac{\text{Pressure decrease}}{\text{Inertia}} = \frac{\Delta\rho}{\rho u^2}. \tag{6.35}$$

Due to the predominant viscous frictional forces in the micro-range, pressure decrease $\Delta\rho$ is often large, whereas inertia is negligible. EULER number is therefore expected to be very large for microfluidic systems.

Surface tension between two fluid phases (fluid or gaseous fluid) plays an important role for the equilibrium of forces. The ratio of inertia and surface tension is called WEBER number:

$$\mathrm{We} = \frac{\text{Inertia}}{\text{Surface tension}} = \frac{\rho u^2 D_\mathrm{h}}{\sigma}. \tag{6.36}$$

Here, σ is the surface tension between the two fluid phases. Surface tension can be compared to friction. The number of capillarity Ca describes the ratio of frictional forces and surface tension. At the same time, this number expresses the ratio of WEBER number and REYNOLDS number:

$$\mathrm{Ca} = \frac{\text{Surface force}}{\text{Frictional force}} = \frac{\mathrm{We}}{\mathrm{Re}} = \frac{\mu u}{\sigma}. \tag{6.37}$$

There is no spatial variable in the equation defining the number of capillarity. Therefore, this parameter is not affected by miniaturization.

Molecular diffusion and convection determine transport effects in the micro-range. Diffusion is determined by random BROWNian motion of the molecule, whereas convection is caused by the flow. PÉCLET number describes the ratio of convection and diffusion:

$$\mathrm{Pe} = \frac{\text{Convection}}{\text{Diffusion}} = \frac{u L_{\mathrm{ch}}}{D}. \tag{6.38}$$

Here, D is the diffusion coefficient and L_{ch} the characteristic length of the transport process. A small PÉCLET number indicates a transport process dominated by diffusion.

STROUHAL number characterizes the relationship between the advection of a periodic turbulence and the flow:

$$\mathrm{St} = \frac{\text{periodic disturbance}}{\text{Advection}} = \frac{f L_{\mathrm{ch}}}{u}. \tag{6.39}$$

Here, f is the frequency of the periodic turbulence.

Microfluidic systems can be modeled on different levels: on a molecular level, on a physical level and on a system level. The molecular level looks at single molecules. The deterministic method of molecular dynamics (MD) is used to describe the motion of the molecules. The large number of molecules requires an extreme calculation effort. Instead of simulating each individual molecule, a large number of molecules can be modeled as a particle. The interaction of molecules within a particle is statically determined. This static method is called Direct Simulation Monte Carlo (DSMC).

On a physical level, the fluid is considered as a continuum and all characteristics as continuous variables. The description of microfluidic systems and components is called continuum model. For its description, we use an equation set for the conservation of mass, the conservation of angular momentum and the conservation of energy. The conservation equations are often partial differential equations which – for simple cases – can be solved analytically. For issues involving complex geometry and coupled fields, numeric methods have to be applied to solve a continuum model.

The continuum model assumes that the fluid's characteristics are defined continuously through space. Fluid characteristics such as viscosity and density are assumed to be material characteristics. Miniaturization of fluidic components, however, touches the borderline of molecular dimensions where fluids cannot be considered to be a continuum any longer. An important dimensionless number for determining the continuum conditions is KNUDSEN number:

$$\mathrm{Kn} = \frac{\text{Free path}}{\text{Characteristic geometry}} = \frac{\lambda}{D_{\mathrm{h}}}. \tag{6.40}$$

Table 6.5 Similarity between fluid mechanics and electrical engineering.

Fluidic units		Electric units	
Mass flow $\dot{m} = \mathrm{d}m/\mathrm{d}t$	$[\dot{m}] = \mathrm{kgs}^{-1}$	Electric current I	$[I] = \mathrm{A}$
Pressure p	$[p] = \mathrm{kgm}^{-1}\mathrm{s}^{-2}$	Voltage V	$[V] = \mathrm{V}$
Pressure decrease Δp	$[\Delta p] = \mathrm{kgm}^{-1}\mathrm{s}^{-2}$	Voltage decrease ΔV	$[\Delta V] = \mathrm{V}$
Fluidic resistance $R_{\mathrm{fluid}} = \Delta p/\dot{m}$	$[R_{\mathrm{fluid}}] = m^{-1}\mathrm{s}^{-1}$	Electric resistance $R = \Delta V/I$	$[R] = \mathrm{VA}^{-1}$
Fluidic inertia $\Delta p = L_{\mathrm{fluid}} d\dot{m}/dt$	$[L_{\mathrm{fluid}}] = \mathrm{m}^{-1}$	Inductivity $\Delta V = L\mathrm{d}I/\mathrm{d}t$	$[L] = \mathrm{VsA}^{-1}$
Fluidic capacity $\dot{m} = C_{\mathrm{fluid}}\mathrm{d}p/\mathrm{d}t$	$[C_{\mathrm{fluid}}] = \mathrm{ms}^2$	Capacity $I = C\mathrm{d}V/\mathrm{d}t$	$[C] = \mathrm{AsV}^{-1}$

Here, λ is the mean free path of the molecules and D_{h} the hydraulic diameter or another characteristic length. KNUDSEN number increases with miniaturization ($D_{\mathrm{h}} \downarrow$) or for diluted gases ($\lambda \uparrow$). In general, it is possible to use KNUDSEN number to determine different flow ranges and models:

- $\mathrm{Kn} < 10^{-3}$: continuum model, NAVIER-STOKES equations, no-slip boundary conditions,
- $10^{-1} < \mathrm{Kn} < 10$: continuum model, NAVIER-STOKES equations, slip boundary conditions,
- $10^{-3} < \mathrm{Kn} < 10^{-1}$: transition between continuum model and molecular model,
- $\mathrm{Kn} > 10$: free molecular flow.

On a system level, the behavior of microfluidic components is summarized in a macro-model. The behavior can be deducted from either the models on the physical level, the molecular level or directly from experimental results. The macro-models are integrated into a network and are simulated. The behavior of a microfluidic system using macro-models resembles that of a network simulation in electrical engineering. Therefore, the same tools can be used for modeling a microfluidic system. In relation to the physical level, on a system level the calculation effort is largely reduced. Table 6.5 illustrates the similarity between fluid mechanics and electrical engineering.

Example 6.4 Fluidic parameters in micro-mixers

Average flow speed in a micro-mixer is 1 mm/s. The micro-mixer has two inlets and one mixing channel. Two fluids flow parallel to each other from the inlet into the mixing channel. The width of the mixing channel is $B = 1$ mm and its height $H = 100\ \mu$m. Viscosity, density and diffusion coefficient of the fluids are $\mu = 1 \cdot 10^{-3}$ Pa · s, $\rho = 11 \cdot 10^3$ kg/m^3 and $D = 1 \cdot 10^{-9} \mathrm{m}^2/\mathrm{s}$. The task consists in determining REYNOLDS and PÉCLET number of the mixer.

The characteristic length for REYNOLDS number is the hydraulic diameter. For a flat channel ($B \gg H$), the hydraulic diameter can be assumed to approximate channel height:

$$D_{\mathrm{h}} = 4A/U \approx H. \tag{6.41}$$

REYNOLDS number becomes:

$$Re = \frac{\rho u D_\mathrm{h}}{\mu} = \frac{1 \cdot 10^3 \cdot 1 \cdot 10^{-3} \cdot 100 \cdot 10^{-6}}{1 \cdot .10^{-3}} = 0.1. \qquad (6.42)$$

REYNOLDS number is much smaller than the critical REYNOLDS number. The flow is therefore laminar and the mixing process is based on diffusion.

For PÉCLET number, the characteristic length is channel width B. PÉCLET number becomes:

$$\mathrm{Pe} = \frac{u L_\mathrm{ch}}{D} = \frac{uB}{D} = \frac{1 \cdot 10^{-3} \cdot 1 \cdot 10^{-3}}{1 \cdot 10^{-9}} = 1000. \qquad (6.43)$$

The large PÉCLET number means that in this case convection is dominant, and not diffusion. There is a rule of thumb that can be used for channel length. The mixing channel is supposed to have a length of $L = \mathrm{Pe} \cdot B = 1000 \cdot 1\ \mathrm{mm} = 1\ \mathrm{m}$ in order to achieve homogeneous mixing at the end of the mixing channel.

6.2.2 Element Types

(a) Fluidic resistance

A microchannel with an incompressible laminar flow constitutes a linear fluidic resistor. Pressure decrease over a straight microchannel is proportional to mean flow speed:

$$\Delta p = f_\mathrm{F}\mathrm{Re}(2L\mu/D_\mathrm{h}^2)u. \qquad (6.44)$$

Here, f_F is FANNING friction factor and L channel length. For a laminar flow, the product of FANNING friction factor and REYNOLDS number is constant and only related to the geometry of the channel cross-section. The mass flow is determined using cross-section area A:

$$\dot{m} = \rho A u \qquad (6.45)$$

According to Table 6.5, the fluidic resistance of a straight microchannel with laminar flow is calculated as follows:

$$R_\mathrm{fluid} = \frac{\Delta p}{\dot{m}} = f_\mathrm{F} Re \frac{2L\mu}{\rho A D_\mathrm{h}^2} = f_\mathrm{F} Re \frac{2L\nu}{A D_\mathrm{h}^2}. \qquad (6.46)$$

With miniaturization, fluidic resistance increases with the third order. Systems with long microchannels will show strong damping characteristics. Table 6.6 shows typical cross-sectional forms and the calculation of product $f_F\mathrm{Re}$. Non-linear fluidic resistances such as for turbulent flows of non-NEWTONian fluids can be represented as pressure-related functions. The resistance of a real fluidic component, such as a passive valve, can be described using a table with values derived from experiments.

Table 6.6 Product of FANNING frictional factor and REYNOLDS number.

Cross-sectional form	Product of FANNING frictional factor and REYNOLDS number
Circle	$f_F\text{Re} = 16$
Parallel plates	$f_F\text{Re} = 24$
Rectangle	$f_F\text{Re} = 24(1{-}1.3553\alpha + 1.9467\alpha^2{-}1.7012\alpha^3{+}0.9564\alpha^4{-}0.2537\alpha^5)$ with $\alpha = H/W$
Rectangle with round corners	$f_F\text{Re} = 24 - 23.616\alpha + 22.346\alpha^2{-}4.724\alpha^3{-}3.0672\alpha^4 + 1.0623\alpha^5)$ with $\alpha = H/W$ and $0 \leq \alpha \leq 1$

A fluid resistor can only be applied in unsteady-state conditions, when mass flow and pressure decrease are identical. The so-called dynamic REYNOLDS number represents the ratio of dynamic excitation and viscous damping:

$$Re_d = \frac{\text{Dynamic excitation}}{\text{Viscousdamping}} = \frac{D_h}{2}\sqrt{\frac{2\pi f \rho}{\mu}} \tag{6.47}$$

Here, f is the exciting frequency. Small dynamic REYNOLDS numbers mean that the viscous damping exceeds the dynamic excitation. For numbers below the critical dynamic REYNOLDS number, the fluidic resistors can be used for unsteady-state problems. The critical REYNOLDS number of a circular channel is 2.45, for instance.

Example 6.5 Modeling of a cylindric capillary

Determine the maximum excitation frequency of a cylindric capillary with a diameter of 100 µm, in order to be able to use a fluidic resistor for modeling the capillary! Viscosity and density of the fluids in the capillary are $\mu = 1 \cdot 10^{-3}$ Pa · s and $\rho = 1 \cdot 10^3$ kg/m^3.

Using Equation (6.47), the exciting frequency for the critical dynamic REYNOLDS number $\mathrm{Re}_{d,k}$ can be determined as follows:

$$f = \frac{2\mu}{\pi\rho}\left(\frac{\mathrm{Re}_d}{D_h}\right)^2 = \frac{1 \cdot 10^{-3}}{3 \cdot 14 \cdot 1 \cdot 10^3}\left(\frac{2.45}{100 \cdot 10^{-6}}\right)^2 = 382\ \mathrm{Hz}. \tag{6.48}$$

(b) Fluidic capacity

Similar to electric capacity, fluidic capacity is a fluidic element for storing energy. The storage of potential energy is determined by the compressibility of the fluid or the elastic properties of a microfluidic component. The mass flow generated by compressibility is

$$\dot{m}u = m_0\gamma \mathrm{d}p/\mathrm{d}t, \tag{6.49}$$

where m_0 is the mass in its initial state and γ the compressibility factor of the fluid. According to Table 6.5, it results that

$$C_{\mathrm{fluid}} = m_0\gamma. \tag{6.50}$$

Increasingly, polymer materials such as silicon rubber (PDMS, polydimethylsiloxane) are used for manufacturing microfluidic components and systems. The fluidic capacity results then from the elastic channel wall consisting of silicon rubber. Due to the interaction between structure and fluid, elastic mechanical components, such as valve flaps or pump membranes, also constitute a fluidic capacity. Fluidic capacity caused by elasticity can be determined either through direct measuring or coupled simulation.

(c) Fluidic inertia

Table 6.5 defines the fluidic inertia of a microfluidic component. The kinetic energy of the fluid is stored in the fluidic inertia. Thus, the fluidic inertia of a microchannel is

$$L_{\mathrm{fluid}} = L/A. \tag{6.51}$$

6.2.3 Fluidic Interfaces

As opposed to other microsystems, microfluidic systems also require – in addition to interfaces with macrosystems for information and energy flow – connections for the material flow. Depending on the kind of coupling nature between micro- and macrosystem, there are three different forms of fluidic interconnects:

- press-fit interconnects;
- substance-to-substance interconnects; and
- positive interconnects.

Press-fit interconnects can be carried out using springs and seal rings. Spring structures are designed including mechanical basic elements and can be integrated into microfluidic systems. Systems consisting of polymer materials, such as silicon rubber, are self-sealing and do not require any sealing. It is also possible to use polymer microtechnology in order to integrate seal rings into systems consisting of harder materials. Press-fit interconnects have a long life and a comparatively high compressive strength.

For substance-to-substance interconnects, the connection is carried out using gluing, eutectic bonding, glass soldering and anodic bonding. Here it is particularly important to take into consideration the resistance of joints to aggressive fluids as well as the detachment of bonding material. Gluing is a fast and cost-efficient method for microfluidic lab experiments.

Positive interconnects can directly use silicon. In addition, polymer interconnects are used. The interconnects can be manufactured using micro injection molding. After mounting, heat is used to soften the polymer interconnects, in order to ensure a better positive coupling.

Many microfluidic systems are used in a biological environment. In addition to the material selection for the system, bio-compatibility of the interfaces is an important design aspect. The response of the fluidic interface to its environment and the response of the biological environment to the interface are decisive for a continuous operation of the system, especially for implanted systems. Microfluidic systems can be tested for their bio-compatibility either *in vitro* in the test tube or *in vivo* in living organisms.

6.2.4 Design of Microfluidic Elements and Components

The design process of fluidic elements and components follows the rules of microsystem technology. The design process can use either a top-down or a bottom-up model. As opposed to digital electronics, microfluidics does not dispose of a sufficient number of function elements for structuring complex microfluidic systems. Even though there are microfluidic systems consisting of hundreds of switching elements, such as microvalves and micropumps, microfluidics will – in the near future – not be able to reach the degree of integration used in microelectronics. The top-down model is more suitable for designing individual microfluidic components.

(a) Design of microfluidic elements

The design option for microfluidic elements is largely determined by the manufacturing technology. In the following, we will use the example of designing a microchannel of a electrokinetic fluid network for capillary electrophoresis. This method can be used, for instance, for analyzing DNA (desoxyribonucleic acids) fragments.

Figure 6.10 illustrates the principle of capillary electrophoresis. The fluid network consists of two crossing microchannels: a short injection channel and a large separation channel. When applying electric voltage via the injection channel, electro-osmosis will be used to introduce the sample fluid into the injection channel. Electro-osmosis is the flow of an ionized fluid in an electric field in relation to a charged channel surface. In addition to the channel surface, the velocity of electro-osmosis is related to the strength

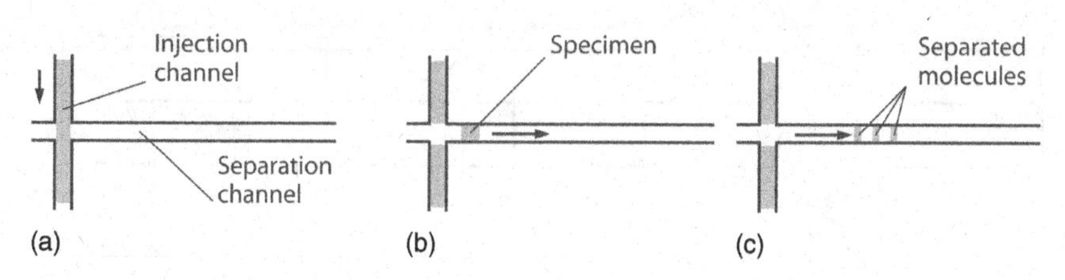

Figure 6.10 Schematic principle of capillary electrophoresis: (a) dosing; (b) injection; (c) separation

of the electric field and the dynamic viscosity of the fluid. The geometry of the channel cross-section does not have the same importance as the pressurized flow. When applying a voltage via the separation channel, a small sample volume is introduced. Together with the electro-osmotic flow, the molecules of the sample separate due to electrophoresis. Electrophoresis is the motion of charged particles in relation to the surrounding fluid under the impact of an electric field. The speed of the electrophoresis only depends on the surface charge, the strength of the electric field, and the dynamic viscosity of the surrounding fluid. Different molecules in the sample can be separated due to their different charges. The separated molecules can be visually detected at the end of the separation channel. Electro-osmosis and electrophoresis constitute reversible electrokinetic transducer effects, which can be considered two-port transducer according to Section 7.1. The following framework has to be taken into consideration:

- The substrate should be electrically insulating and optically transparent. Only glass and polymer fulfill this condition.
- Even though the form of the cross-section is irrelevant, the cross-section area should be sufficiently small in order for the electric resistance of the fluid network to become sufficiently large.
- A large electric resistance produces heat when a high voltage is applied. The increased temperature reduces separation quality. This requires good heat dissipation.

The design of the fluid channeling has to take into consideration the external treatment of the sample fluid and the optical detection. In principle, there are different possible channel forms. For a glass substrate, only form 1 in Figure 6.11a is suitable due to isotropic glass etching. Polymer techniques such as hot embossing can be used to produce forms 1 to 3 in a polymer substrate. GAUSSian profile in form 4 can be generated using direct laser writing. The combination of material selection and channel form is evaluated in accordance with the criteria included in the task. In the process ranging from formulating the task to a final solution, a number of compromises (e.g. between production costs and the good heat dissipation capacity of glass and polymer) have to be made. The given preconditions and the relationships determining the function will be used during the optimization process. Figure 6.11b and Figure 6.11c show a typical capillary electrophoresis separator and the corresponding separation result. Each peak in Figure 6.11c represents a DNS fragment.

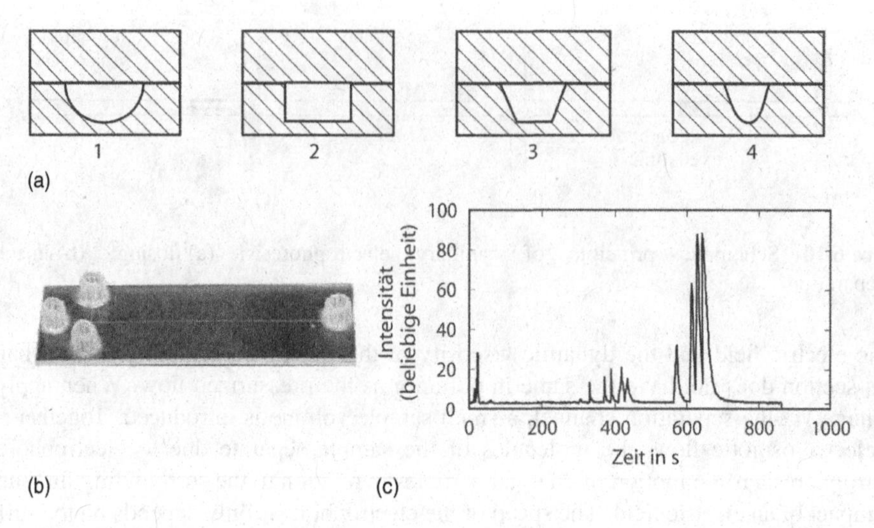

Figure 6.11 Design of microchannels for capillary electrophoresis: (a) channel forms; (b) typical capillary electrophoresis separator; (c) typical signal for optical detection at the end of the separation channel

(b) Design of microfluidic components

The available elements can be used to design diverse microfluidic components. In the following, we will present design examples of polymer microvalves and micropumps. Depending on whether the microvalves will be used as actuating elements or not, they are divided into active and passive microvalves. Whereas active valves are categorized according to their driving principles, passive valves are grouped according to their structural form. Passive valves are rectifying fluidic components. The simplest versions are microchannels with rectifying characteristics, such as diffusor/nozzle or TESLA elements (Figure 6.12). Movable parts or valve flaps are used to achieve a better fluidic rectifying capacity. Figure 6.12c shows different polymer flap valves. Valve characteristics are determined by material properties and the geometry of the spring structure. The design of the desired behavior can therefore be based on the design of mechanical elements (Section 6.1).

On a system level, microvalves can be modeled as nonlinear fluidic resistors and nonlinear fluidic capacitors. Assuming uncompressible fluids and a hard channel wall, rectifying channel structures such as diffusor/nozzle or TESLA element can be exclusively assumed to be non-linear resistors; the fluidic capacity is negligible (Figure 6.13).

As opposed to simple rectifying channel structures, the spring structure in a flap valve stores potential energy. The effect of spring stiffness on the behavior of the flap valve becomes obvious when we look at the characteristic curve of mass flow – pressure decrease in Figure 6.13. On the system level, a flap valve can be modeled as a nonlinear fluidic resistor which is coupled in parallel to a nonlinear fluidic capacitor.

Similar to microvalves, micropumps can be divided into mechanical (with movable parts) and nonmechanical (without movable parts) ones. The pump membrane is moved by an actuator. The actuator can be based on diverse transducer concepts (Chapter 7). On a system level, the pump membrane represents a fluidic capacity. The actuator can

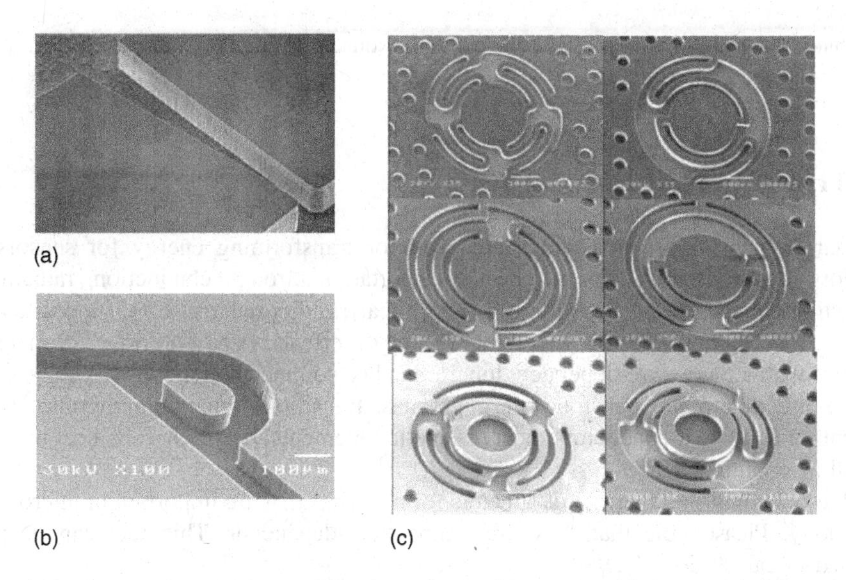

Figure 6.12 Design forms of polymer microvalves without movable parts; (a) diffusor/nozzle-element, (b) TESLA element; and with movable parts: (c) flap valves. Reproduced by permission from Nanyang Technological University, Singapore

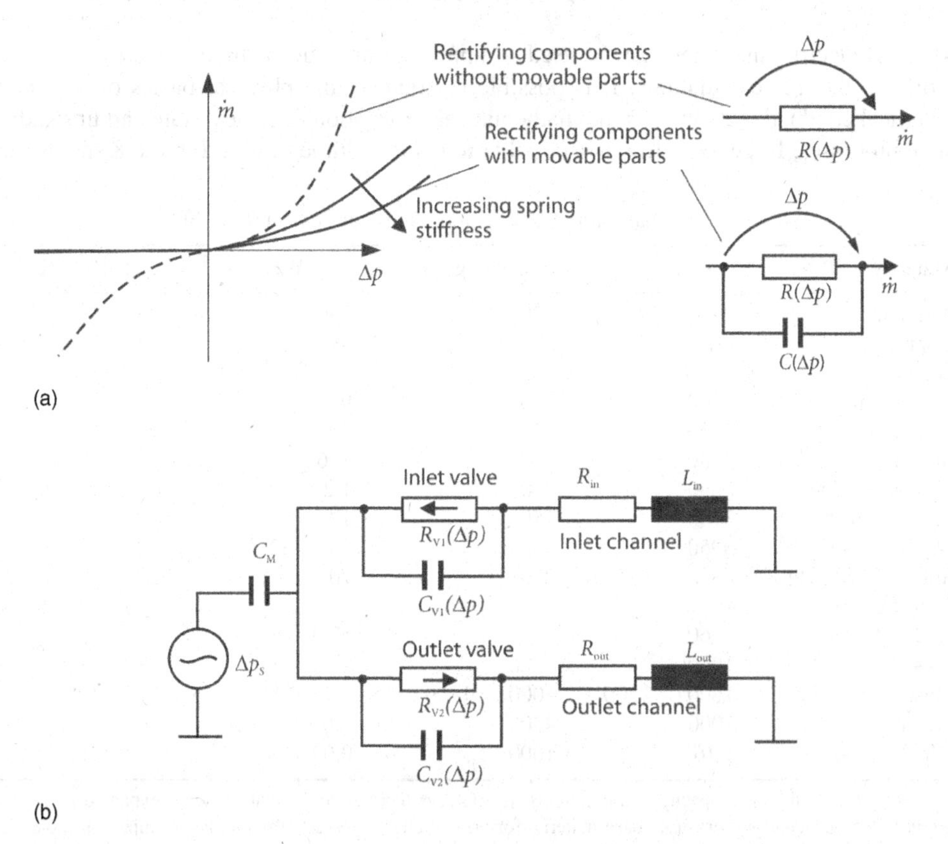

Figure 6.13 Macromodels of microfluidic components: (a) microvalves; (b) micropumps

be considered to be a time-dependent pressure source. Figure 6.13 shows the simplified macromodel of a micropump.

6.3 THERMAL ELEMENTS

Thermal elements in microsystems are used for transforming energy for sensors and actuators (see Section 7.1) or for heat transportation through conduction, radiation or convection. In the first group, there are e.g. heaters, thermal resistors, diodes, as well as transistors, thermal-electric generators (SEEBECK effect), PELTIER cooler, bi-materials, elements with temperature-dependent longitudinal or volume changes which can be caused by both direct thermal expansion or through phase transition. In the second group, we find elements such as heat conductors, heat dissipation elements, heat radiators, heat insulators as well as cooling surfaces.

Table 6.7 comprises thermal parameters for materials that are important in microsystem technology. Please note that they are temperature-dependent. This fact can be partly included in calculations VDI04.

6.3.1 Thermal-electric Analogies

Thermal-electric analogies can be used to apply the analytic methods of electrical networks to thermal calculations. It is possible to compare the physical basics of heat and electrical conduction as well as the mathematical description of steady-state and unsteady-state processes. In general, it is not possible to compare time scales. Table 6.8 shows the

Table 6.7 Thermal parameters of important materials at $T = 300$ K.

Material	ρ in kgm^{-3}	c in Wskg^{-1}K^{-1}	λ in Wm^{-1}K^{-1}	α in 10^{-6} K^{-1}
Aluminium	2700	920	230	23
Copper	8900	390	390	17
Gold	19300	125	314	15
Chromium	6900	440	95	6.6
Platinum	21500	133	70	9.0
Silicon	2330	710	156	2.3
Silicon dioxide	2660	750	1.2	0.3
Silicon nitride	3100	750	19	2.8
Al_2O_3 ceramics[a]	3950	900	30...32	6.6
Solder (63Sn37Pb)	8340	200	70	21
Lead-free solder (SnAg3.5)	7360	219	57.3	22
FR–4[b]	1900	1500	0.2...0.3	12...16
Plastics[c]	1000...2500	1000...1400	0.1...0.5	10...100
Water	1000	4200	0.6	260[d]
Air	1.16	1000	0.026	3400[d]

ρ density; c specific heat capacity; λ thermal conductivity; α thermal coefficient of linear expansion; [a]values are heavily dependent on porosity; [b]base material for printed circuit boards; [c]thermoplastic (injection molded); duroplast (transfer molded); [d]thermal coefficient of volume expansion.

Table 6.8 Thermal-electric analogy.

Thermal units		Electric units	
Heat flow Φ	$[\Phi] = \mathrm{W}$	Current I	$[I] = \mathrm{A}$
Heat flow density q	$[q] = \mathrm{Wm^{-2}}$	Current density J	$[J] = \mathrm{Am^{-2}}$
Thermal energy Q	$[Q] = \mathrm{Ws}$	Charge Q	$[Q] = \mathrm{As}$
Temperature T	$[T] = \mathrm{K}$	Potential φ	$[\varphi] = \mathrm{V}$
Temperature difference $\Delta T, \vartheta_e$	$[\Delta T, \vartheta_e] = \mathrm{K}$	Voltage V	$[V] = \mathrm{V}$
Thermal resistance R_{th}	$[R_{th}] = \mathrm{KW^{-1}}$	Resistance R	$[R] = \mathrm{VA^{-1}}$
Heat capacity C_{th}	$[C_{th}] = \mathrm{WsK^{-1}}$	Capacity C	$[C] = \mathrm{AsV^{-1}}$
Thermal conductivity λ	$[\lambda] = \mathrm{Wm^{-1}K^{-1}}$	Specific conductivity κ	$[\kappa] = \mathrm{AV^{-1}m^{-1}}$
Heat transmission coefficient α	$[\alpha] = \mathrm{Wm^{-2}K^{-1}}$	Area conductivity $R^{-1}\mathrm{A}^{-1}$	$[R^{-1}A^{-1}] = \mathrm{AV^{-1}m^{-2}}$

most commonly used analogies where the similarities between heat transport and electrical transport are reflected by the practical identity of thermal and electric units. This fact becomes at once obvious when looking at steady-state thermal conduction and the corresponding 'OHM's law of thermal conduction'. In this case, thermal flow is replaced by electric current, temperature difference by voltage and thermal resistance by electric resistance. The relationships that are known from electrical engineering regarding the coupling of resistors in series and in parallel can be applied in an analogous way to thermal resistors.

6.3.2 Basic Equations for Heat Transport

The SI unit of temperature T is Kelvin ($[T] = \mathrm{K}$). The absolute zero point is $0\ K = -273.15\ ^\circ\mathrm{C}$. The relation to the CELSIUS scale of temperature can be calculated as follows:

$$T/\mathrm{K} = \vartheta/^\circ\mathrm{C} + 273.15 \qquad (6.52)$$

Often, we do not look at absolute temperature T, but rather at temperature difference ΔT in relation to reference temperature (e.g. ambient temperature T_{amb}). This temperature difference is also called excess temperature ϑ_e:

$$\Delta T = \vartheta_e = T(t) - T_{amb}. \qquad (6.53)$$

Heat transportation in solid bodies, gases or fluids can occur through heat conduction, convection or heat radiation. In solid bodies, heat transport is generally carried out through heat conduction. In most cases, heat conduction contributes the paramount part of heat transport. Often it is also the process determining the time. FOURIER's law on one-dimensional heat conduction states that a heat flow Φ_{thcd} flows towards the colder region if there is a temperature difference $\Delta T = T_g - T_d$ at the front faces of a cuboid body (Figure 6.14):

$$\Phi_{thcd} = \frac{\mathrm{d}Q}{\mathrm{d}t} = -\lambda A \frac{\mathrm{d}T}{\mathrm{d}x}. \qquad (6.54)$$

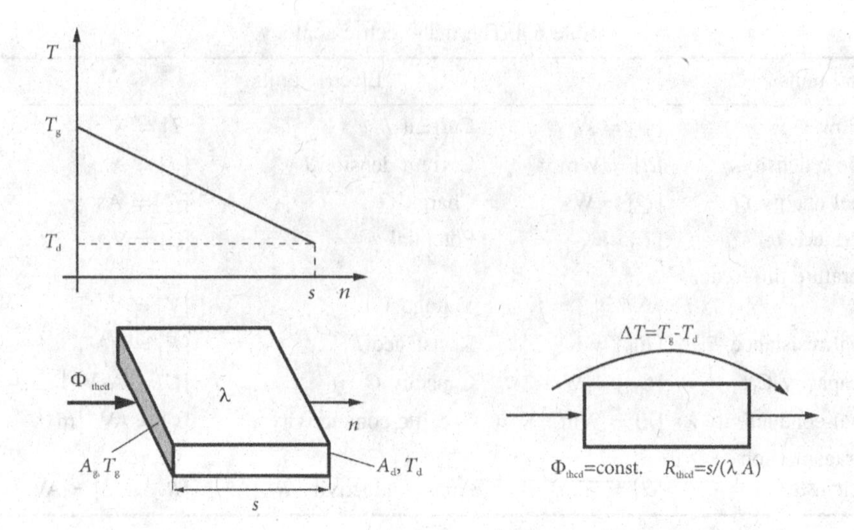

Figure 6.14 One-dimensional heat conduction in a cuboid body: Φ_{thcd} heat flow, λ thermal conductivity, A_g, A_d, T_g, T_d heat-generating and heat-dissipating areas and their temperatures

The ratio of temperature difference and heat flow is defined as the thermal resistance (OHM's law of thermal conduction):

$$R_{\text{thcd}} = \Delta T / \Phi_{\text{thcd}} = s/\lambda \cdot A. \tag{6.55}$$

The following results from Equation (6.54):

- Heat conduction occurs in the material, without mass transport.
- The heat flow is direct proportional to the thermal conductivity of the material, the cross-section area that the heat passes as well as the temperature difference; it is reciprocal to the length of the cross-section that the heat flows through.
- Thermal resistance shows a behavior that is analogous to electric resistance.

Heat conduction does not only occur in solid bodies, but also in fluids and gases. This is of interest for technical applications, such as for elements for generating gas bubbles or for heat dissipation in elements where heat transportation in solid bodies is low due to the small area of cross section A (e.g. element in Figure 6.2d). In fluids and gases, the thermal conductivity (a variable of the material) is usually smaller than that of metals by several orders of magnitude (Table 6.7). For gases, it is independent of pressure unless the mean free path of the molecules is not sufficiently small in comparison to the recipient's dimensions (KNUDSEN number Kn, see Section 6.2). For a pressure of 1 bar, the mean free path of air is about 0.06 μm; for 10 mbar, it is already ca. 6 μm. This means that for micromechanical dimensions of gaps or cavities and for low pressures, thermal conductivity is not unrelated to pressure any longer, but directly proportional to density or absolute pressure, respectively. There is no simple model for heat conduction in fluids. Thermal conductivity is related to temperature and largely dependent on phase transitions.

For water, thermal conductivity λ decreases in a range between $T = 273$ K (ice) and $T = 400$ K (steam) from 2.2 $\text{Wm}^{-1}\text{K}^{-1}$ to 0.03 $\text{Wm}^{-1}\text{K}^{-1}$.

At convection, heat transmission occurs from a solid body (wall) with area A to a fluid (or gas). Convection is caused by the motion of gas or fluid, respectively. It is strongly related to thermal and fluidic conditions in general. There are analytical approaches for estimating the generated heat flow. However, exact calculation is difficult. A distinction is made between free convection and forced convection (cooling fan, water cooling). Similar to heat conduction, also thermal resistance of convection (heat transmission resistance) is defined by the ratio of temperature difference and heat flow:

$$R_{\text{thcv}} = \Delta T/\Phi_{\text{thcv}} = 1/\alpha_{\text{cv}}A. \tag{6.56}$$

Heat transmission coefficient α_{K} is related, among others, to the flow rate of the fluid, wall temperature, fluid density and height of the wall. For free convection, values of $\alpha_{\text{cv}} = 5\ldots 25$ $\text{Wm}^{-2}\text{K}^{-1}$ are used for calculations; for forced convection, values lie in the range of $\alpha_{\text{cv}} = 10\ldots$ 120 $\text{Wm}^{-2}\text{K}^{-1}$. Thermal resistance during convection is particularly important for the efficiency of cooling bodies. Heat transmission by evaporation or condensation constitutes a special form of convection. For microsystems operating in vacuum, it is important that there is no convection-related heat flow.

- Convection is connected to mass transport in the fluid.
- Convection is characterized by heat transmission coefficient α_{cv}.

Heat radiation is heat transport by adsorption or emission of electromagnetic waves. As opposed to heat conduction and convection, this mechanism is not connected to matter at all. STEFAN-BOLTZMANN law describes the radiated heat flow in relation to temperature and radiator surface:

$$\Phi_{\text{thr}} = \text{d}Q/\text{d}t = \varepsilon \cdot \sigma \cdot A_{\text{r}}(T_{\text{r}}^4 - T_{\text{amb}}^4). \tag{6.57}$$

It contains the dimensionless number emissivity (corrective factor for the proportion of reflection in the radiator material in relation to the black, non-reflecting radiator), σ STEFAN-BOLTZMANN constant ($5.67 \cdot 10^{-8}$ Wm^{-2} K^{-4}), A_{r} the radiator area, T_{r} the temperature of the radiator, and T_{amb} ambient temperature. For $T_{\text{r}} - T_{\text{amb}} \ll T_{\text{amb}}$, i.e. for radiator temperatures that only slightly exceed the reference temperature, heat radiation can be estimated using linear approximation:

$$T_{\text{r}}^4 \approx T_{\text{amb}}^4 + 4T_{\text{amb}}^3(T_{\text{r}} - T_{\text{amb}}). \tag{6.58}$$

Equation (6.57) thus becomes

$$\Phi_{\text{r}} \approx \varepsilon \cdot \sigma \cdot A_{\text{r}} \cdot 4T_{\text{amb}}^3(T_{\text{r}} - T_{\text{amb}}). \tag{6.59}$$

Heat radiation can only be exchanged between two areas that 'face' each other. Visibility factor F_{12} of area A_1 in relation to a second area A_2 describes the proportion of the radiation emitted by radiating area A_1 and absorbed by area A_2:

$$\Phi_{\text{thr}} = \varepsilon \cdot \sigma \cdot A_1 \cdot F_{12} \cdot T_{\text{r}}^4. \tag{6.60}$$

Also for heat radiation, the basic equation of heat flow $\Phi = \alpha A \Delta T$ can be used to define thermal resistance:

$$R_{\mathrm{thr}} = \Delta T/\Phi_{\mathrm{thr}} = 1/(\alpha_{\mathrm{r12}} \cdot A_{\mathrm{r}}). \qquad (6.61)$$

Here, the heat transmission coefficient is

$$\alpha_{\mathrm{r12}} = f(x)\rho_{\mathrm{eff}} \cdot \sigma \cdot T_{\mathrm{r}}^3 \qquad (6.62)$$

with $f(x) = 1 + x + x^2 + x^3$ and $x = T_{\mathrm{amb}}/T_{\mathrm{r}}(T_{\mathrm{amb}} < T_{\mathrm{r}})$. The effective absorption coefficient ρ_{eff} is a factor that takes into account the areas of radiator and receiver as well as their degrees of absorption:

$$\rho_{\mathrm{eff}} = \frac{1}{\frac{1}{\rho_1} + \frac{A_1}{A_2}\left(\frac{1}{\rho_2} - 1\right)} \qquad (6.63)$$

Absorbencies ρ_1 and ρ_2 of areas 1 and 2 are related to temperature and surface structure. For the common case $A_2 \gg A_1$, $\rho_{\mathrm{eff}} = \rho_1$.

It results from Equation (6.60):

- Heat radiation is proportional to the fourth power of radiator temperature (T^4-law).
- Only for high radiator temperatures does the heat flow that can be dissipated through heat radiation become a relevant portion of heat transmission.

In relation to a specific element, heat change Q and temperature difference ΔT are proportional to each other. Heating an element with mass m from an initial temperature T_{amb} to temperature T_1 requires a heat input of

$$Q = mc(T_1 - T_{\mathrm{amb}}). \qquad (6.64)$$

Here, c is the specific heat capacity of the material (see Table 6.7). Often, the power required for heating is of interest

$$P = \frac{\mathrm{d}Q}{\mathrm{d}t} = mc\frac{\mathrm{d}T(t)}{\mathrm{d}t} = C_{\mathrm{th}}\frac{\mathrm{d}T(t)}{\mathrm{d}t}. \qquad (6.65)$$

The heat capacity

$$C_{\mathrm{th}} = mc = V \cdot \rho \cdot c \qquad (6.66)$$

is a measure of the change of the heat stored in the element, in relation to the corresponding change in temperature. Applying the temperature difference in relation to a reference temperature, such as in Equation (6.53), and not the absolute temperature, we can use Equation (6.65) which results in

$$P = C_{\mathrm{th}}\frac{\mathrm{d}\vartheta_{\mathrm{e}}(t)}{\mathrm{d}t}. \qquad (6.67)$$

Equation (6.67) clearly shows that the temperature of an element cannot jump as this would require an infinitely large power. It also shows the following: If a thermally ideally insulated element with heat capacity C_{th} is from time $t = 0$ supplied with power $P = \text{const.}(= \text{heat flow } \Phi)$, this will result due to

$$\frac{\mathrm{d}\vartheta_e(t)}{\mathrm{d}t} = \frac{P}{C_{th}} \neq 0 \tag{6.68}$$

in an unsteady-state behavior. For the time-dependent heating of the element, it applies that

$$\vartheta_e(t) = \vartheta_e(t = 0) + 1/C_{th} \int_0^t P dt. \tag{6.69}$$

In practice, elements are not thermally ideally insulated, which means that part of the supplied power will be transmitted via thermal resistors R_{th} to the adjacent media:

$$P = C_{th} \frac{d\vartheta_e}{dt} + P_a. \tag{6.70}$$

The power transmitting to the adjacencies P_a (or heat flow Φ_a, respectively) can be calculated using the equations for heat conduction, convection and heat radiation, with $\Delta T(t) = \vartheta_e(t) = T(t) - T_{amb}$ and R_{th} resulting from the corresponding heat transport mechanism. For the time-dependent heating of the element, solving the linear differential Equation (6.70) and applying the conditions $P(t) = \text{const.}$ and $\vartheta_e(0) = 0$ will result in

$$\vartheta_e(t) = P R_{th}(1 - e_{th}^{-1/\tau}) \tag{6.71}$$

with thermal time constant $\tau_{th} = C_{th} R_{th}$.

For $t \gg \tau$, the steady-state condition will be reached:

$$\mathrm{d}\vartheta_e/\mathrm{d}t = 0 \text{ and } \Delta T = \vartheta_e = P R_{th}. \tag{6.72}$$

The thermal compensation processes have died; the heat capacities are 'charged' and the entire supplied power is transmitted to the surrounding media. The element reaches its excess temperature. The smaller the thermal time constant τ_{th}, the faster this state will be reached. Thermal elements in microsystem technology have a substantially smaller thermal time constant than in the macroworld. It is therefore possible to reach frequencies in the kilohertz range for heating and cooling cycles.

6.3.3 Equivalent Circuits

Thermal equivalent circuits consist of heat flow and temperature sources (Φ or ΔT, respectively) as well as networks containing thermal resistors R_{th} and thermal capacitors C_{th} coupled in series or in parallel. Using the analogies provided in Table 6.8 and thermal junction and loop rules, we can carry out calculations that are analogous to network calculations in electrical engineering:

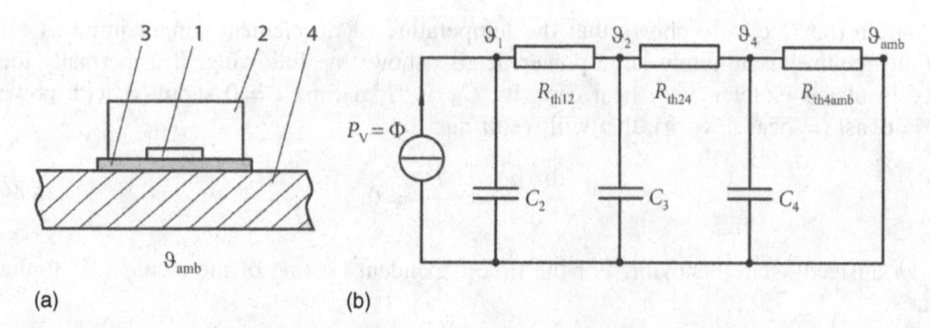

Figure 6.15 Thermal equivalent circuit of a thermally insulated semiconductor component: 1: heat-generating chip (power loss P_V); 2: component housing; 3: insulating disc; 4: heat sink

- The thermal junction rule states that the sum of supplied power (heat-generating component), dissipated power (heat exchange with the environment) and stored power (in the element of a cooling device) is zero. Power (heat flow) behaves analogous to electric power.
- The thermal loop rule states that the sum of the temperature differences along a closed loop is zero. Temperature differences can be compared to voltage differences.
- During unsteady-state temperature behavior, the heat flow occurs from the heat source to the environment thus charging the heat capacities.
- Maximum power P_{max} that can be dissipated after reaching the steady-state condition results from Equation (6.72) as the ratio of maximum admissible temperature rise and total thermal resistance between source and environment

Figure 6.15 shows the example of the thermal equivalent circuit of an semiconductor component in package which generates power loss P_v and is mounted on a cooling body (heat sink).

The internal thermal resistance R_{th12} of the semiconductor component depends on the internal structuring and bonding technique used and is commonly included in data sheets. Resistance R_{th24} characterizes heat transmission from the component housing to the heat sink. It comprises the heat transmission between housing and insulating disc, the heat transition through the insulating disc and the heat transmission from the insulating disc to the heat sink. The insulating disc has a high thermal conductivity. Total transmission resistance R_{th24} depends on the thermal coupling between component housing and heat sink, e.g. on the thickness of the insulating disc, contact pressure, surface quality and heat conducting paste. Resistance R_{th4amb} characterizes the heat transmission from heat sink to the environment (convection resistance). Convection resistance R_{th2amb}, describing the direct heat transmission from component housing to environment, would have to be included as dissipation resistance opposite to the mass potential (ambient temperature T_{amb} or ϑ_a, respectively). Its magnitude is very large compared to the other thermal resistance and it can therefore be neglected. Heat capacities $C_{th} = mc$ always refer to the potential of the ambient temperature. The heat quantity required to 'charge' them amounts to $Q = C_{th}\Delta T$.

Example 6.6 Calculation of an air flow sensor

Air flow can be measured using a sensor that contains a heating element and a temperature probe (Figure 6.16). The air flow dissipates heat resulting in a temperature difference $\Delta T = T_1 - T_{amb}$ that is proportional to the air stream. Temperatures T_{amb} and T_1 are measured with integrated temperature sensors. Temperature difference ΔT between heater and point T_{amb} is controlled in order to remain constant. The air flow around the sensor causes increased heat dissipation and an increase of the required heating power. The supplied heating power is the measure of the flow velocity of the air. A silicon chip with dimensions $10 \times 2 \times 0.5\ \text{mm}^3$ is used as a thermal transducer element. The silicon has a thickness of 0.5 mm, a dioxide thickness of 2 μm, and a nitride thickness of 0.5 μm. The chip contains a heater structure with an area of $2 \times 1.5\ \text{mm}^2$.

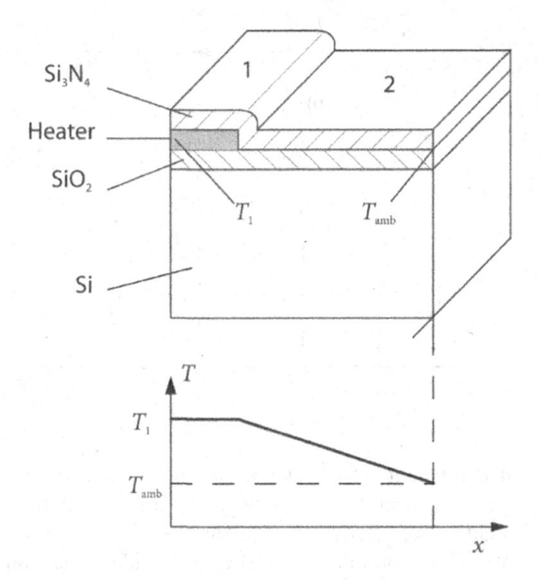

Figure 6.16 Air flow sensor (not in scale)

The sputtered platinum heater is located on the SiO_2 layer, its height is negligible. Due to its structural arrangement in the environment, the right-hand side of the silicon chip operates as heat sink. Heat dissipation through free convection and radiation can be neglected as it is very small. When calculating thermal resistances it is important to know that heat flows in the heater range (1) move vertically to the heater surface, whereas in the remaining area (2), they flow in parallel direction to point T_{amb}. Each partial heat flow reaches the entire cross-sectional area of all three layers. Subsequently, we will discuss the following questions and their solutions:

(a) What does the thermal equivalent circuit look like in the steady-state case (no air flow)? What simplifications can be made? How large is the heating power required for maintaining a temperature difference of $\Delta T = 90$ K along the entire sensor chip?

(b) What does the thermal equivalent circuit look like in the unsteady-state case (the element is heated from switching on until reaching the steady-state condition)? What is the heating time till the steady-state condition is reached?

Figure 6.17 presents the thermal equivalent circuit. Thermal resistors and heat capacitors can be calculated using Equations (6.51) and (6.62). The dimensions and material sizes provided in Table 6.7 are applied to arrive at the values for the thermal resistances and capacities of the individual layers given in Table 6.9.

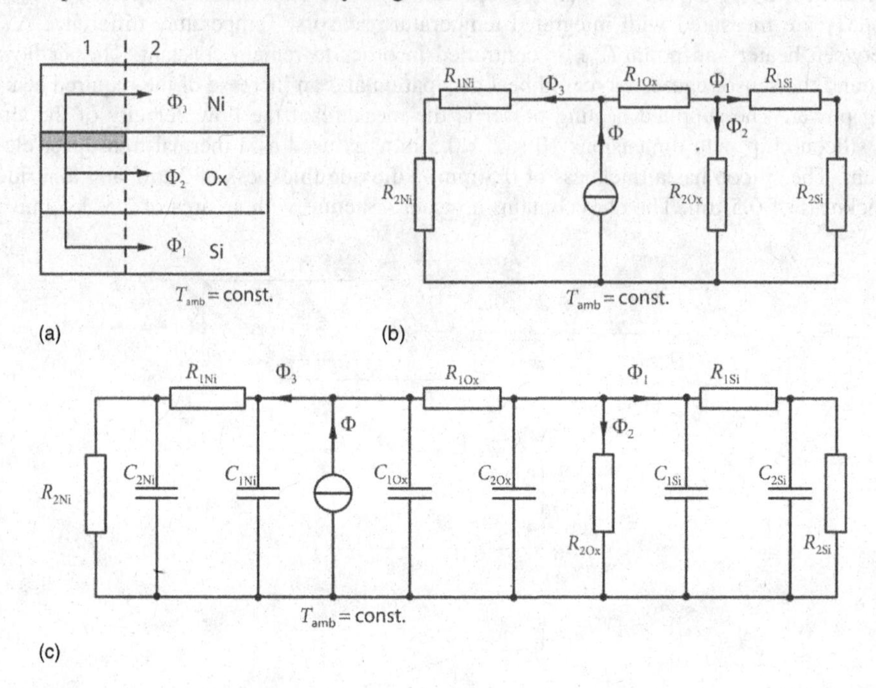

Figure 6.17 Thermal equivalent circuit of the air flow sensor according to Figure 6.16: (a) layer stack of heater and heat flows in the layers 1,2; (b) thermal equivalent circuit for the steady-state condition; (c) thermal equivalent circuit for an unsteady-state condition (heating of the element). Ni silicon nitride, Ox silicon dioxide, Si silicon

Table 6.9 Thermal resistances and thermal capacities related to Example 6.6.

Thermal resistance	R_{th} in KW^{-1}	Thermal capacity	C_{th} in WsK^{-1}
R_{1Ox}	0.56	C_{1Ox}	$12.0 \cdot 10^{-6}$
R_{1Ni}	0.009	C_{1Ni}	$3.5 \cdot 10^{-6}$
R_{1Si}	1.07	C_{1Si}	$2.5 \cdot 10^{-3}$
R_{2Ox}	$1.77 \cdot 10^{6}$	C_{2Ox}	$67.8 \cdot 10^{-6}$
R_{2Ni}	$0.45 \cdot 10^{6}$	C_{2Ni}	$19.8 \cdot 10^{-6}$
R_{2Si}	54.5	C_{2Si}	$14.2 \cdot 10^{-3}$

(a) Steady-state case

As the heat flows in the range of the heater (1) in an approximately vertical direction, whereas in the remaining range (2) they flow in a longitudinal direction, it has to be taken into account that there is a total of three air flows Φ_1, Φ_2, Φ_3 that are directed from the heater to point T_0; and each flows through ranges (1) and (2). On their way, the heat flows encounter the following thermal resistances (Figure 6.17a):

$$\Phi_1 : R_{1\text{Ox}}, R_{1\text{Si}}, R_{2\text{Si}};\ \Phi_2 : R_{1\text{Ox}}, R_{2\text{Ox}};\ \Phi_3 : R_{1\text{Ni}}, R_{2\text{Ni}}.$$

$(\Phi_1 + \Phi_2)$ flows through resistance $R_{1\text{Ox}}$ resulting in a thermal equivalent circuit for the steady-state case according to Figure 6.17b.

It is possible to assume a number of simplifications: The numeric values show that due to $R_{1\text{Ni}}, R_{1\text{Si}}, R_{1\text{Ox}}, R_{2\text{Si}} \ll R_{2\text{Ni}}, R_{2\text{Ox}}$ heat flows Φ_2 and Φ_3 are very small. Basically, almost the entire heat flow is represented by Φ_1. The effective total thermal resistance thus corresponds to $R_{1\text{Ox}}, R_{1\text{Si}}, R_{2\text{Si}}$, coupled in series, i.e. $R_{\text{thL}} = (0.56 + 1.07 + 54.5)\ \text{KW}^{-1} = 56.13\ \text{KW}^{-1}$. For a further simplification, also $R_{1\text{Ox}}$ and $R_{1\text{Si}}$ can be neglected.

In order to keep a constant temperature difference $\Delta T = 90$ K along the entire sensor chip, heating power

$$P_{\text{heat}} = \Phi = \frac{\Delta T}{R_{\text{thL}}} = \frac{90K}{56.13\ KW^{-1}} = 1.6\ W \tag{6.73}$$

has to be supplied.

(b) Unsteady-state case

The thermal capacities are to include in the equivalent circuit according to Figure 6.17c. They belong to the corresponding thermal resistances and always relate unilaterally to reference temperature T_{amb}.

We can use the following simplifications: Thermal capacities $C_{1\text{Si}}$ and $C_{2\text{Si}}$ are substantially larger than all the others. Therefore, the total thermal capacity can be calculated using the approximation

$$C_{\text{th tot}} = C_{1\text{Si}} + C_{2\text{Si}} = 16.7 \cdot 10^{-3}\text{WsK}^{-1}. \tag{6.74}$$

Thus, the time constant becomes

$$\tau_{\text{th}} = C_{\text{th}} R_{\text{th}} = 0.94\ \text{s}. \tag{6.75}$$

Applying $t \geq 3\tau_{\text{th}}$ to calculate the time until the steady-state condition is reached, results in $t \cong 3$ s.

Using the approximate values for R_{th} and C_{th}, the equivalent circuit can be simplified to a degree that it only consists of a heat source and a parallel coupling of R_{thcv} with $C_{\text{th\_tot}}$.

EXERCISES

6.1 The task is to design a capacitive acceleration sensor with a measuring range of $a_N = 50\ \text{ms}^{-2}$ using silicon micromachining. The eigenfrequency of the micromechanical element (flexible capacitor plate, coupled to the thick rim via a cantilever spring) is $f_e = 1.4$ kHz (limit frequency $f_g = 1$ kHz). How large is displacement y of the flexible capacitor plate which can be used for the capacitive signal conditioning at nominal value?

6.2 The task is to calculate for the silicon micromechanical torsion mirror in Figure 6.18 according to Figure 6.2d the heat flow that moves due to heat conduction from the mirror plate to the frame. The mirror plate is used to deflect laser beams. Part of the radiation is absorbed resulting in a temperature of 80 °C on the mirror plate. The frame has a temperature of 20 °C. Spring cross-section and length are $30 \times 60\ \mu m^2$ and 1000 μm. Thermal conductivity of silicon is $\lambda_{Si} = 156$ W/Km. Heat dissipation through convection is negligible.

6.3 What effect do relative changes $\Delta h/h$ of structure thickness h of micromechanical oscillators have on their eigenfrequency f_e, if they are manufactured using

(a) silicon bulk micromachining
(b) silicon surface micromachining?

Spring and oscillator mass have the same thickness. The spring has a rectangular cross-section. For bulk structures, motions of the structures occur in the direction of the wafer normal and for surface structures in the wafer plane.

6.4 How can the damping of flexible micromechanical elements be influenced? Does a changed damping affect other parameters of the oscillatory system?

6.5 The tactile system of a coordinate measuring device in Figure 6.19 is to be used for measuring contours in industrial metrology. The tactile head consists of a silicon

Figure 6.18 Micromechanical torsional mirror made of silicon. Reproduced by permission from the Center for Microtechnologies, Chemnitz University of Technology, Germany

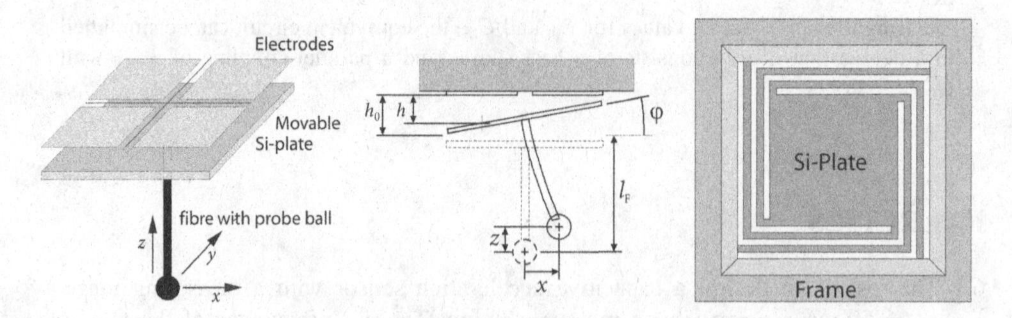

Figure 6.19 Micromechanical tactile system

plate that is flexibly suspended on a silicon spring according to g; a glass fiber with a probe ball is fixed to the plate in direction of the surface normal. The probe ball moves along the surface of a work piece in direction $x - y$ and is displaced when doing this. The displacement is transmitted to the silicon plate and can be capacitively detected. The following values are given:

Maximum displacement of probe ball $s_{x,y\,max} = 20\ \mu m$

Glass fibre diameter $d_F = 18$, 40 or 50 μm

Probe ball diameter $d_B = 33$ or 100 μm

Glass fibre length $l_F = 1$, 2 or 3 mm

Fibre material (quartz) $E_F = 75$ GPa; $\rho_F = 2200\ kgm^{-3}$.

Stiffness of silicon spring $c_{Sz} = 4.61$ N/m; $c_{Srot\ x,y} = 15.3 \times 10^{-6}$ Nm

Silicon plate: area $3 \times 3\ mm^2$, thickness 30 μm, $\rho_{Si} = 2300\ kg/m^3$.

(a) When moving in direction x, y, how does stiffness $c_{Tx,y}$ acting on the probe ball depend on the stiffnesses c_F of the glass fibre and c_ς of the silicon plate (sensor element)? The glass fiber is exposed to bending stress, and the silicon spring to torsional stress.
(b) How large do spring stiffnesses c_F of the glass fibre and effective spring stiffnesses $c_{Tx,y}$ become for the given fiber length and diameter?
(c) What fiber lengths and diameters are feasible for a contact force at s_{max} in the range of 30...250 μN?
(d) How large is static sagging z_{stat} of the silicon plate due to its own weight (the weight of the glass fiber is negligible)?
(e) How large is eigenfrequency f_z of the sensor element?

6.6 Determine Weber number for a capillary flow with a water/air boundary layer at a velocity of 100 μm/s! The capillary is cylindrical and has a diameter of 100 μm. Surface tension or the water/air boundary layer amounts to 72 N/m. It is assumed that the capillary is totally hydrophile (contact angle is 0°).

6.7 The mean free path of air molecules under ambient conditions (25 °C, 1 bar) is $6.11 \cdot 10^{-8}$ m. Determine KNUDSEN number for air in a cylindrical capillary with a diameter of 10 μm.

6.8 Will the thermal time response calculated in Example 6.6 under (b) change if the sensor is operated?

6.9 What temperature does the lower side of the silicon substrate in the range of the heater structure reach if we use the heating power calculated in Example 6.6 under (a)?

REFERENCES

[MEHNER00] Mehner, J. (2000) *Entwurf in der Mikrosystemtechnik* (Design in Microsystem Technology; in German). Dresdner Beiträge zur Sensorik (Dresden Contributions to Sensorics). Bd. 9. Dresden University Press.

[NEXUS02] European Commission, Network of Excellence in Multifunctional Microsystems (NEXUS), *Market Analysis for Microsystems 1996–2002*.

[NGUYEN02] Nguyen, N. T, Wereley, S. (2002) *Fundamentals and Application of Microfluidics*. Artech.

[NGUYEN04] Nguyen, N. T. (2004) *Mikrofluidik – Entwurf, Herstellung und Characterization*; in German). (Microfluidics – Design, Manufacturing, Characterization). Stuttgart: Teubner.

[VDI04] *VDI-Wärmeatlas* (VDI Heat Atlas), 9th edn, Berlin, Heidelberg, New York: Springer, 2004.

7

Sensors and Actuators

Microsystems complement microelectronics – which are dedicated mainly to the processing of analog and digital electric signals as well as to signal and data storage – with sensoric and actuating functions. They thus constitute a link to non-electric environments of technical systems with predominantly non-electric state variables. The direct coupling of electric or electronic components with non-electric (mechanical, thermal, optical) components is typical for sensors and actuators.

The name sensor is used for components that convert a (preferably non-electric) measurand into an electric measuring signal. As opposed to this, an actuator is a component that converts an electric drive energy into mechanical work. In the schematic presentation of a technical or natural control circuit (Figure 7.1), sensors can be considered to be 'input transducers' that convert the measurand into an electric measuring signal and transport it to a processing unit (processor, brain).

The processing unit generates in accordance with a given objective function the triggering signals for the actuators that – as 'output transducers' – affect a technical process or the environment.

Thus, sensors and actuators are transducers that convert the energy of one form into that of another form. As energy conversion, in principle, can occur in opposite transduction directions, it is often possible to use the transduction principle both for actuators and (inversely) for sensors. Examples are the piezoelectric effect used for both piezoactuators in linear precision drives and piezoelectric oscillation absorbers or acceleration sensors; or electrostatic transducers that are applied in electrostatic drives as well as in capacitive sensors. Figure 1.6, for instance, presents a yaw rate sensor that measures the yaw rate via the CORIOLIS effect using a capacitive acceleration sensor (Figure 1.8d). The interdigital finger structure can be used in separate areas for both capacitive distance measuring and electrostatic force compensation which moves the movable part of the electrode back into steady state [KOVACS98]. This allows to operate such a sensor as a force compensating sensor in feedback-loop mode which results in extremely precise measuring [KHAZAN94]. A precondition for such applications is the fact that transduction mechanisms are reversible. In the following, we will therefore at first look at principle characteristics of transducers for sensors and actuators before presenting individual transducer principles.

Introduction to Microsystem Technology: A Guide for Students Gerald Gerlach and Wolfram Dötzel

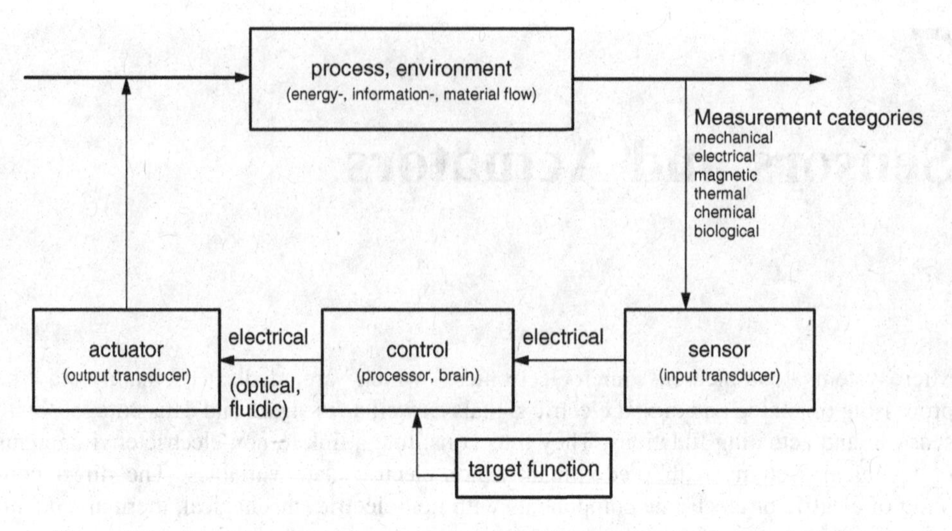

Figure 7.1 Sensors and actuators as input and output transducers in a control circuit

7.1 REVERSIBLE AND PARAMETRIC TRANSDUCERS

7.1.1 Reversible Transducers

Transducers are assumed to be closed systems with an internal energy W_{i} consisting of electrical energy W_{el} and mechanical energy W_{mech}, for instance:

$$W_{\mathrm{i}} = W_{\mathrm{el}} + W_{\mathrm{mech}} \tag{7.1}$$

If the system (sensor, actuator) is supposed to show constant properties over longer periods, the internal energy of the systems as a state condition has to be constant:

$$\mathrm{d}W_{\mathrm{i}} = 0. \tag{7.2}$$

At the same time, transducers are supposed to have a reversible function. This means that the contour integral

$$\oint W_i = 0 \tag{7.3}$$

has to be path-independent.

For a general electromechanic system according to Figure 7.2, it results that

$$\mathrm{d}W_i = V\mathrm{d}Q + F\mathrm{d}s, \tag{7.4}$$

with voltage V, charge Q, force F and displacement s.

If we have an energy-conserving system, i.e. there is no energy flow (e.g. diffusion or Ohm's current, heat transport), we have a static transducer and Equation (7.2) can easily

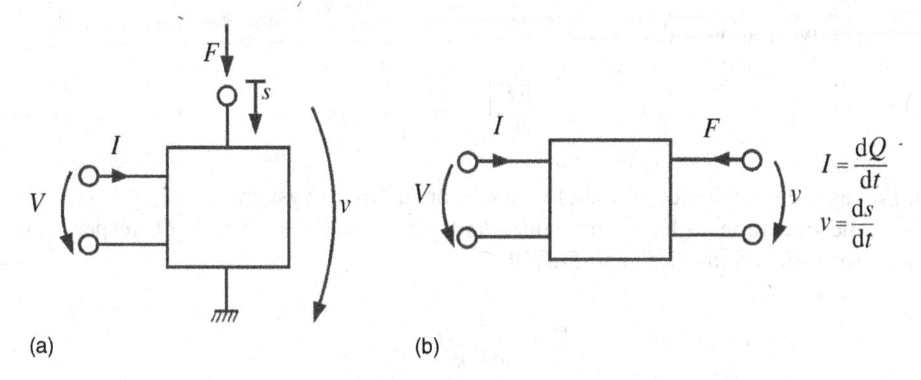

Figure 7.2 General electromechanic transducer, (a) principle representation; (b) transducer two-port

be applied. Electrostatic transducers with only a displacement current and no conducting current are such transducers.

Voltage $V = V(Q, s)$ and force $F = F(Q, s)$ are dependent on charge Q and displacement s. Two of the four coordinates are sufficient to completely describe the state of the transducer. With the conditions in Equation (7.3) and using Equation (7.4), it is valid that

$$\oint(V\mathrm{d}Q + F\mathrm{d}s) = \oint[V(Q, s)\,\mathrm{d}Q + F(Q, s)\,\mathrm{d}s] = 0 \tag{7.5}$$

It results from this that the integral in Equation. (7.5) has to be a complete differential:

$$\begin{aligned}\Delta V &= \left.\frac{\partial V}{\partial Q}\right|_{s_0}\Delta Q + \left.\frac{\partial V}{\partial s}\right|_{Q_0}\Delta s\\ \Delta F &= \left.\frac{\partial F}{\partial Q}\right|_{s_0}\Delta Q + \left.\frac{\partial F}{\partial s}\right|_{Q_0}\Delta s\,\cdot\end{aligned} \tag{7.6}$$

It is valid that

$$\left.\frac{\partial V}{\partial s}\right|_{Q_0} = \left.\frac{\partial F}{\partial Q}\right|_{s_0} = A_0 = \text{const.}, \tag{7.7}$$

with s_0 and Q_0 being constant values for displacement s and charge Q and with changes Δs and ΔQ occuring around them.

The coefficients in Equations (7.5) to (7.7) have specific physical meanings [LENK01]: Parameter A_0 from Equation (7.7) is the transducer constant or the reciprocity parameter. C_K is the capacitance of the mechanically clamped (short-circuited: $s_0 = \text{const.}$) system.

$$\left.\frac{\partial V}{\partial Q}\right|_{s_0} = \frac{1}{C_K}. \tag{7.8}$$

Subsequently, it applies that

$$\left.\frac{\partial F}{\partial s}\right|_{Q_0} = \frac{1}{n_\ell}, \tag{7.9}$$

with n_ℓ as the compliance of the electrically open-circuit system.

For the transition to harmonic values and with $\underline{v} = j\omega\underline{s}$ or $\underline{I} = j\omega\underline{Q}$, respectively, it results from Equations (7.5) and (7.6) that

$$\begin{aligned} \underline{V} &= \underbrace{\frac{1}{j\omega C_K}\underline{I}}_{\underline{V}_C} + \underbrace{\frac{A_0}{j\omega}\underline{v}}_{\underline{V}_W} \\ \underline{F} &= \underbrace{\frac{A_0}{j\omega}\underline{I}}_{\underline{F}_W} + \underbrace{\frac{1}{j\omega n_\ell}\underline{v}}_{\underline{F}_n}. \end{aligned} \tag{7.10}$$

The lumped equivalent circuit in Figure 7.3 can be used to represent the system of Equation (7.10). The following statements can be derived from Equation (7.10) or Figure 7.3a, respectively:

The electromechanical transducer is reversible. When electric power is supplied, force $\underline{F}$ and velocity $\underline{v}$ change. This way, the transducer operates as an actuator. In the reversed case, a mechanical load with $\underline{v}$ or $\underline{F}$, respectively, causes changes in current $\underline{I}$ and voltage $\underline{V}$. In this case, the transducer operates as a sensor.

When operating a transducer as an actuator, the input is electric energy. Part of the electric power is stored in capacitor C_K, the remaining part is converted into mechanical

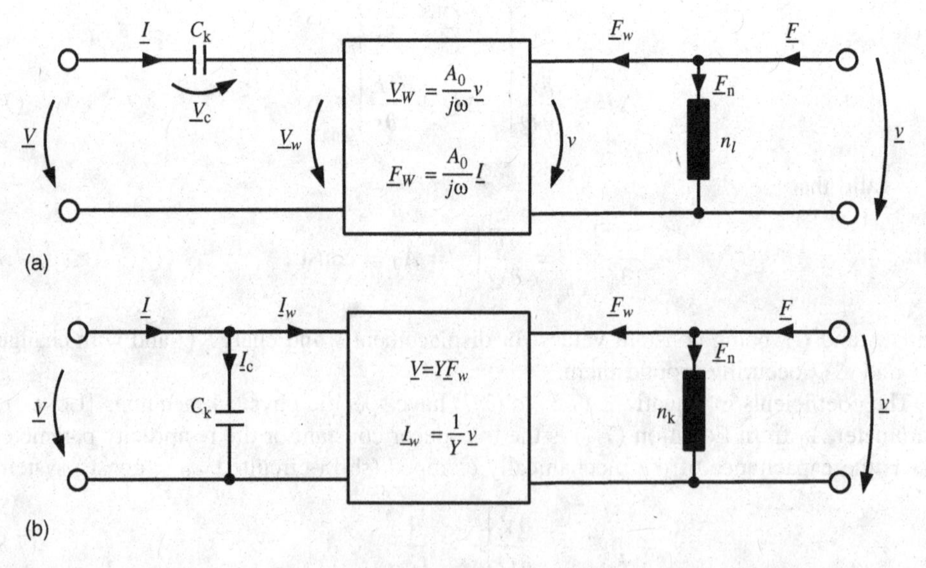

Figure 7.3 Equivalent diagram of a reversible static transducer: (a) with imaginary transformer; (b) with a gyrator as coupling two-port

energy. However, the converted mechanical energy is not entirely at free disposal, as part of it is utilized for the elastic deformation of the mechanical component (spring n_ℓ).

For mechanoelectrical sensors, mechanical energy is diverted from the measuring process. This leads partially to the deformation of the spring n_ℓ. The remaining energy is converted into electric energy, a part of which is stored in capacitor C_K.

The system of Equation (7.10) and the lumped equivalent circuit in Figure 7.3a include a transformer-like coupling two-port which is, however, completely imaginary. Therefore Equation (7.10) is mainly written in the following form:

$$\begin{aligned} \underline{I} &= \underbrace{\frac{1}{Y}\underline{v}}_{\underline{I}_W} + \underbrace{j\omega C_0 \underline{V}}_{\underline{I}_C} \\ \underline{F} &= \underbrace{\frac{1}{j\omega n_K}\underline{v}}_{\underline{F}_n} + \underbrace{\frac{1}{Y}\underline{V}}_{\underline{F}_W} \end{aligned} \quad \text{with} \quad \begin{cases} n_K = n_\ell \| n_C = \dfrac{n_\ell n_C}{n_\ell + n_C} \\ n_C = \dfrac{1}{A_0^2 C_K} \\ Y = \dfrac{1}{A_0 C_K} \end{cases} \tag{7.11}$$

The corresponding equivalent circuit now includes a transducer two-port that converts effort variables ($\underline{V}$, $\underline{v}$) into flow variables ($\underline{I}_W$, $\underline{F}_W$) and vice versa. Such a convertor is called gyrator. The imaginary transformer in Figure 7.3a has thus been transformed into a real gyrator.

In general, Equation (7.11) can be written as follows:

$$\begin{aligned} \underline{I} &= \gamma_{11}\underline{V} + \gamma_{12}\underline{v} \\ \underline{F} &= \gamma_{21}\underline{V} + \gamma_{22}\underline{v}. \end{aligned} \tag{7.12}$$

Coeffient γ_{11} describes electric input conductance $\underline{I}/\underline{V}$ for mechanical clamping ($\underline{v} = 0$), γ_{22} mechanical admittance $\underline{F}/\underline{v}$ for short circuit $\underline{V} = 0$. Coefficients γ_{12} and γ_{21} are coupling constants that are identical:

$$\gamma_{12} = \gamma_{21}. \tag{7.13}$$

When changing the direction of current $\underline{I}$ or force $\underline{F}$ in the system, γ_{12} and γ_{21} would have the same quantity, but with opposite sign.

Example 7.1 Electrostatic transducer

Figure 7.4 presents an electrostatic transducer with a fixed lower capacitor electrode. The upper electrode can be shifted parallel to the lower one. In normal state, the overlapping area of the electrodes is $a \cdot b$ and the distance is d. In this normal state, the spring has compliance n.

An applied voltage V generates force F_{el}, which results – for displacement $s = \Delta b$ – from the changing electrostatic field energy in the capacitor:

$$F_{el} = \frac{dW_{el}}{db} = \frac{d\left(\dfrac{Q^2}{2C}\right)}{db} = \frac{Q^2}{2}\frac{d\left(\dfrac{1}{C}\right)}{db}. \tag{7.14}$$

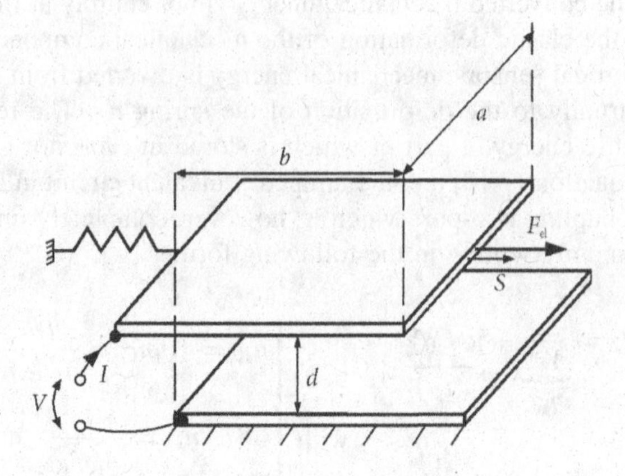

Figure 7.4 Electrostatic transducer

Here, C is the capacitance and Q the charge on the electrodes. With $C = \varepsilon ab/d$, it results that

$$F_{\mathrm{el}} = \frac{1}{2}V^2\varepsilon\frac{a}{d}. \tag{7.15}$$

Similar to Equations (7.5) and (7.6), we can determine that:

$$\left.\frac{\partial Q}{\partial V}\right|_{s=0} = C = \varepsilon\frac{ab}{d}, \tag{7.16}$$

$$\left.\frac{\partial F}{\partial s}\right|_{V=V_0} = \frac{1}{n} \tag{7.17}$$

and from Equation (7.15)

$$\left.\frac{\partial F}{\partial V}\right|_{s=0, V=V_0+\Delta V} = \varepsilon\frac{a}{d}\cdot V_0 = \left.\frac{\partial Q}{\partial s}\right|_{V=V_0} \tag{7.18}$$

Thus it results that

$$\begin{aligned} \Delta Q &= C\Delta V + \varepsilon_0\frac{a}{d}V_0 s \\ \Delta F &= \varepsilon\frac{a}{d}V_0\Delta V + \frac{1}{ns} \end{aligned} \tag{7.19}$$

or after the transition to sinusoidal variables ($\underline{I} = \mathrm{j}\omega\underline{Q}$, $\underline{v} = \mathrm{j}\omega\underline{s}$), respectively:

$$\underline{I} = \underbrace{j\omega C\underline{V}}_{\underline{I}_C} + \underbrace{\varepsilon\frac{a}{d}V_0\underline{v}}_{\underline{I}_W}$$

$$\underline{F} = \underbrace{\varepsilon \frac{a}{d} V_0 \underline{V}}_{\underline{F}_W} + \underbrace{\frac{1}{j\omega n} \underline{v}}_{\underline{F}_n}. \tag{7.20}$$

Equation (7.20) corresponds exactly to the equivalent diagram in Figure 7.3b with $n = n_K$, $C = C_K$ and $Y = \varepsilon a V_0/d$. The operation of the electrostatic transducer requires electric bias voltage V_0, which is superimposed by exciting voltage $\underline{V}$.

Example 7.2 Coupling factor of an electrostatic transducer

Transducers transform energy of one form into that of another form. However, only a part of the input energy is converted; another part is stored in the storage elements of the original physical domain.

We want to calculate coupling factor K^2 for the electrostatic transducer in Figure 7.4, i.e. the ratio of converted mechanical energy and the supplied electric energy.

Coupling factor K^2 of a reversible electromechanical transducer is

$$K^2_{\text{mech}} = \frac{\text{mechanical energy}}{\text{supplied electrical energy}} = \frac{W_{\text{mech}}}{W_{\text{el}}}, \tag{7.21}$$

when operated as an actuator, and

$$K^2_{\text{el}} = \frac{\text{electrical energy}}{\text{supplied mechanical energy}} = \frac{W_{\text{el}}}{W_{\text{mech}}}, \tag{7.22}$$

when operated as a sensor. For reversible transducers

$$K^2_{\text{mech}} = K^2_{\text{el}} = K^2. \tag{7.23}$$

For steady-state processes, it is possible to replace the energies by the corresponding powers P_{mech} and P_{el}.

In case the electrostatic actuator in Figure 7.2 is driven by sinusoidal voltage $\underline{V}_0$ and output force is $\underline{F} = 0$, it results that (Figure 7.3b, Equation (7.20))

$$P_{\text{el}} = \underline{V}\,\underline{I} \tag{7.24}$$

and

$$P_{\text{mech}} = \underline{v} \cdot \underline{F}_n = \underline{V} \cdot \underline{I}_W. \tag{7.25}$$

The latter equation results from the fact, that $\underline{V} \cdot \underline{I}_W$ is the electric power which is transformed from the left-hand electric side in Figure 7.3b to the right-hand mechanical side. Using the transformation equations of the gyrator, it follows that ($\underline{F} = 0$, $\underline{F}_W = -\underline{F}_n$)

$$\frac{\underline{I}_W}{\underline{V}} = \frac{1}{Y^2} \cdot \frac{\underline{v}}{\underline{F}_W} = \frac{1}{Y^2} \cdot \frac{\underline{v}}{\underline{F}_W} = \frac{1}{Y^2} j\omega n = j\omega C_n \tag{7.26}$$

with $C_n = n/Y^2$.

The spring on the mechanical side becomes the additional capacitor C_n on the electric side and is located parallel to C_K. This means that the deformation of spring n leads to an increase in the measurable input capacitance of the actuator. Applying Equation (7.26),

$$P_{\text{mech}} = \underline{V} \cdot \underline{I}_W = \underline{V}^2 j\omega C_n, \quad P_{\text{el}} = \underline{V} \cdot \underline{I} = \underline{V}^2 \mathrm{j}\omega (C_K + C_n), \tag{7.27}$$

resulting in a coupling factor

$$K^2 = \frac{P_{\text{mech}}}{P_{\text{el}}} = \frac{C_n}{C_K + C_n}. \tag{7.28}$$

This way, the coupling factor must be smaller than unity. If C_K is substantially smaller than C_n, we have a good energy conversion.

7.1.2 Parametric Transducers

So far, we have been looking at reversible transducers. When using them as actuators or sensors, the main interest lies in keeping the characteristics of the transducer constant in order to achieve a unique relationship of the state variables. For the general case in Equation (7.12), for instance, this means that coefficients γ_{ij} have to be constant in the coefficient matrix. This condition does not always have to be applied. For an illustration, we will provide another extensive presentation of Equation (7.12):

$$\begin{aligned} \underline{I} &= \gamma_{11}\underline{V} + \gamma_{12}\underline{v} = \left.\frac{\partial I}{\partial V}\right|_{v_0} \cdot \underline{V} + \left.\frac{\partial I}{\partial v}\right|_{V_0} \cdot \underline{v} \\ \underline{F} &= \gamma_{21}\underline{V} + \gamma_{22}\underline{v} = \left.\frac{\partial F}{\partial V}\right|_{v_0} \cdot \underline{V} + \left.\frac{\partial F}{\partial v}\right|_{V_0} \cdot \underline{v}. \end{aligned} \tag{7.29}$$

If – in a specific case – we cause selected variables $\underline{I}$, $\underline{V}$, $\underline{F}$ or $\underline{v}$ to become zero and do not keep parameters γ_{ij} constant any longer, but allow changes in the geometrical or material parameters in γ_{ij}, we obtain a parametric transducer.

Example 7.3 Capacitive displacement sensor as a parametric transducer

Figure 7.5 shows a capacitive displacement sensor. The sensor is fed with sinusoidal current I. Output voltage V is supposed to change in relation to displacement $s = \Delta b$. This arrangement corresponds to the electrostatic transducer in Figure 7.4. The movable upper electrode is not fixed to a spring n, which means that there is no mechanical counterforce F_n. According to Equation (7.20), it is valid for the coefficients in Equation (7.29) that

$$\gamma_{11} = \mathrm{j}\omega\varepsilon\frac{ab}{d} \qquad \gamma_{12} = \gamma_{21} = \varepsilon\frac{a}{d}V_0 \qquad \gamma_{22} = \frac{1}{\mathrm{j}\omega n} \tag{7.30}$$

As no bias voltage V_0 is applied, $\gamma_{12} = \gamma_{21} = 0$. The missing spring can be represented as $n \rightarrow \infty$ or $\gamma_{22} = 0$, respectively.

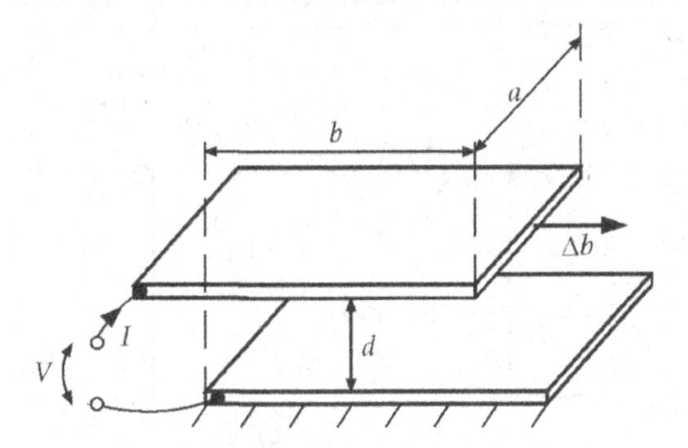

Figure 7.5 Capacitive displacement sensor

Thus, Equation (7.29) is reduced to:

$$\underline{I} = \mathrm{j}\omega\varepsilon\frac{ab}{d}\underline{V} = \mathrm{j}\omega C\underline{V}. \tag{7.31}$$

Displacement s of the upper electrode loses its physical importance as a state variable which it had in Example 7.1. It now acts as a changeable parameter $b = b_0 + \Delta b = b_0 + s$. This means that the capacitive displacement sensor is not a reversible transducer any longer, as – due to the missing counterforce of the spring – a change in voltage V would not result in a defined change of displacement s.

According to Equation (7.31), the arrangement only operates as a parametric transducer, with a capacitance change being achieved by changing all parameters:

$$\frac{\Delta C}{C} \approx \frac{\Delta\varepsilon}{\varepsilon} + \frac{\Delta a}{a} + \frac{\Delta b}{b} - \frac{\Delta d}{d}. \tag{7.32}$$

This way, the sensor in Figure 7.5 could also be used as a level sensor where a rising dielectric fluid causes a change in dielectric constant ε between the capacitor plates.

Reversible and parametric transducers can be distinguished according to Figure 7.6:

- For reversible transducers, energy W_{in} of the input variable carries the measuring signal (for sensors). Energy W_{out} of the output signal is directly generated by energy conversion during the measuring process.
- Parametric transducers are used as sensors. The measuring signal changes parameters γ_{ij} of an energy transducer. The input energy is the feeding energy which is available through feeding voltage or current. It does not carry the measuring signal. Parametric effects are also called modulation effects.

In general, it applies that:

- All resistive, capacitive and inductive sensors are parametric transducers.
  When restoring forces are integrated in form of spring elements, capacitive parametric transducers can be turned into reversible electrostatic transducers (see Examples 7.1

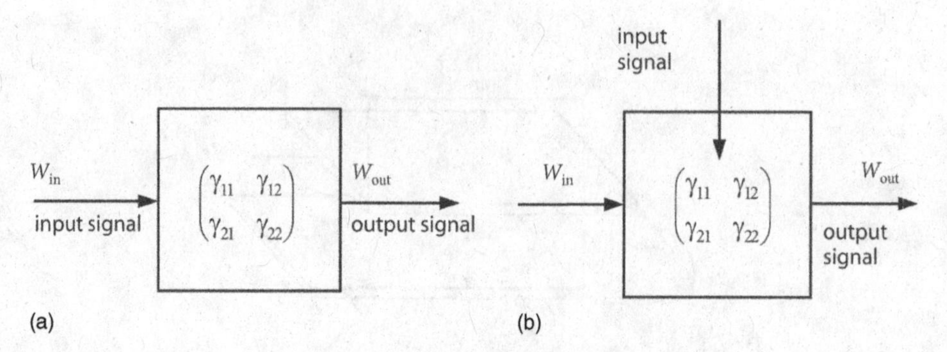

Figure 7.6 (a) Reversible and (b) parametric transducers (according to [MIDDELHOEK98])

and 7.2). The same applies to parametric inductive and reversible electrodynamic transducers (see Figure 7.15).

- Actuators always apply reversible transducer principles as they focus on the conversion of electrical into mechanical energy.

7.1.3 Stationary Reversible Transducers

A large number of sensoric and actuating conversion mechanisms is based on transportation processes of charge quantity Q (electric current density $J_{\mathrm{el}} = \mathrm{d}^2q/\mathrm{d}t\,\mathrm{d}A$), of particles (particle flow density $J_{\mathrm{T}} = \mathrm{d}^2n/\mathrm{d}t\,\mathrm{d}A$) or of thermal energy W_{th} (heat flow density $J_{\mathrm{th}} = \mathrm{d}^2\ W_{\mathrm{th}}/\mathrm{d}t\,\mathrm{d}A$). The mentioned currents and flows are caused by gradients in potential ∇U, concentration ∇n and temperature ∇T where it is valid that

$$\nabla = \begin{pmatrix} \partial/\partial x \\ \partial/\partial y \\ \partial/\partial z \end{pmatrix}. \tag{7.33}$$

This process is connected to an energy transport through space, which means that we cannot originally assume that the thermodynamic equilibrium applies to this system. Therefore, we cannot call these transducers static, as the ones discussed in Section 7.1.1. However, according to ONSAGER's theory, it is still possible to apply the principle of the microscopic reversibility of thermodynamic processes [ONSAGER31]. It is a precondition though, that it is possible to find effort gradients X_i for flow variables J_i that comply with ONSAGER's criterion according to which product J_iX_i has to correspond to an entropy generation (S entropy):

$$J_iX_i = \frac{1}{V}\frac{\mathrm{d}S}{\mathrm{d}t}. \tag{7.34}$$

Electric drift current: OHM current leads to JOULE heating

$$\mathrm{d}Q = T\,\mathrm{d}S = (V \cdot I)\mathrm{d}t \tag{7.35}$$

where Q is the change in heat and $(VI)\,dt$ the change in electric energy. With $VI = EJ_{\text{el}}V$ (E electric field strength, J_{el} electric current density, V volume), it results that

$$T\mathrm{d}S = EJ_{\text{el}}V\mathrm{d}t$$

or, respectively

$$\left(\frac{1}{V}\frac{\mathrm{d}S}{\mathrm{d}t}\right)_{\text{el}} = J_{\text{el}} \cdot \frac{E}{T} = J_{\text{el}}X_{\text{el}}. \qquad (7.36)$$

$X_{\text{el}} = E/T$ complies with ONSAGER's criterion from Equation (7.34). In the same way, J_{el}/T could be chosen as flow variable and ∇T as the effort gradient.

Heat flow: Heat Q flows from place x_{A} to x_{B}, if $T(x_{\text{B}}) < T(x_{\text{A}})$. Assuming small changes in temperature $\Delta T = T(x_{\text{A}}) - T(x_{\text{B}}) \ll T(x_{\text{A}})$, it applies to the entropy change:

$$\Delta S = \Delta Q\left(\frac{1}{T(x_{\text{A}})} - \frac{1}{T(x_{\text{B}})}\right) \approx -\Delta Q\frac{\Delta T}{T^2}. \qquad (7.37)$$

With heat flow density $J_{\text{th}} = \mathrm{d}^2Q/\mathrm{d}t\,\mathrm{d}A \approx \Delta Q/A\Delta t$, it results that

$$\left(\frac{1}{V}\frac{\mathrm{d}S}{\mathrm{d}t}\right)_{\text{th}} = \frac{1}{V}\cdot\lim_{\Delta t\to 0}\frac{\Delta S}{\Delta t} = \lim_{\Delta t\to 0}\left(-\frac{1}{V}J_{\text{th}}A\frac{\Delta T}{T^2}\right)$$

$$= \lim_{\Delta t\to 0}\left(-\frac{A\Delta x}{V}J_{\text{th}}\frac{\Delta T}{\Delta x}\frac{1}{T^2}\right) = J_{\text{th}}\frac{-\nabla T}{T^2} = J_{\text{th}}X_{\text{th}};\ X_{\text{th}} = -\frac{\nabla T}{T^2}. \qquad (7.38)$$

Pairs of flow variables J_i and effort gradients X_i can be assumed to be reversible if they fulfil ONSAGER's criterion from Equation (7.34). Such a system can thus be represented as follows:

$$J_1 = L_{11}X_1 + L_{12}X_2$$
$$J_2 = 7L_{21}X_1 + L_{22}X_3. \qquad (7.39)$$

L_{ij} denominates the transportation coefficients. Compliance with ONSAGER's criterion leads to $L_{12} = L_{21}$. The coupling factor is

$$K^2 = \frac{L_{12}L_{21}}{L_{11}L_{22}} = \frac{L_{12}^2}{L_{11}L_{22}} < 1.$$

Example 7.4 Electrothermal transducers

Electrothermal transducers convert electric energy into thermal energy and vice versa. This process includes electric current and heat transmission. According to [KWAAITAAL93], transducer Equation (7.40) can be found for $J_1 = J_{\text{el}}$, $J_2 = J_{\text{th}}$, $X_1 = E/T$ and $X_2 = -\nabla T/T^2$:

$$\begin{pmatrix} J_{\text{el}} \\ J_{\text{th}} \end{pmatrix} = \begin{pmatrix} \frac{1}{\rho}T & \frac{1}{\rho}\alpha_s T^2 \\ \frac{1}{\rho}\alpha_s T^2 & \lambda T^2 + \frac{1}{\rho}\alpha_s T^3 \end{pmatrix}\begin{pmatrix} E/T \\ -\nabla T/T^2 \end{pmatrix} \qquad (7.40)$$

Here, ρ is the specific resistivity, α_S the SEEBECK coefficient, and λ the thermal conductivity.

Transducer Equation (7.40) includes several physical laws:

For constant temperature $\nabla T = 0$, OHM's law follows from Equation (7.40a)

$$J_{el} = \frac{1}{\rho} E. \tag{7.41}$$

According to SEEBECK's effect, a temperature gradient ∇T between two points of a conductor results in potential difference ΔV and thus an electric field strength E. For $J_{\mathrm{el}} = 0$, Equation (7.40) becomes:

$$E = \alpha_s \nabla T. \tag{7.42}$$

If no current flows through the transducer ($J_{\mathrm{el}} = 0$; the currentless state is the experimental condition for measuring thermal conductivity) and a temperature gradient ∇T is given, FOURIER's law of heat conduction applies. Inserting Equation (7.42) for $J_{\mathrm{el}} = 0$ in Equation (7.40b) results in:

$$J_{\mathrm{th}} = \frac{1}{\rho} \alpha_S^2 T \nabla T - \lambda \nabla T - \frac{1}{\rho} \alpha_S^2 T \nabla T = -\lambda \nabla T. \tag{7.43}$$

Conversely, an electric current produces a heat flow that can be used for electrothermal cooling (PELTIER effect). This results from coupling of Equations (7.40a) and (7.40b) for $\nabla T = 0$:

$$J_{\mathrm{th}} = \alpha_S \cdot T \cdot J_{\mathrm{el}}. \tag{7.44}$$

The two thermoelectric transducer effects correspond to SEEBECK and PELTIER effect according to Equations (7.42) and (7.44).

Resistive sensors function according to Equation (7.41). Applying $I = J_{\mathrm{el}} A$ and $V = E\ell$, resistance R becomes

$$R = \frac{V}{I} = \rho \cdot \frac{\ell}{A}. \tag{7.45}$$

Specific resistivity ρ, resistor length ℓ and resistor cross-section A are the parameters that the sensor measurands can change. Therefore, resistive sensors are not reversible, but parametric transducers, as was described in Section 7.1.2.

7.2 TRANSDUCERS FOR SENSORS AND ACTUATORS

Table 7.1 lists the most important transducer principles of microsystem technology. Energy conversion can comprise one or several steps: one-step conversion for electrostatic and piezoelectric transducers, for instance; multi-step for conversion chains electric-thermal-electric, for instance. In addition to the actual transducer elements for sensor and actuator functions, practical applications require further components for processing signals and for power supply as well as interfaces to the macro-environment.

Table 7.1 Important transducer principle in microsystem technology

	Actuator		Sensor	
Mechanical variable $F, \Delta l$	electrostatic	⟷	capacitive	Electric variable $V, I, \Delta R, \Delta C$
	piezoelectric (inverse)	⟷	piezoelectric	
	electrodynamic	⟷	electrodynamic	
	electromagnetic	⟷	inductive	
			piezoresistive[a]	
	thermomechanical[b]			
	shape memory effect[c]			

[a, b, c] parametric, see Section 7.1

7.2.1 Electrostatic Transducers

Electrostatic transducers can be assumed to be electrode systems with variable capacitance $C(x)$. The variable capacitance is most frequently based on plate capacitors with movable plates whose moving direction x results from their suspension in a fixed frame (Section 6.1). Table 7.2 presents the three most important forms of capacitive transducers. Using the equation for the capacitance of plate capacitors

$$C = \frac{\varepsilon A}{d} \tag{7.46}$$

variable capacitance $C(x)$ can be reached by changing the basic distance of the electrodes $d = d(x)$ or by changing electrode area $A = A(x)$, with x being the corresponding moving direction (operating direction) of the movable capacitor plate and ε the permittivity.

The basic types can be modified by using the movable electrodes as part of a differential capacitor and by applying the distance or area variation not only to an individual capacitor, but to multi-cell arrays of capacitors (comb structures, Table 7.2).

Here, the electrode areas of the comb arms are electrically connected, and eventually only two electrodes remain as effective capacitance (or in the case of differential capacitors, four electrodes for two capacitors).

Capacitance function $C(x)$ can be used to describe the electromechanical effects of the movable electrodes. It is the coupling variable in the equations that describes sensor and actuator effects of electrostatic transducers as well as connecting the electrical and mechanical subsystem. For all electrostatic transducer types, it is important to determine the parameters in Equations (7.6) or (7.12), respectively, for specific arrangements, i.e. for the corresponding capacitance function $C(x)$.

For comb structures with distance variations according to Table 7.3, a comb motion in the given direction produces two different capacitance functions for each comb arm and its fixed counter-electrodes (differential capacitor):

$$C_1(x) = \frac{\varepsilon ab}{d - x},$$

$$C_2(x) = \frac{\varepsilon ab}{d + x}. \tag{7.50}$$

Table 7.2 Basics types of electrostatic transducers

Variation of		Diagram	Capacitance function and properties
Basic distance	Parallel	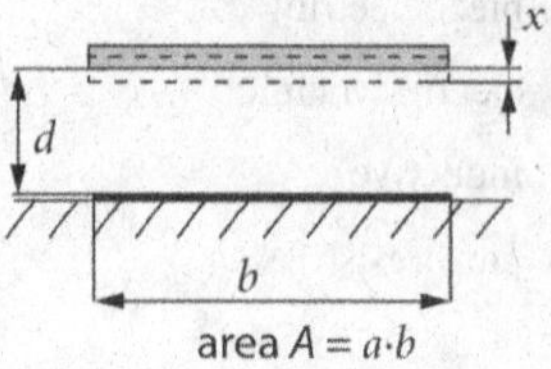 area $A = a \cdot b$	$C(x) = \dfrac{\varepsilon A}{d - x} = \dfrac{\varepsilon ab}{d - x}$ (7.47) • Relatively large capacitance change • Nonlinear characteristic curve $C(x)$ • Pull-in effect has to be considered
	Angle	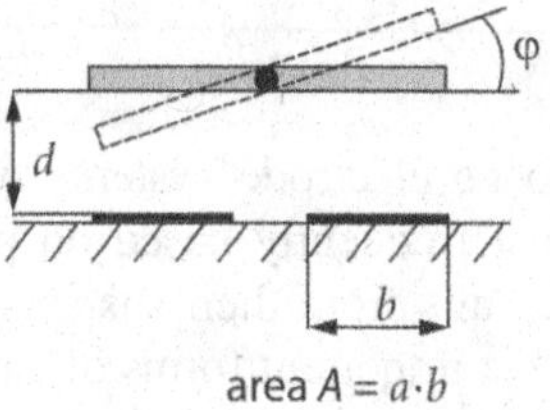 area $A = a \cdot b$	$C(\varphi) = \dfrac{\varepsilon a}{\tan\varphi} \cdot \ln\left(\dfrac{b\tan\varphi}{d} + 1\right)$ (7.48) (for one electrode) • Relatively large capacitance change • Nonlinear characteristic curve $C(x)$ • Pull-in effect has to be considered
Area		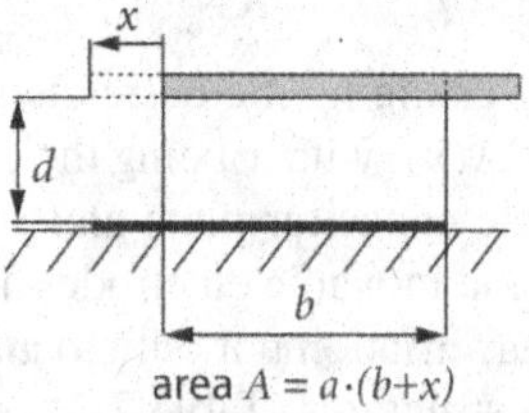 area $A = a \cdot (b+x)$	$C(x) = \dfrac{\varepsilon a(b+x)}{d}$ (7.49) • Relatively small capacitance change • Linear characteristic curve $C(x)$ • No pull-in effect

The first derivation of these capacitance function is nonlinear, e.g.

$$\frac{dC_1(x)}{dx} = \frac{\varepsilon ab}{(d - x)^2}. \tag{7.51}$$

Total capacitance C_{tot} of the comb structure with distance variation results from multiplying the capacitance change in Equations (7.50a) or (7.50b), respectively, with number n of the comb arms. For the comb structure with area variation according to Table 7.3b, for instance, the resulting capacitance function is

$$C(x) = \frac{2n\varepsilon a(b + x)}{d}. \tag{7.52}$$

Its first derivation is constant and does not contain the moving direction x:

$$\frac{dC(x)}{dx} = \frac{2n\varepsilon a}{d}. \tag{7.53}$$

Table 7.3 Comb structures as modified basic types of electrostatic transducers

Distance variation	Area variation	
(a)	(b)	(c)

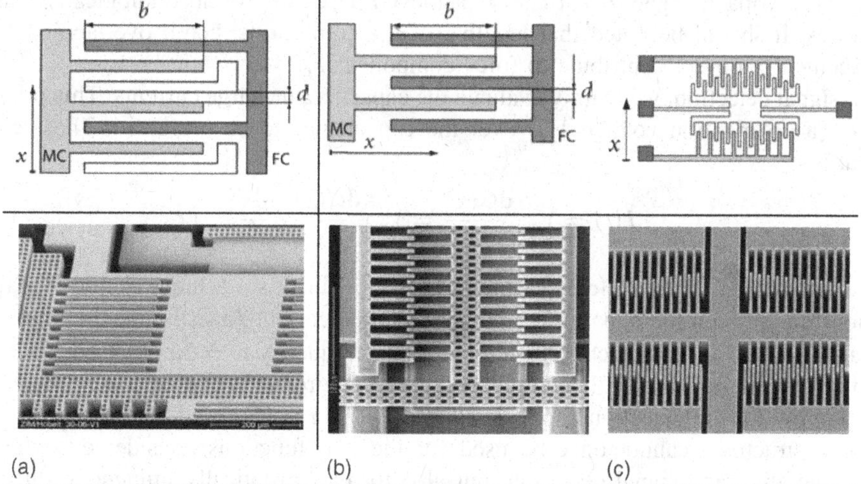

MC movable combs, FC fixed combs, *x* moving direction, *a* comb depth (not represented in the figure), *b* overlapping of combs, *d* electrode distance

Reproduced by permission from the Center for Microtechnologies, Chemnitz University of Technology, Germany

According to the principle of virtual displacement, we can determine the electrostatic force acting between the two electrodes of a plate capacitor with area A and distance d using electrostatic energy W_{el} (see Equation 7.14). That means that the virtual displacement of a movable capacitor plate by $\mathrm{d}x$ produced by electric force F_{el} results in a change in electric energy $\mathrm{d}W_{\mathrm{el}}$:

$$\mathrm{d}W_{\mathrm{el}} = F_{\mathrm{el}} \cdot \mathrm{d}x. \tag{7.54}$$

The electric energy stored in a plate capacitor is

$$W_{\mathrm{el}} = \frac{CV^2}{2} = \frac{\varepsilon A V^2}{2d}. \tag{7.55}$$

With Equation (7.54), the resulting electrostatic force is

$$F_{\mathrm{el}} = \frac{\mathrm{d}W_{\mathrm{el}}}{\mathrm{d}x} = \frac{V^2}{2} \cdot \frac{\mathrm{d}C(x)}{\mathrm{d}x}, \tag{7.56}$$

where we use $C(x)$ for the component-specific capacitance function. Using Equation (7.51), for the comb structure with area variation given in Table 7.3b, it is valid that

$$F_{\mathrm{el}} = \frac{V^2}{2} \cdot \frac{2n\varepsilon a}{d}. \tag{7.57}$$

Equation (7.57) shows that the electrostatic driving force that can be achieved by comb drives is mainly determined by number n of comb arms as well as by the values of comb depth a and capacitor gap d that can be achieved depending on technologically feasible tolerances. It should be noted that length l of the comb arms, comb overlapping b and displacement x do not contribute to force component F_{el}.

For signal detection, we mainly evaluate the capacitive recharge currents. This principle requires a polarization voltage V_{pol} over the capacitance to be evaluated. The recharge current is

$$I(t) = V_{pol}\frac{dC(x)}{dt} = V_{pol}\frac{dC(x)}{dx} \cdot \frac{dx}{dt}. \tag{7.58}$$

Due to $dx/dt = \hat{X}\omega\cos\omega t$, for measurand $x(t) = \hat{X}\sin\omega t$ we achieve the desired linear relationship between the maximum value of recharge current $\hat{I}$ and the maximum oscillation amplitude $\hat{X}$ to be detected, if $dC(x)/dx$ and frequency $\omega = 2\pi f$ are constant. The comb structures in Table 7.3b comply with this requirement. For an ideal situation, it should apply for the detection capacities that $dC(x)/dx = \text{const.}$ and $d^2C(x)/dx^2 = 0$.

Comb structures cannot only be used for the two functions considered so far, i.e. force generation and signal detection, but also for electrostatically influencing the stiffness of elastic suspensions (electrostatic softening). As opposed to Equations (7.52) and (7.53), for this application, it has to be valid for the second derivation of the capacitance function that $d^2C(x)/dx^2 = \text{const.}$ The result is an electrostatic tuning force that is independent of displacement and, consequently, we have a linear characteristic curve. The capacitance function increases quadratically to the displacement. Comb structures with stepwise combs, as shown in Table 7.3c, can approximately fulfil this requirement. A quadratic capacitance function can be achieved if the comb arms successively enter the area of the fixed counter-electrodes. Table 7.4 provides a comparison of advantages and disadvantages of electrostatic transducer principles.

Table 7.4 Advantages and disadvantages of the electrostatic transducer principle

Advantages	Disadvantages
• Simple construction and technology: Plate capacitor with movable plate • Sensor and actuator functions work with the same electrode arrangement (advantageous for closed control loops) • High resolution and high dynamics (small mass) • Small temperature impact • Simple multi-cell capacitor arrays (comb structure) • Monolithic integration of micromechanics and microelectronics • Option of different frequency ranges and static signals • Favourable scaling behavior for actuating applications	• Parasitic and stray capacitance occur • Often nonlinear characteristics • Stability issues may arise (pull-in effect, see Chapter 8) • Sensitive to humidity and foreign particles • Inspite of small plate distance comparatively large voltages required for actuating applications • Often screening necessary

Example 7.5 Electrostatic softening (continuation of Example 6.3)

The spring stiffness of the elastic suspension of comb structures can be changed by electrostatic softening. Figure 7.7 shows a typical comb structure for electrostatic softening. This way it is possible to vary the resonance frequencies of oscillators within a given range, for instance. For the electrostatic softening, we use tuning voltage V_T that generates displacement-dependent electrostatic forces and causes a reduction of mechanical spring stiffness c_{mech} by electric spring stiffness c_{el}. For variable resonance frequency f_e^* and electric spring stiffness c_{el}, the following relationships apply:

$$f_e^* = \frac{1}{2\pi}\sqrt{\frac{c_{tot}^*}{m}} = \frac{1}{2\pi}\sqrt{\frac{c_{mech} - c_{el}}{m}}, \tag{7.59}$$

$$c_{el} = \frac{dF_{el}(x)}{dx} = \frac{V_T^2}{2} \cdot \frac{d^2C(x)}{dx^2}. \tag{7.60}$$

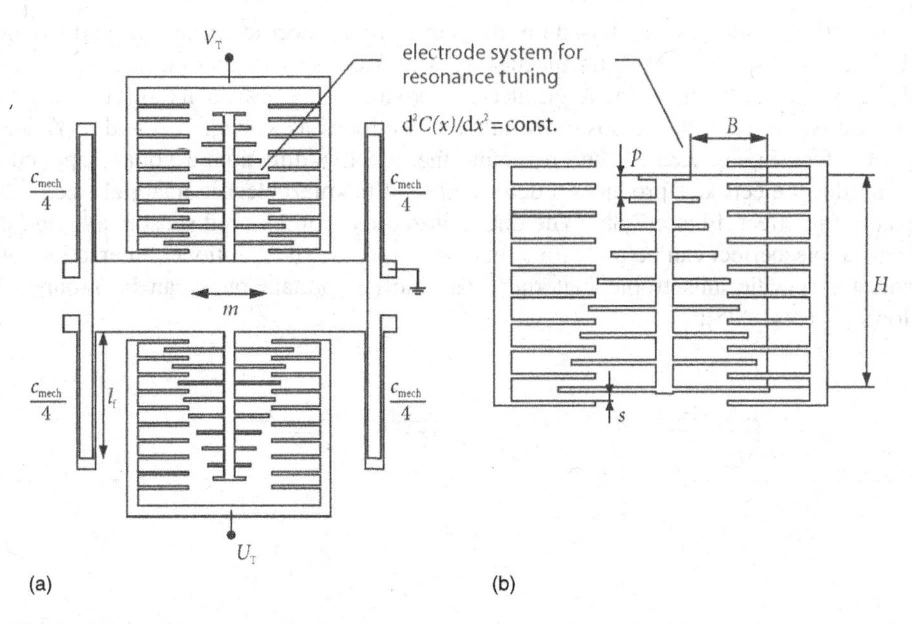

Figure 7.7 Oscillators with linearly stepwise arranged comb elements for electrostatic softening

In practice, we choose an electrode arrangement with a quadratic curve of $C(x)$, such as the linearly stepwise arranged combs in Table 6.3c. It applies for a symmetric oscillator half according to Figure 7.7b and to [LEE98] that:

$$\frac{d^2C(x)}{dx^2} = \frac{4\varepsilon hH}{psB}. \tag{7.61}$$

The height of the combs is not given in the figure; it is $h = 50$ µm. Other dimensions are $p = 10$ µm, $s = 2$ µm, $B = 40$ µm and $H = 400$ µm. Permittivity is $\varepsilon = 8.85 \cdot 10^{-12}$ As/Vm. In order to reduce the resonance frequency by 5 %, for instance, electrostatic softening is used to reduce the mechanical spring stiffness from $c_{\text{mech}} = 7.13$ Nm^{-1} to 6.43 Nm^{-1}, according to Example 6.3. That means that we have to provide an electric spring stiffness of $c_{\text{el}} = 0.7$ Nm^{-1}. The required tuning voltage V_{T} for two symmetric oscillator halfs results from

$$V_{\text{T}} = \sqrt{\frac{2c_{\text{el}}}{\mathrm{d}^2C(x)/\mathrm{d}x^2}} = \sqrt{\frac{2c_{\text{el}}}{2\dfrac{4\varepsilon \cdot h \cdot H}{p \cdot s \cdot B}}} = \sqrt{\frac{2c_{\text{el}} \cdot p \cdot s \cdot B}{2 \cdot 4\varepsilon \cdot h \cdot H}} = 28 \text{ V}. \tag{7.62}$$

7.2.2 Piezoelectric Transducers

Piezoelectric transducers[1] are based on the interaction of electric variables (field strength and dielectric displacement) with mechanical variables (stress and strain). If a mechanical force $F(t)$ acts on a piezotransducer, it produces charges on its surface that can be measured between the opposite metallized electrodes as voltage $V(t)$ (direct piezoelectric effect, Figure 7.8a). When reversing the effective direction, a voltage applied to the transducer electrodes produces a deformation of the piezoelectric material (reciprocal piezoelectric effect, Figure 7.8b). The direct piezoeffect can be used sensorically and the reciprocal piezoeffect can be used for actuating purposes. The mentioned interaction only occurs for specific anisotropic materials. The most important ones can be grouped as follows [FRÜHAUF05]:

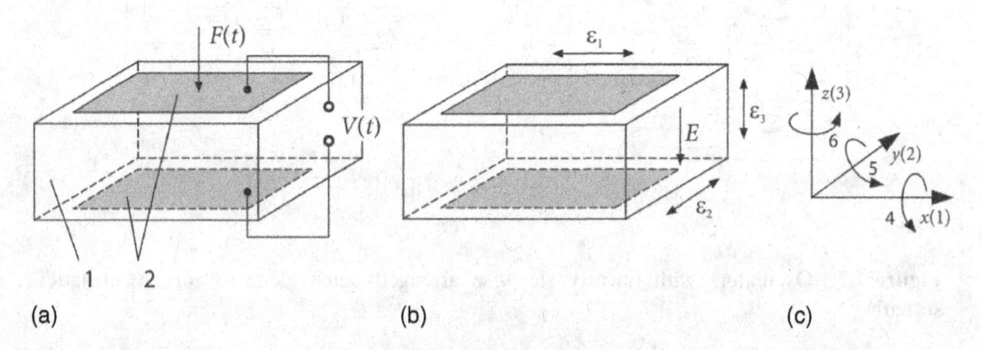

Figure 7.8 (a) Direct and (b) reciprocal piezoelectric effect; 1: piezomaterial; 2: electrodes; $F(t)$ force; $V(t)$ force-independent voltage; E electric field; ε_i strain

[1]Mainly referred to in its abbreviated form: piezotransducers; not to be confused with piezoresistive transducers!

- piezocrystals
- piezoceramics
- piezopolymers.

State equations are used to describe the piezoelectric effect. Depending on the independent variable selected, they can have different representations. Assuming mechanical stress and electric field strength to be independent variables, the state equations become

$$D_i = d_{in}\sigma_n + \varepsilon_{im}^{\sigma} E_m \text{ (direct piezoelectric effect)} \tag{7.63}$$

$$\varepsilon_j = s_{jn}^{E}\sigma_n + d_{jm} E_m \text{ (reciprocal piezoelectric effect)} \tag{7.64}$$

with $i, m = 1\ldots 3$ and $j, n = 1\ldots 6$.

Here, D_i represents the components of the dielectric displacement vector, σ_n the components of the mechanical stress tensor, E_m the components of the electric field strength vector, ε_j the components of the strain tensor, d_{in} or d_{jm} the piezoelectric coefficients, ε_{im} the permittivity and s_{jn} the elastic constants. Superscript indices mean that the parameter has to be kept constant in order to be able to determine the corresponding coefficient. Indexing of the parameters with numbers 1, 2 and 3 corresponds to the directions of the coordinate system in Figure 7.8c, where axis 3 usually represents the polarization direction of the piezoelectric material. Numbers 4, 5 and 6 provide the shearing on axes 1, 2 and 3, i.e. mechanical stress that acts tangentially to the areas comprised by the coordinate system. Piezoelectric coefficients d_{in} or d_{jm}, respectively, relate components σ_n of the mechanical stress state and the resulting dielectric displacement D_i to each other or, respectively, components E_m of the electric field strength to the resulting strain ε_j. If we only look at the directional correlations of commonly used stress situations, we can distinguish longitudinal, transversal and shearing effects. According to the main loading type, d_{11} and d_{33} describe longitudinal effects (d_L), d_{31} and d_{32} transversal effects (d_Q) as well as d_{14} and d_{15} shearing effects (d_S). The piezoelectric coefficient d_{15}, for instance, relates a shearing on axis 5 to the resulting dielectric displacement towards axis 1. Table 7.5 provides parameters of important piezoelectric materials.

Table 7.5 Piezoelectric materials and their piezoelectric coefficients

Material	Piezoelectric coefficients d_{in} in 10^{-12} C/N		
Quartz	$d_{11} = 2.31$	$d_{14} = -0.727$	
Lithium niobate (LN; $LiNBo_3$)	$d_{15} = 69.2$	$d_{22} = 20.8$	$d_{33} = 6.0$
Zink oxide (ZnO)	$d_{15} = -8.3$	$d_{31} = -5$	$d_{33} = 12.4$
Barium titanate ($BaTiO_3$)	$d_{15} = 550$	$d_{31} = -150$	$d_{33} = 374$
Lead zirconate titanate (PZT)	$d_{15} = 265\ldots765$	$d_{31} = -60\ldots-270$	$d_{33} = 380590$
Aluminum nitride (AlN)		$d_{31} = 1.9$	$d_{33} = 4.9$
Polyvinylidene difluoride (PVDF)	$d_{31} = 18\ldots28$	$d_{32} = 0.9\ldots4$	$d_{33} = -20\ldots-35$

After production or deposition on a substrate, piezoceramics and piezopolymers have to be polarized in an electric field in order to impress the piezoelectric properties. Piezoceramics PZT and barium titanate show a high piezoelectric effect and are therefore particularly suitable for actuating applications. Piezocrystals, as single crystalline bulk material, are used as oscillator quartz in watches, for instance. Thin layers of ZnO, PZT or AlN are used for micromechanical sensors and actuators. They are generated through sputtering, MOCVD or sol-gel techniques. They are not standard processes of semiconductor technology and are often applied as backend processes. The achieved piezoelectric coefficients, e.g. for PZT, are smaller as those for compact ceramics.

For actuating applications, an unloaded piezotransducer produces – according to Equation (7.63) with $\sigma_n = 0$ and with the electric field direction given in Figure 7.8b – a mechanical strain of

$$\varepsilon_3 = d_{33} \cdot E_3. \tag{7.65}$$

With $\varepsilon_3 = \Delta z/z$ and $E_3 = V/z$ (z is the dimension of the piezotransducer in the direction z), it results that

$$\Delta z = d_{33} \cdot V. \tag{7.66}$$

Equation (7.66) shows that for a z-orientation of the electric field, the length change Δz in field direction is independent of the dimensions of the piezotransducer in direction z.

For length changes Δx in direction x as well as an initial length of x and a z-orientation of the electric field

$$\varepsilon_1 = d_{31} \cdot E_3 \tag{7.67}$$

it results from $\varepsilon_1 = \Delta x/x$ and $E_3 = V/z$

$$\Delta x = x d_{31} \cdot V/z. \tag{7.68}$$

In addition to piezoelectric coefficient d, coupling factor K^2 provided in Equations (7.21) and (7.22) constitutes an important parameter. The coupling factor can be calculated from the material tensors provided in Equations (7.63) or (7.64) respectively. It applies, for instance, for the coupling factor of the free thickness oscillator that

$$k_{33} = \frac{d_{33}^2}{s_{33} \cdot \varepsilon_{33}}. \tag{7.69}$$

Actuating applications require piezomaterials with large k_{33}. Table 7.6 presents several parameters, equivalent circuits and coupling Equations (7.70) to (7.72) that are important for piezotransducers.

Table 7.6 Parameters of piezoelectric transducer types

Function	Length mode oscillators	Thickness mode oscillators	Shear mode oscillators
Mechanical and electric boundary conditions	$\sigma_2 \ldots \sigma_6 = 0$ $F = \sigma_1 l_2 l_3;\ S = \varepsilon_1 l_1$	$\varepsilon_1, \varepsilon_2, \varepsilon_4 \ldots \varepsilon_6 = 0$ $F = \sigma_3 l_1 l_2;\ S = \varepsilon_3 l_3$	$\sigma_1, \ldots \sigma_3, \sigma_4, \sigma_6 = 0$ $F = \sigma_5 l_1 l_2;\ S = \varepsilon_5 l_3$
	$D_1 = D_2 = 0;\ Q = D_3 l_1 l_2;\ E_1 = E_2 = 0;\ V = E_3 l_3;\ I = \mathrm{x}Q/\mathrm{d}t$		
$A_{\mathrm{el}}, l_{\mathrm{el}}$	$A_{\mathrm{el}} = l_1 l_2;\ l_{\mathrm{el}} = l_3$		
$A_{\mathrm{mech}}, l_{\mathrm{mech}}$	$A_{\mathrm{mech}} = l_2 l_3;\ l_{\mathrm{mech}} = l_1$	$A_{\mathrm{mech}} = l_1 l_2;\ l_{\mathrm{mech}} = l_3$	
K^2	$\dfrac{d_{31}^2}{\varepsilon_{33}^{\sigma} S_{11}^E}$	$\dfrac{1}{1 + c_{33}^E \varepsilon_{33}^{\varepsilon} / e_{33}^2}$	$\dfrac{d_{35}^2}{\varepsilon_{33}^{\sigma} S_{55}^E}$
e	d_{31}/s_{11}^E	e_{33}	d_{35}/S_{55}^E
ε	$\varepsilon_{33}^{\sigma}(1 - k^2)$	$\varepsilon_{33}^{\varepsilon}$	$\varepsilon_{33}^{\sigma}(1 - k^2)$
c	$1/S_{11}^E$	C_{33}^E	$1/S_{55}^E$
Lumped equivalent circuit	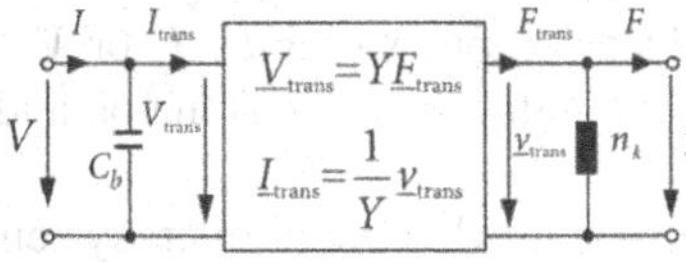		

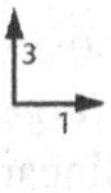

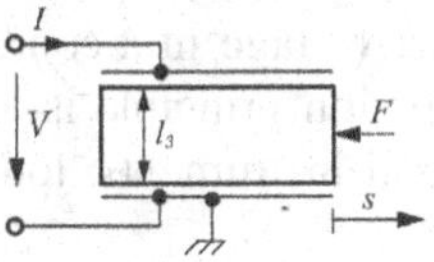

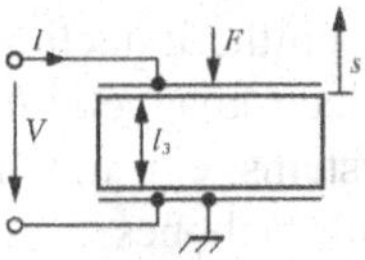

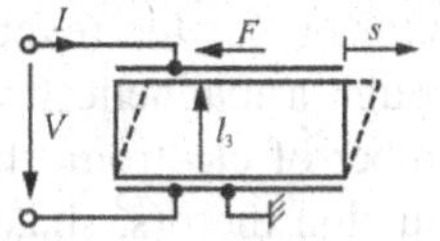

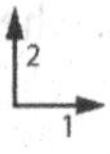

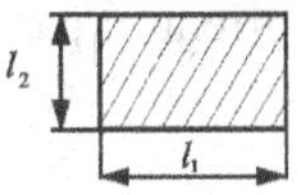

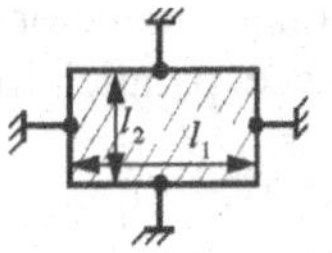

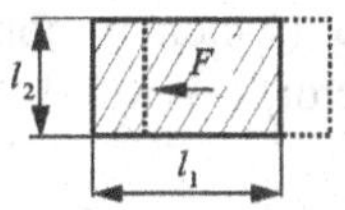

$$\begin{pmatrix} I \\ F \end{pmatrix} = \begin{pmatrix} j\omega C_{\mathrm{b}} & 1/Y \\ 1/Y & -1/\omega n_{\mathrm{k}} \end{pmatrix} \begin{pmatrix} V \\ v \end{pmatrix} \tag{7.70}$$

$$C_{\mathrm{b}} = \varepsilon \frac{A_{\mathrm{el}}}{l_{\mathrm{el}}};\ Y = e \frac{A_{\mathrm{el}}}{l_{\mathrm{mech}}} = e \frac{A_{\mathrm{mech}}}{l_{\mathrm{el}}};\ n_{\mathrm{k}} = \frac{l_{\mathrm{mech}}}{c A_{\mathrm{mech}}} \tag{7.71}$$

$$K^2 = \frac{1}{1 + c\varepsilon/e^2} = \frac{1}{1 + Y^2 C_{\mathrm{b}}/n_{\mathrm{k}}};\ \frac{1}{Y} = \frac{k}{\sqrt{1-k^2}}\sqrt{C_{\mathrm{b}}/n_{\mathrm{k}}} \tag{7.72}$$

Index b: mechanically clamped system; Index k: system in the electric short-circuit.

7.2.3 Electrodynamic Transducers

Electrodynamic transducers are based on the force of a magnetic field acting on a conductor or, for the reverse effect, on the induction of a voltage in a conductor that moves through a magnetic field. The electromagnetic conversion principle is applied in a large number of electromechanical systems, e.g. moving-coil instruments, loudspeakers, linear induction motors, shakers and microphones.

With LORENTZ force F acting on a conductor with length l, that is located in magnetic field B (which is constant in space parallel to l) and current I passing through the conductor, it is valid that

$$\vec{F} = I(\vec{l} \times \vec{B}) \text{ or, respectively, } F = IBl \cdot \sin(\vec{l}, \vec{B}) \tag{7.73}$$

with $\vec{F}, \vec{l}$ and $\vec{B}$ as vector quantities.

According to the principle of effect and counter-effect, the same force applies to the carrier of the magnetic field, e.g. the permanent magnet, the pole areas of the magnetic circuit or the current-conducting wires. In practice, we often use a permanent magnet for inducing the magnetic field and the conductors are cylindric or planar coils. For the design of systems, it is desirable that $\sin(\vec{l}, \vec{B}) = \sin 90^\circ = 1$. Thus, Equation (7.73) becomes

$$F = I \cdot B \cdot l, \tag{7.74}$$

which is schematically represented in Figure 7.9 for a classical plunger-type coil.

For sensor applications, the conductor is moved with the velocity v in a magnetic field with magnetic flux density B, accordingly the voltage V is induced in the conductor. In analogy to Equation (7.74), it applies to the voltage

$$V = vBl. \tag{7.75}$$

For homogeneous magnetic fields, transducer output variables F or V, respectively, are linear to input variables I or v. This is advantageous for actuator and sensor applications.

When applying the electrodynamic transducer principle in microsystem technology, the movable transducer element can be realized using the basic forms of mechanical elements presented in Section 6.1. Two structural forms can be used. They are schematically presented in Figure 7.10. We will use the element in Figure 6.2d as an example of a movable transducer element.

For the structure according to Figure 7.10a, the critical issues are power supply via movable joints, structuring the planar coil on the movable element as well as – very importantly – its temperature load due to the current. In addition, planar coils require a large number of windings and thus a large actuator area as well as a comparatively large OHM's resistance (power consumption $P_{el} = I^2R$). For the structure in Figure 7.10b, the mass of the permanent magnet on the movable transducer element causes a reduction of its dynamics. Neither microassembly of permanent magnets nor their direct structuring on movable transducer elements through microtechnology (microgalvanics, screen printing, sputtering) are currently established techniques in microsystem manufacturing. Direct

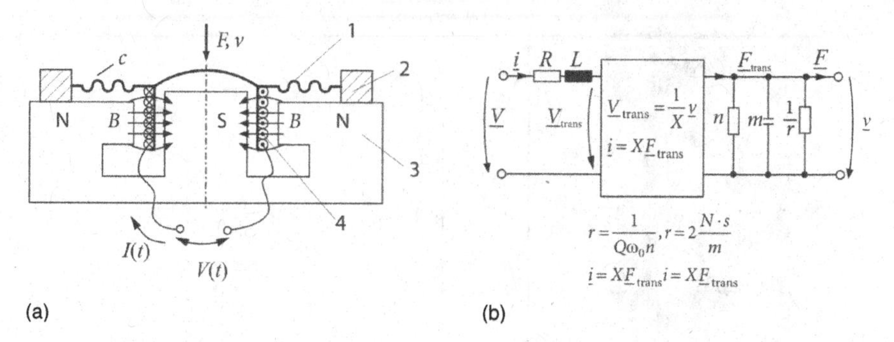

Figure 7.9 Electrodynamic transducer according to the plunger-type coil principle: (a) structure; (b) lumped equivalent circuit; 1: suspension; 2: frame; 3: magnetic circuit with ring-shaped air gap; 4: moving coil; N north pole, S south pole, F force, v velocity, r frictional impedance, m mass, n mechanical compliance, $I(t)$ current, $V(t)$ voltage, X transducer constant ($X = 1/Bl$)

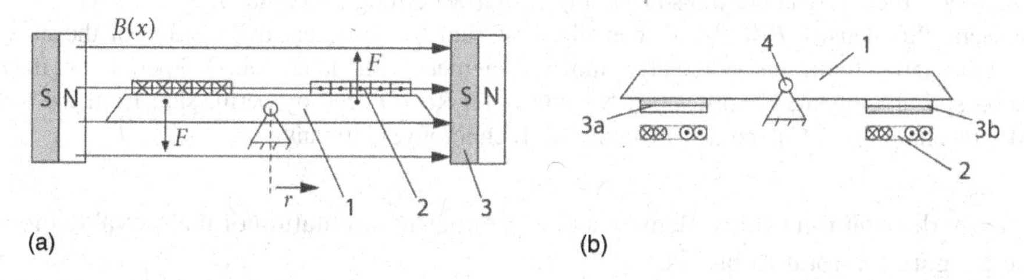

Figure 7.10 Designs of electrodynamic transducers in microsystem technology: (a) planar coil on movable silicon element, permanent magnet fixed on frame; (b) permanent magnet on movable silicon element, coil fixed on frame; 1: movable element; 2: planar coil; 3: permanent magnet; 4: torsion spring element; r distance to rotational axis; F force; $B(x)$ magnetic flux density, x coordinate direction, N north pole, S south pole

structuring through microtechnology requires high temperatures and provides only small magnetic characteristics.

Microassembly of permanent magnets is very costly and requires specific, nonmagnetizable equipment (adjusting, mounting).

Figure 7.11a shows an electrodynamic transducer produced according to Figure 7.10a. Due to its form and the structural size of the movable transducer element, magnetic flux density B over the relatively large air gap is not constant. Instead of Equation (7.74), we have to use modified analytic approaches or numeric solutions for calculating the resulting force.

The torque, for instance, can be determined analytically by summing up all partial torques M_k that have been generated by the electrodynamic effect of the individual conductors k ($k = 1 \ldots n$). According to Figure 7.10a, it applies to the partial torques that

$$M_k = F_k r_k = I B_k l_k r_k. \tag{7.76}$$

Here, r_k presents the distance of the conductors to the rotational axis of the movable element. B_k is the value of magnetic flux density and l_k the effective conductor length perpendicular to the magnetic flux, always in relation to coordinate x of k. Flow density is assumed to be constant along the conductor and across the conductor width. For a

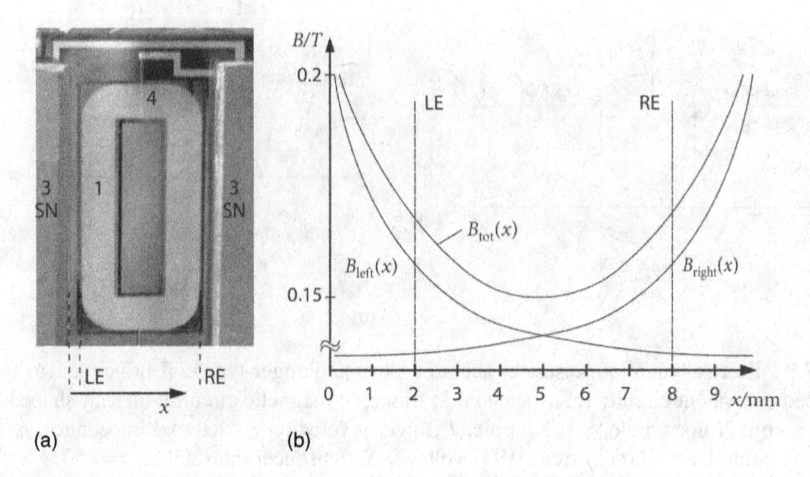

Figure 7.11 Electrodynamic transducer: (a) design according to Figure 7.10a; (b) typical curve of magnetic flux density B in the air gap (lines LE and RE represent the position of the movable element in relation to the x-coordinate). 1 movable element with planar coil; 2, 3 permanent magnet; 4 torsion spring element; N north pole; S south pole. Reproduced by permission from the Center for Microtechnologies, Chemnitz University of Technology, Germany

symmetric distribution of flow density and a symmetric orientation of the movable element in the air gap, the total torque is

$$M_{\text{tot}} = 2 \cdot \sum_{k=1}^{n} M_k, \tag{7.77}$$

with n being the number of the conductor loops in the planar coil.

For actuating applications, achievable transducer force F or, for the example in Figure 7.10, torque M are important parameters. The maximum achievable torque is mainly dependent on magnetic flux density B that can be achieved in the air gap and on maximum current $I_{\max}$. As an approximation, it is possible to use quotient M^2/P_{el} where M can be determined according to Equation (7.77) and P_{el} as

$$P_{\text{el}} = I^2 R = I^2 \rho \frac{l_{\text{PC}}}{A} = I^2 \rho \frac{l_{\text{PC}}}{bh}. \tag{7.78}$$

In Equation (7.78), P_{el} is the electric power consumed by the transducer, ρ is the specific resistivity of the conductor material, l_{PC} the conductor length of the planar coil and bh the cross-sectional area of its conductor.

Applying Equations (7.76) to (7.78),

$$\frac{M_{\text{tot}}^2}{P_{\text{el}}} = \frac{4bh}{\rho l_{\text{PC}}} \left(\sum_{k=1}^{n} B_k l_k r_k \right)^2. \tag{7.79}$$

Equation (7.79) uses the example in Figure 7.11 to show which parameters affect the design of electrodynamic transducers and how many degrees of freedom they have. The electric power P_{el} that the conductor can consume is limited by the allowed temperature load (current density) of the planar coil.

Example 7.6 Displacement angle of an electrodynamic transducer

Conservation of angular momentum $M_{el} = M_{mech}$ can be used to determine the displacement angle φ of an electrodynamic transducer according to Figure 7.10a in relation to current I. The characteristic curve $\varphi(I)$ results from Equations (6.23), (7.76) and (7.77)

$$\varphi = I \cdot \frac{l \cdot \sum_{k=1}^{n} B_k l_k r_k}{G \cdot I_t}. \tag{7.80}$$

where G is the shear modulus, I_t the torsional resistance of the cross-section and l the length of a torsional spring element. Equation (7.80) shows that, for an electrodynamic transducer, displacement angle φ has a linear relationship to current I. As opposed to this, the characteristic curve $\varphi(V)$ of the electrostatic transducer in Table 7.2 with an angle-dependent capacitance function according to Equations (7.48) and (7.56) is quadratic.

Electrodynamics transducers according to the principle in Figure 7.10a or Figure 7.11a, respectively, i.e. with a planar coil on a movable silicon element and a fixed permanent magnet, have been increasingly publicized over the past years as actuators for MEMS scanners and have be commercialized by the company Microvision. [2,3,4]

Only permanent magnets of very small volumes can be used for miniaturized transducers. It is necessary to use magnetic material with a very high energy density $(BH)_{max}$ in order to reach a sufficient force, nonetheless. Sintered permanent magnets based on rare earths such as samarium-cobalt (SmCo) and neodymium magnets (NIB neodymium, iron, and boron or NdFeB, respectively) are particularly suitable for this. Table 7.7 provides illustrative examples of magnetic parameters. In cuboid-shape structures, the smallest available magnets have a thickness of 1 mm and an edge length of 2 mm.

7.2.4 Thermomechanical Transducers

Thermomechanical transducers are based on a two-step energy conversion. At first, electric or light energy is transformed into heat and then, subsequently, the heat expansion of solid bodies, fluids or gases carries out mechanical work. Commonly, the current flowing through a heating resistor is used to generate thermal energy. The supplied or dissipated thermal energy can also produce phase transitions solid ↔ fluid or fluid ↔ gaseous. Such thermodynamic processes are combined with large volume changes and are therefore very suitable for actuating effects. Also the shape-memory effect, that is not included in this

[2]Sprague, R. *et al.* (2005) Bi-axial magnetic drive for scanned beam display mirrors. *Proc. of SPIE*, Vol. 5721 (SPIE Bellingham), pp. 1–13.

[3]Yan, J. *et al.* (2003) Magnetic actuation für MEMS scanners for retinal scanning displays. *Proc. of SPIE*, Vol. 4985 (SPIE San Jose), pp. 115–20.

[4]Urey, H. *et al.* (2003) MEMS raster correction scanner for SXGA resolution retinal scanning display. *Proc. of SPIE*, Vol. 4985 (SPIE San Jose), pp. 106–14.

Table 7.7 Magnetic parameters of permanent magnets at an ambient temperature of 20 °C

Magnetic material	Samarium cobalt (Sm_2Co_{17})		Neodymium magnets ($Nd_2Fe_{14}B$)	
	VACOMAX 225 HR	VACOMAX 240 HR	VACODYM 655 HR	VACODYM 745 HR
Energy density $(BH)_{max}$ in kJ/m^3	190...225	200...240	280...315	370...400
Coercivity H_{cJ} in kA/m	1590...2070	640...800	1670...1830	1115...1195
Remanence B_r in T	1.03...1.10	1.05...1.12	1.22...1.28	1.40...1.44
Maximum permanent temperature in °C	350	300	150	70

Source: Vacuumschmelze GmbH&Co. KG Hanau, VAC, Germany.

book, is based on a thermally generated phase transition with crystal structures of specific alloys changing between an austenite and a martensite phase.

An important feature of all thermomechanical transducer designs is their time behavior. The heating time required for reaching the necessary minimum temperature depends on the supplied heat flow Φ and thermal time constant $\tau_{th} = R_{th}C_{th}$. The cooling-down time is determined by the heat transmission to the environment (see Section 6.3). Miniaturization of components can substantially increase the velocity of thermal processes such as heating and cooling-down: Thermal capacitance C_{th} scales like a volume. Low heat capacity means little energy and short heating times. Heat transmission scales with surface area, i.e. for a miniaturization of transducer dimensions it increases relatively to heat capacity, which promotes a faster heat dissipation. In spite of the favorable scaling tendencies, thermomechanical transducers are relatively slow compared to other principles. Response times typically lie in the range of double-digit milliseconds. Another disadvantage is the comparatively large heating power. Advantages include simple design and technology as well as the large forces and displacements that can be achieved.

(a) Thermal expansion of solid bodies

For thermomechanical transducers that utilize length or volume changes, the following equations apply

$$\Delta l/l = \alpha \Delta T \text{ where } \Delta l = l(T_1) - l(T_0) \tag{7.81}$$

$$\Delta V/V = \gamma \Delta T \text{ where } \Delta V = V(T_1) - V(T_0). \tag{7.82}$$

The length or volume values at temperature T_0 are the corresponding reference values, α or γ are the coefficients of longitudinal or volume expansion. For volume expansion coefficient γ, a good approximation for solid bodies is $\gamma = 3\alpha$ (if $\Delta l/l \ll 1$).

Figure 7.12a shows a micromechanical beam of length l, mass m and the linear thermal coefficient of expansion α. The beam is stiffly attached to a frame and contains an integrated heat resistor R_H, that produces thermal power $P = I^2R_H$ when a current is applied. Assuming that the temperature in the entire beam is the same (even distribution

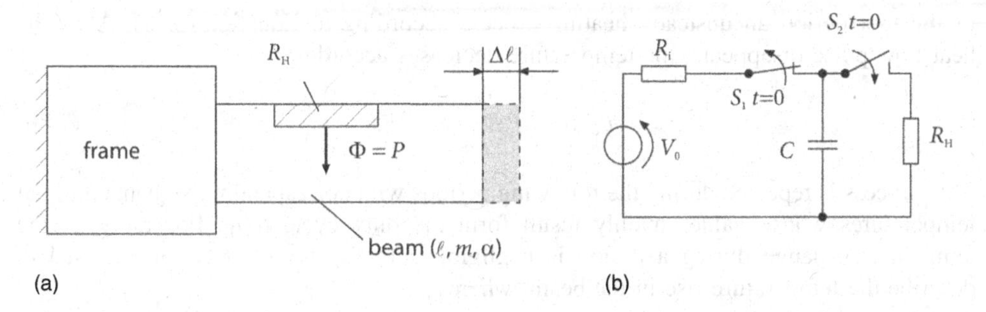

Figure 7.12 Thermal expansion in a beam: (a) schematic; (b) generation of heating power through capacitor discharging. m beam mass, α linear coefficient of thermal expansion, l beam length, Δl length expansion, $\Phi = P$ heat flow, Vvoltage, R_i internal resistor, R_H heating resistor, C capacitor

of the heating power over the beam volume due to very good thermal conductivity of the beam or due to current flowing through the entire beam), the temperature in the beam is only a function of time. Using Equation (6.61), it applies to the temperature rise that

$$\vartheta_e(t) = PR_{th}(1 - e^{-t/\tau_{th}}).$$

Here, $\vartheta_e(t)$ is the temperature difference to the environment, R_{th} the thermal resistance between beam and environment (convection resistance), $\tau_{th} = R_{th}C_{th}$ the thermal time constant and $C_{th} = mc$ the heat capacity of the beam. According to Equation (7.81), the thermal expansion of the beam is proportional to the temperature rise:

$$\Delta l(t) = l \cdot \alpha \cdot \Delta T(t) = l \cdot \alpha \cdot P \cdot R_{th}(1 - e^{-t\tau_{th}}) \tag{7.83}$$

with $\Delta l_{max} = l \cdot \alpha \cdot P \cdot R_{th}$. Equation (6.47) can be applied to determine the heat resistance.

If the heat flow in the beam is not generated by a constant heating power, but by the discharging of a capacitor with capacitance C, such as in Figure 7.12b for instance, and voltage V, the heat flow in the beam caused by a capacitor discharging $V(t) = V_0 \cdot e^{-t/\tau}$ is

$$\Phi = \Phi(t) = P(t) = \frac{V_0^2}{R_H} \cdot e^{-2t/\tau} \text{ with } \tau = R_H \cdot C. \tag{7.84}$$

At $t = 2\tau$, this heat flow is reduced to about 2 % of its initial value.

For periodical capacitor discharges with period duration t_P, the resulting pulsed heat flow $\Phi(t)$ can be determined from $t_I = 2\tau$ or, respectively,

$$\Phi_I = \frac{V_0^2}{4R} \tag{7.85}$$

using a simplified representation as a sequence of rectangular pulses, with pulse duration t_I and pulse height Φ_I.

Equation (7.85) assumes that the rectangular pulse and the pulse of the capacitor charge have an identical energy content. Such a pulse frequency causes during the pulse duration of the first period an unsteady heating process according to Equation (6.61). After the heat flow pulse disappears, the temperature decreases according to

$$\vartheta_{\mathrm{e}}(t) = P R_{\mathrm{th}} \cdot \mathrm{e}^{-t/\tau_{\mathrm{th}}}. \tag{7.86}$$

This process is repeated during the following periods with correspondingly changed initial temperatures whose values mainly result form the duty cycle $t_{\mathrm{I}}/t_{\mathrm{P}}$. For $\tau_{\mathrm{th}} \gg t_{\mathrm{P}}$, the temperature change during a period is negligible and Equation (6.61) can be used to describe the temperature rise in the beam, where

$$t = n \cdot t_{\mathrm{p}} \text{ and } \Phi = \Phi_{\mathrm{m}} = \Phi_{\mathrm{I}} \frac{t_{\mathrm{I}}}{t_{\mathrm{P}}} \tag{7.87}$$

have to be inserted. Φ_{m} corresponds to the mean heat flow.

If the power loss produced by the heating resistor is not evenly distributed over the beam volume, the temperature in the beam becomes a function of time and locus. At first, the heat flow produced by the heat resistor heats the different beam regions one after the other. A lumped thermal equivalent circuit can be used to model them as small disks with the corresponding heat resistance and heat capacity. The time- and locus-dependent heating of the beam are then superposed.

(b) Thermal expansion of bimaterials

Thermal expansion of bimaterials is commonly used as an actuating transducer principle for temperature-dependent electric switches. Classical thermal bimetals are standardized according to the German standard DIN 1715. Micromechanical designs often have the form of a cantilever beam (Figure 7.13), consisting of a compound of two materials with different thermal expansion coefficients.

A temperature increase from T_1 to T_2, which is usually generated by an electric heating resistor, causes a different expansion in the two material layers. This causes stress with opposite signs and consequently a bending of the beam by radius r.

Force and torque equilibrium in the beam's cross-section can be used to calculate the displacement of the unattached beam end [MEHNER94]. Here, the expansion in the contact

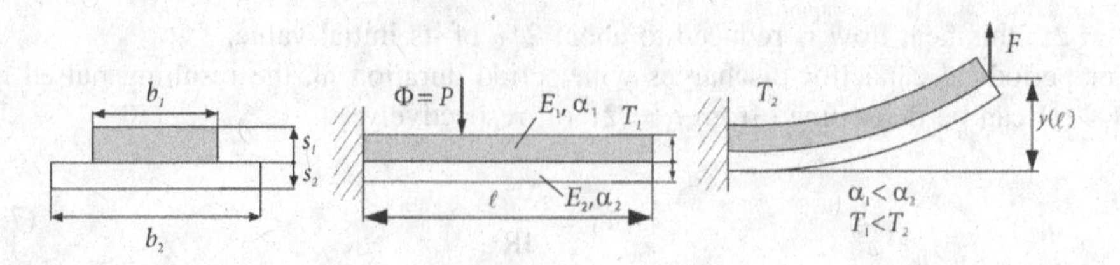

Figure 7.13 Bending of a two-layer beam at temperature change $T_1 \rightarrow T_2$. $\Phi = P$ heat flow, l beam length, b layer width, s layer thickness, α thermal expansion coefficient, $y(l)$ displacement, F force, T temperature

range of the two material layers has to be continuous. It applies that

$$y(l) \cong \frac{l^2}{2r}, \tag{7.88}$$

with the bending radius being determined from

$$r = \frac{(b_1 E_1 s_1^2)^2 + (b_2 E_2 s_2^2)^2 + 2b_1 b_2 E_1 E_2 s_1 s_2 (2s_1^2 + 3s_1 s_2 + 2s_2^2)}{6 \cdot \Delta\alpha \cdot \Delta T \cdot b_1 b_2 E_1 E_2 s_1 s_2 (s_1 + s_2)} \tag{7.89}$$

Here, $\Delta\alpha = \alpha_2 - \alpha_1$ and $\Delta T = T_2 - T_1$.

For identical layer widths $b_1 = b_2 = b$, the bending radius becomes

$$r = \frac{s_1 + s_2}{6} \cdot \frac{3(1 + s_1/s_2)^2 + \left(1 + \frac{s_1 E_1}{s_2 E_2}\right)\left[(s_1/s_2)^2 + \frac{s_2 E_2}{s_1 E_1}\right]}{\Delta\alpha \cdot \Delta T (1 + s_1/s_2)^2}. \tag{7.90}$$

Displacement $y(l)$ is affected to a larger degree by the variation of dimensions and expansion coefficients than by the variation of YOUNG's modulus. The bending radius of the transducer element becomes small if, for similar YOUNG's modulus values, the dimensions of the two layers are similar. Assuming for an approximation of the displacement in Equation (7.90) an identical thickness of the two layers $s_1 = s_2 = s$ and YOUNG's modulus $E_1 = E_2 = E$, it results that

$$y(l) = 3 \cdot l^2 \Delta\alpha \cdot \Delta T / (8 \cdot s)\,. \tag{7.91}$$

Equation (7.91) shows how the achievable displacement can be influenced by the selection of design parameters l, s, $\Delta\alpha$. Miniaturized dimensions reduce response time and the required heating power.

Force F at the unattached beam end is

$$F = 3EI \cdot y(l)/l^3, \tag{7.92}$$

where the rotational stiffness of the beam can be determined using

$$EI = \frac{(b_1 E_1 s_1^2)^2 + (b_2 E_2 s_2^2)^2 + 2b_1 b_2 E_1 E_2 s_1 s_2 (2s_1^2 + 3s_1 s_2 + 2s_2^2)}{12(s_1 b_1 E_1 + s_2 b_2 E_2)} \tag{7.93}$$

For identical layer widths $b_1 = b_2 = b$

$$F = \frac{3b}{4l} \cdot \frac{s_1 + s_2}{\frac{1}{s_1 E_1} + \frac{1}{s_2 E_2}} \cdot \Delta\alpha \cdot \Delta T. \tag{7.94}$$

Similar to the thermal expansion of solid bodies, the thermal expansion of bimaterials can be combined with the dynamic effect of loaded membranes or beams, resulting in a snap effect that can be used for switching functions.

Example 7.7 Thermomechanically operated valve [NGUYEN02]

Figure 7.14 shows a micromechanical valve with a fixed square valve seat 6 that has an edge length of 500 μm. Four cantilever beams 1 to 4 are suspended at a frame 5. The cantilever beams consist of a bimaterial and have an integrated heater. When a current is flowing through the heater, the bimaterial effect causes the opening and closing of the valve seat. All cantilever beams have a length of 500 μm and a width of 200 μm. The bimaterial consists of a silicon-aluminum-compound with layer thicknesses $s_{Si} = 10$ μm and $s_{Al} = 2$ μm. YOUNG's moduli are $E_{Si} = 170$ GPa and $E_{Al} = 70$ GPa, the thermal coefficients of expansion are $\alpha_{Si} = 2.3 \cdot 10^{-6}$ K^{-1} and $\alpha_{Al} = 23 \cdot 10^{-6}$ K^{-1}. The aluminum layer is vapor-deposited at a temperature of 400 °C. During normal conditions, the valve is supposed to be closed at 25 °C. Therefore the gap between valve seat 6 and valve opening 7 has to be smaller than 10 μm. This value can be determined as follows:

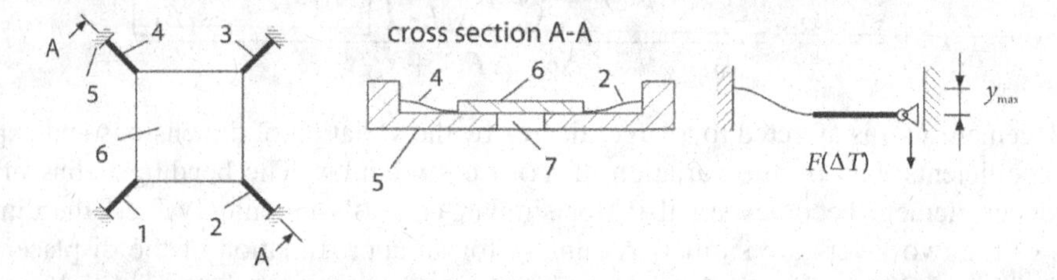

Figure 7.14 Thermomechanical valve with bimorphous actuator: 1 to 4 cantilever beams, 5 frame, 6 valve seat, 7 valve opening, y_{max} displacement, $F(\Delta T)$ force

Assuming that the bimorphous layers are load-free for an evaporation at 400 °C, it applies for the force at the unattached end of the cantilever beam at 25 °C according to Equation (7.94): $F = 3.615 \cdot 10^{-3}$ N. According to Equation (7.93), their rotational stiffness amounts to $EI = 3.774 \cdot 10^{-9}$ Pa · m^4 each. Due to the double-sided attachment, the bending line of the cantilever beams corresponds to that of an S-type spring. According to Equation (6.29), their spring stiffness results from $c = 12\ EI/l^3$. Due to $c_{tot} = 4c$ and $F_{tot} = 4F$, the displacement of the valve seat based on the bimorphous effect amounts to

$$y_{max} = F_{tot}/c_{tot} = Fl^3/(12EI) = 10\ \mu\text{m}.$$

(c) Thermal expansion of gases and fluids

In comparison to the thermal expansion of solid bodies, the thermal expansion of gases and fluids is large. When supplied with thermal energy, thermomechanical transducers based on gases and fluids can produce comparatively large forces and strokes.

- If heat is supplied to a fluid in a closed cavity that is covered with a membrane, it reacts according to Equation (7.82) with an increase in volume. For water (thermal volume expansion coefficient $\gamma = 0.26 \cdot 10^{-3}$ K^{-1}), heating by 20 K causes a volume increase of ca. 0.5 %. Due to the incompressibility of fluids, this would result in a displacement of the membrane, even for large counterforces.

- If there is gas in a closed cavity, a heat supply will result in a pressure increase due to the compressibility of the gas:

$$p_1(T) = p_0(1 + \gamma \Delta T). \tag{7.95}$$

Here, p_0, T_0 are the initial values of pressure or temperature, respectively, and ΔT the temperature increase due to the heat supply.

- If there is a fluid-gas mix in a closed cavity, the evaporation of the fluid can lead to a particularly large pressure increase. In the cavity, there will result an equilibrium of liquid and gaseous phase. The increase in pressure in relation to temperature is

$$p(T) = p_c \cdot e^{\frac{L_0}{R}\left(\frac{1}{T_c} - \frac{1}{T}\right)}, \tag{7.96}$$

where R is the gas constant, p_c the critical pressure of the applied fluid, L_0 its evaporation heat and T_c its critical temperature. It applies that $T_c > T$. It results from Equation (7.96) that for a thermodynamical equilibrium at the phase boundary between liquid and gaseous phase, there is an exponential increase in pressure if the temperature rises from T_1 to T_2 [KITTEL01]:

In order to increase the pressure from its initial value p_1 to a value p_2, a temperature increase of

$$\Delta T = T_2 - T_1 = \frac{RT_1^2 \ln p_2/p_2}{L_0 - RT_1 \ln p_2/p_1} \tag{7.97}$$

is required. Fluids with a low boiling point are most suitable for evaporation.

Deriving Equation (7.96) for the temperature

$$\frac{\mathrm{d}p(T)}{\mathrm{d}T} = \frac{p_c L_0}{RT^2} e^{\frac{L_0}{R}\left(\frac{1}{T_c} - \frac{1}{T}\right)} \tag{7.98}$$

shows that critical pressure p_c has to be large and critical temperature T_c as well as evaporation heat L_0 have to be small in order for a small temperature increase to cause a maximum pressure increase. The temperature increase ΔT required to displace a membrane using the thermal expansion of gases and fluids is smallest for closed cavities filled with a fluid-gas mix.

- A common application of the thermomechanical transducer principle with a phase transition liquid → gaseous are the printing heads of ink printers, which are known as bubble jet printing heads. Here the ink is both medium and actuating element (Figure 7.15). Before each drop ejection, a short current pulse heats the heater for a few microseconds. The interface between heater and ink reaches a temperature of more than 300 °C and produces a closed vapor bubble above the heater. Thermal conductivity to the ink has to be good, whereas thermal conductivity to the substrate has to be poor. With bubbles forming, the heat flow from heater to ink is almost completely interrupted. The vapor bubble expands fast and the dynamic effect ejects an ink drop

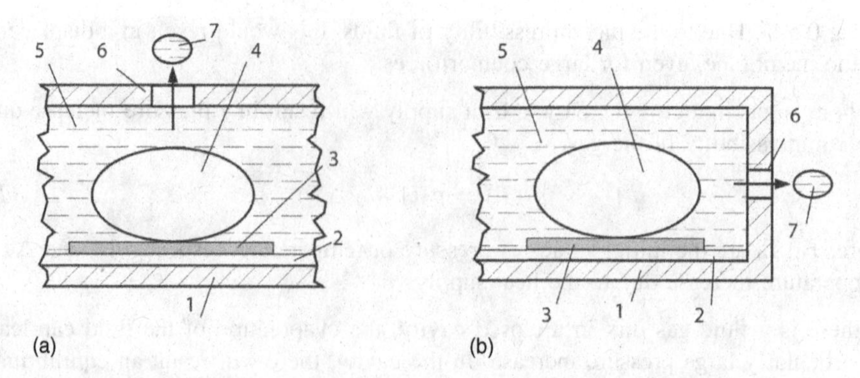

Figure 7.15 Variants of the bubble-jet principle: (a) nozzle frontal; (b) nozzle sidewise; 1: ink chamber; 2: thermal insulation; 3: heater; 4: vapor bubble; 5: ink; 6: nozzle; 7: ink drop.

from the nozzle. The drops reach a speed of over 10 m/s, their mass lies between 50 and 100 ng. A cycle between two consecutive drops comprises the phases heating, film boiling, bubble formation, drop ejection, collapsing of the bubble, drop detachment, capillary refilling of the ink reservoir. Such a cycle takes approximately 200 μs, drop frequency amounts to ca. 5 kHz. The transducer principle that the bubble jet printing head is based on, i.e. the generation and utilization of vapor bubbles as driving elements, can also be used for displacing movable mechanical structures [LIN94].

7.2.5 Piezoresistive Transducers

(a) Piezoresistive Effect

Due to the resistive measuring effect, resistive semiconductor sensors have a simple and robust design and, due to the solid state effect, they are very suitable for miniaturization.

According to Section 7.1.2, resistive sensors are parametric transducers. Example 7.3 showed a derivation where a resistance as parameter according to Equation (7.45) can be used to change specific resistivity ρ as well as length ℓ and cross-sectional area A of the resistor:

$$R = \frac{V}{I} = \rho \cdot \frac{\ell}{A}. \tag{7.99}$$

In the following, we want to look at a resistor with a force acting axially on the resistor. The result is relative expansion or strain $\varepsilon = \Delta\ell/\ell$ (Figure 7.16). The relative change in resistance consists of the change of geometric and material parameters:

$$\frac{\Delta R}{R} = \frac{\Delta\rho}{\rho} + \frac{\Delta\ell}{\ell} - \frac{\Delta A}{A} \approx \begin{cases} \dfrac{\Delta\ell}{\ell} - \dfrac{\Delta A}{A} & \text{for } \dfrac{\Delta\rho}{\rho} \ll \dfrac{\Delta\ell}{\ell} - \dfrac{\Delta A}{A} \\[2ex] \dfrac{\Delta\rho}{\rho} & \text{for } \dfrac{\Delta\ell}{\ell} - \dfrac{\Delta A}{A} \ll \dfrac{\Delta\rho}{\rho} \end{cases} \tag{7.100}$$

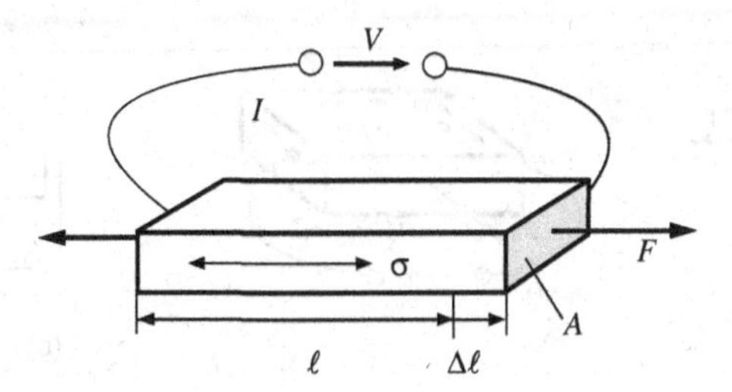

Figure 7.16 Resistance during tensile stress

In the first case, geometric changes dominate. This is called strain-gauge effect and is utilized in strain gauges. As the resistor becomes longer with the expansion and – at the same time – more slender, both portions in Equation (7.101) act in the same direction.

The second case, where the change in resistivity determines the change of resistance during stress, is called piezoresistive effect.

Specific resistivity ρ depends on the number of charge carriers N in the volume $V = A \cdot \ell$ and their mobility μ (e unit charge):

$$\rho = \frac{V}{N \cdot \mu \cdot e}. \tag{7.101}$$

This results in:

$$\frac{\Delta\rho}{\rho} = \frac{\Delta V}{V} - \frac{\Delta(N \cdot \mu)}{N \cdot \mu}. \tag{7.102}$$

With $\varepsilon = \Delta\ell/\ell$ as well as

$$\frac{\Delta V}{V} \approx \frac{\Delta A}{A} + \frac{\Delta\ell}{\ell} \tag{7.103}$$

for small lengths and cross-section changes up to the procent range, it results from Equations (7.101) and (7.103).

$$\frac{\Delta R}{R} = \left[2 - \frac{1}{\varepsilon}\frac{\Delta(N\mu)}{N\mu}\right]\varepsilon = K\varepsilon. \tag{7.104}$$

Here, the resistance change $\Delta R/R$ is proportional to the strain. The proportionality coefficient is the K-factor. The first part of the term in brackets is determined by geometric changes. For a metallic strain gauge, it applies that

$$\left.\frac{\Delta R}{R}\right|_{\text{metal}} = 2\varepsilon. \tag{7.105}$$

For semiconductors, the number of charge carriers and their mobility changes with deformation. This is the result of the deformed band structure due to a distorted lattice structure.

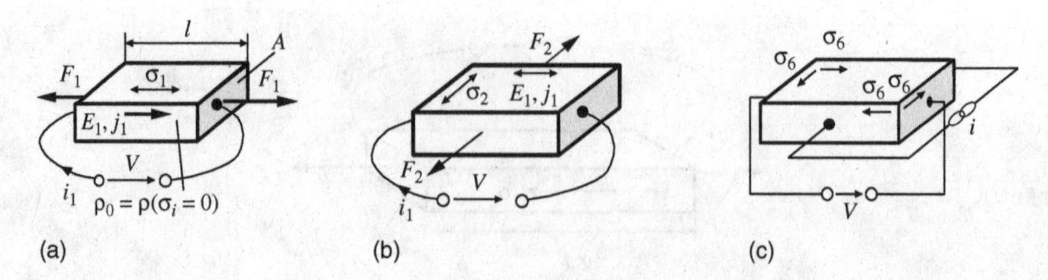

Figure 7.17 Piezoresistive couplings: (a) longitudinal effect; (b) transversal effect; (c) shearing effect

This way the second part of the bracket term in Equation (7.104) becomes dominant

$$\left.\frac{\Delta R}{R}\right|_{\text{semiconductor}} = (20\ldots100)\varepsilon. \tag{7.106}$$

Piezoresistors can be easily integrated into semiconductors (Figures 1.2, 7.17). Then, they are part of the deformation body, e.g. part of a deformable plate or of a cantilever that can be displaced by force or pressure. That means, that a mechanical stress or strain field acts upon them.

Integrated silicon resistors in silicon deformation bodies have both identical stress and strains at the interface between piezoresistor and substrate. Due to that, integrated piezoresistors are described via piezoresistive coefficients π_{ij} (Figure 7.18). The piezoresistive longitudinal effect means that force F_1 acts in the direction of current flow I_1 and generates a stress field σ_1. This changes the relationship of $E_1 = \rho \cdot J_1$ to

$$E_1 = \rho_0(1 + \pi_{11}\sigma_1)J_1. \tag{7.107}$$

Here, π_{11} is the piezoresistive longitudinal coefficient that describes the correlation of the change in resistivity between $\rho(\sigma_1 = 0) = \rho_0$ and $\rho(\sigma_1 \neq 0)$. The relative resistance

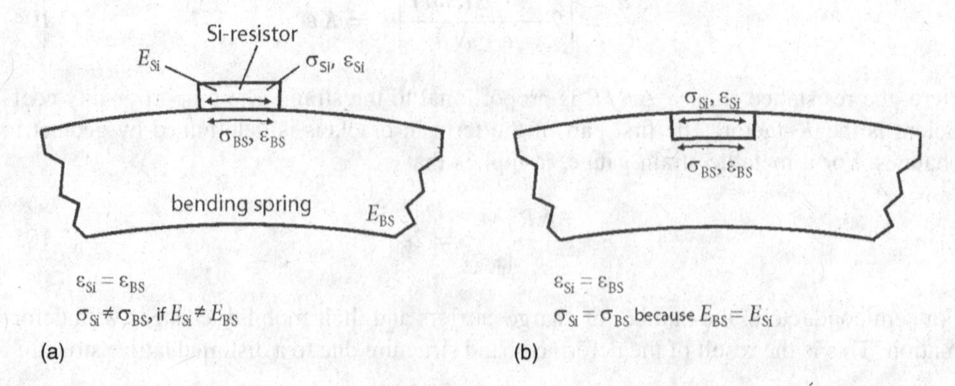

Figure 7.18 Strain or stress transmission for (a) a silicon strain gauge; (b) a silicon piezoresistor

change $\Delta R/R$ due to the longitudinal effect results from Equation (7.107)

$$\frac{\Delta R}{R} = \frac{R(\sigma_1) - R(0)}{R(0)} = \pi_{11}\sigma_1. \tag{7.108}$$

The piezoresistive transversal effect describes the effect of force F_2 and the corresponding stress σ_2 acting perpendicularly to current direction I_1:

$$E_1 = \rho_0(1 + \pi_{12}\sigma_2)J_2 \tag{7.109}$$

or, respectively

$$\frac{\Delta R}{R} = \frac{R(\sigma_2) - R(0)}{R(0)} = \pi_{12}\sigma_2. \tag{7.110}$$

The piezoresistive shearing effect describes that for a shear load (shear stress σ_6 in the 1–2-plane), electric field E_1 will be generated if current I_2 passes perpendicularly through the resistor:

$$E_1 = \rho_0\pi_{66}\sigma_6 J_2. \tag{7.111}$$

As opposed to the longitudinal and transversal effect, the shear effect is not a resistance effect. Due to the similarity to the HALL effect, where magnetic flux density B produces a transversal coupling between current and the voltage acting perpendicularly to it, the shear effect often is called piezoresistive HALL effect. Due to patent issues, the shear effect is basically almost only applied by the company Motorola.

If all six possible stress components $\sigma_1 \ldots \sigma_6$ occur, it follows from Equations (7.107), (7.109) and (7.111) by linear superposition that

$$E_1 = \rho_0[(1 + \pi_{11}\sigma_1 + \pi_{12}\sigma_2 + \pi_{13}\sigma_3)J_1 + \pi_{66}\sigma_6 J_2 + \pi_{55}\sigma_5 J_3]. \tag{7.112}$$

Analog equations apply to E_2 and E_3. Due to the symmetry of the cubic crystal, in silicon there are only three unrelated piezoresistive coefficients: $\pi_{11} = \pi_{22} = \pi_{33}$; $\pi_{12} = \pi_{21} = \pi_{13} = \pi_{31}$ as well as $\pi_{44} = \pi_{55} = \pi_{66}$.

If there is a plane, biaxial stress condition with σ_1 and σ_2, using Equations (7.108) and (7.110), the relative change in resistance r becomes

$$r = \frac{\Delta R}{R} = \pi_{11}\sigma_1 + \pi_{12}\sigma_2. \tag{7.113}$$

Table 7.8 shows the piezoresistive coefficients of silicon. The largest value corresponds to π_{44} in p-conducting silicon. That means that there are two options:

- utilization of the piezoresistive shear effect (see above);
- directing the piezoresistors in the silicon crystal in a way that a large proportion of π_{44} is included in the effective piezoresistive coefficients.

Table 7.8 Piezoresistive coefficients π_{ij} for p- and n-conducting silicon

Type	ρ_0/Ω cm	$\pi_{11}/10^{11}\text{Pa}^{-1}$	$\pi_{12}/10^{11}\text{Pa}^{-1}$	$\pi_{44}/10^{11}\text{Pa}^{-1}$
n-Si	11.7	−102.2	53.4	−13.6
p-Si	7.8	6.6	−1.1	138.1

The latter is applied in the majority of the practical cases. In addition, it is possible to use the fact that a shear deformation can also be achieved by two normal stress components σ_1 and σ_2 if one stress is compressive and the other one tensile (Figure 7.19). In general, for any crystal direction, the resulting piezoresistive longitudinal and transversal coefficients are [LENK75], [MESCHEDER04]

$$\pi_L = \pi_{11} - 2(\pi_{11} - \pi_{12} - \pi_{44})(\ell_1^2 m_1^2 + \ell_1 n_1^2 + m_1^2 n_1^2)$$
$$\pi_Q = \pi_{12} + 2(\pi_{11} - \pi_{12} - \pi_{44})(\ell_1^2 \ell_2^2 + m_1^2 m_2^2 + n_1^2 n_2^2). \quad (7.114)$$

Here, $\ell_1 \ldots n_2$ are the direction cosine components between two coordinate systems, with one system being transformed on to the other by rotation at angles v, ψ and φ:

$$\ell_1 = \cos\psi\cos\varphi - \cos v \sin\psi \sin\varphi; \ \ell_2 = -\cos\psi\sin\varphi - \cos v\sin\psi\cos\varphi$$
$$m_1 = \sin\psi\cos\varphi + \cos v\cos\psi\sin\varphi; \ m_2 = -\sin\psi\sin\varphi + \cos v\cos\psi\cos\varphi$$
$$n_1 = \sin v\sin\varphi; \qquad n_2 = \sin v\cos\varphi$$

For resistors that are rotated, according to Figure 7.20, in the base crystal by 45° ([1 1 0]-or [1$\bar{1}$0]- direction, respectively, in the (1 0 0)-plane), it results that $\nu = 0$, $\psi = 45°$, $\vartheta = 0$ and thus $\ell_1 = -\ell_2 = m_1 = m_2 = 1/\sqrt{2}$ and $n_1 = n_2 = 0$. Equation (7.114) becomes

$$\pi_L = \frac{1}{2}(\pi_{11} + \pi_{12} + \pi_{44}) \qquad \pi_T = \frac{1}{2}(\pi_{11} + \pi_{12} - \pi_{44}) \quad (7.115)$$

and for p-silicon:

$$\pi_L \approx -\pi_T \approx \pi_{44}/2. \quad (7.116)$$

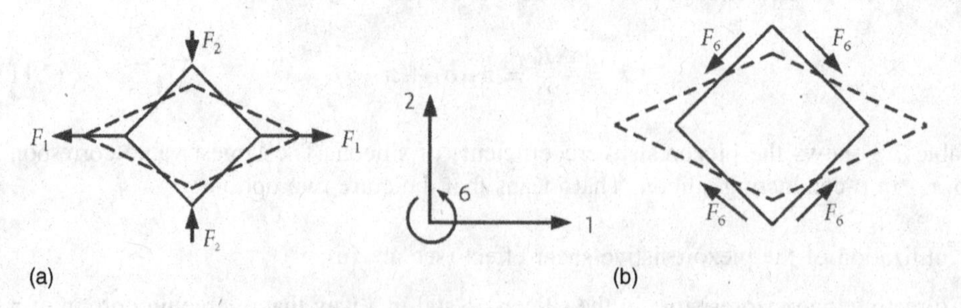

Figure 7.19 Relationship between (a) normal stresses σ_1, σ_2 and (b) shear stress σ_6

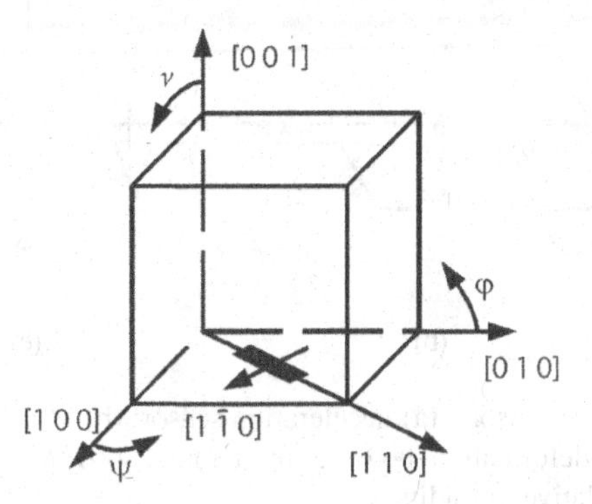

Figure 7.20 Utilization of the π_{44} coefficient by rotating the resistors in the Si base lattice

For two paired variable resistors that are situated perpendicular to each other in a biaxial stress field, with Equation (7.113) the relative change in resistance becomes:

$$r_L = \pi_L \sigma_L + \pi_T \sigma_T \text{ and}$$
$$r_T = \pi_T \sigma_L + \pi_L \sigma_T \quad (7.117)$$

Indices L and T of resistances and mechanical stress can be arbitrarily chosen, whereas π_L and π_T denominate the piezoresistive longitudinal and transversal effect, depending on whether the mechanical stress acts parallel or vertically to the current direction in the resistor.

(b) Piezoresistive Sensors

Piezoresistive pressure sensors operate according to the principle presented in Figure 7.21. Measuring pressure p deforms a pressure plate and generates a mechanical stress or strain field, respectively (σ_L, σ_T or $\varepsilon_L, \varepsilon_T$). At a time, two resistors R_L and R_T are differently affected via the piezoresistive effect. They form a WHEATSTONE full-bridge whose initial stress varies proportionally to measuring pressure p.

The piezoresistive measuring principle can use all measurands that cause a deformation of the deformation body (Figure 7.22). This may be an acceleration, for instance, that displaces via an inert mass a silicon cantilever spring with integrated resistors (Figure 7.22a). For AFM (Atomic Force Microscopes), a force acting on a tip results in a deformation of

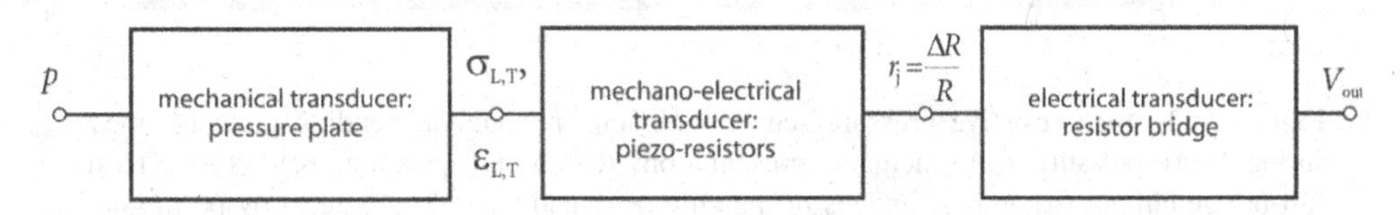

Figure 7.21 Operating principle of piezoresistive pressure sensors: p pressure, σ_i, ε_i stress or strain, r_j relative resistance change, V_{out} sensor output voltage, L longitudinal, T transversal

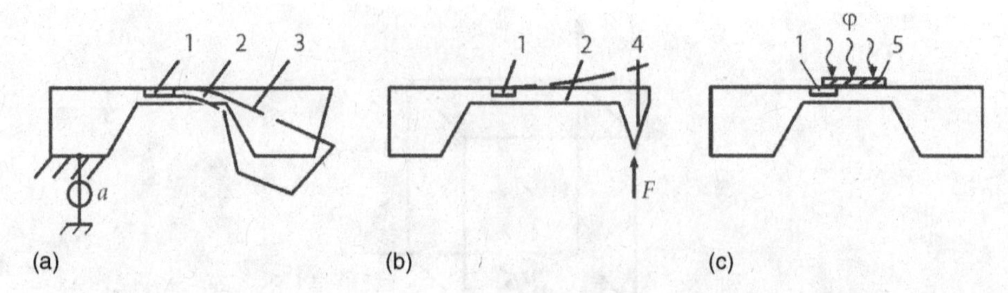

Figure 7.22 Piezooresistive sensors: (a) acceleration sensor; (b) atomic force sensor; (c) humidity sensor; 1 piezoresistor, 2 deformation body, 3 inert mass, 4 AFM tip, 5 swellable polymer, *a* acceleration, *F* force, φ relative humidity

the cantilever spring (Figure 7.22b). In piezoresistive humidity sensors, the humidity-dependent swelling of a polymer layer results in a bimetal-like displacement of the cantilever beam that is locally thinned down in the silicon chip (Figure 7.22c).

Example 7.8 Sensitivity of a piezoresistive silicon pressure sensor

We are looking for the sensitivity of the piezoresistive pressure sensor in Figure 7.23. The sensor chip has a WHEATSTONE full-bridge circuit where the resistances in the

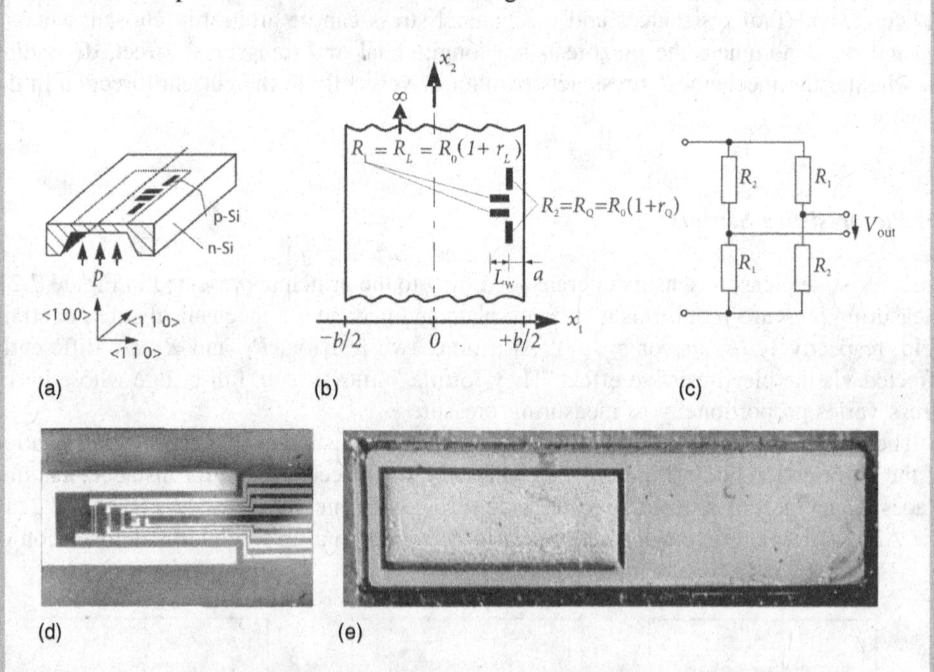

Figure 7.23 Piezoresistive pressure sensor with long rectangular bending plate for measuring heart pressure; (a) schematic presentation; (b) resistor position; (c) WHEATSTONE full-bridge circuit; (d) and (e) images of the chip front and back. Reproduced by permission from the Dresden University of Technology, Germany

bridge change identically for diagonal pairs (r_L, r_T). The following is given:

Longitudinal stress	$\sigma_L = \sigma_1(x_1) = \frac{1}{4}\left(\frac{b}{h}\right)^2 p\left[1 - 3\left(\frac{x_1}{b/2}\right)^2\right]$	(7.118)
Transversal stress	$\sigma_T = \sigma_2(x_1) = \nu \cdot \sigma_1(x_1)$	(7.119)
Piezoresistive coefficients	$\pi_L = -\pi_T = 5 \cdot 10^{-10}\,\text{m}^2/\text{N}$	(7.120)
POISSON's ratio	$\nu = 0.063$	
Bridge supply voltage	$V_0 = 5\,\text{V}$	
Pressure plate dimensions	$b = 400\,\mu\text{m}, h = 10\,\mu\text{m}$	
Full scale pressure	100 kPa (air pressure)	

Output voltage V_{out} results from the difference in voltage over resistors R_1 and R_2 in the two voltage dividers $R_1/(R_1 + R_2)$ and $R_2/(R_1 + R_2)$:

$$\frac{V_{out}}{V_0} = \frac{R_2}{R_1 + R_2} - \frac{R_1}{R_1 + R_2} = \frac{R_2 - R_1}{R_1 + R_2} \tag{7.121}$$

With $R_1 = R_0(1 + r_L)$ and $R_2 = R_0(1 + r_T)$, it results that:

$$\frac{V_{out}}{V_0} = \frac{r_T - r_L}{2 + r_L + r_T} \approx \frac{1}{2}(r_T - r_L) \tag{7.122}$$

The latter approximation can be assumed as the relative changes in resistance r_L and r_T usually have a maximum in the per-thousand range and no non-linearities have to be taken into consideration.

Using the correlations in Equation (7.117) for orthogonally oriented resistors R_L and R_T, it results from Equation (7.121) that

$$\frac{V_{out}}{V_0} = \frac{1}{2}(\pi_L - \pi_T)(\sigma_L - \sigma_T). \tag{7.123}$$

This equation is fundamental to piezoresistive full-bridge structures:

- Longitudinal and transversal resistors have to be oriented in a silicon crystal in a way that difference $\pi_L - \pi_T$ becomes as large as possible. For a (1 0 0)<1 1 0>-orientation, the optimum is reached: $\pi_L \approx -\pi_T \approx \pi_{44}/2$.
- Resistors R_L and R_T have to be located on the deformation body in a way that voltage difference $\sigma_L - \sigma_T$ becomes as large as possible. In the example described here: $(\sigma_L - \sigma_T) = (1 - \nu)\sigma_L$. Due to the extremely small POISSON's ratio for this orientation is $1 - \upsilon \approx 1$.

Shifting R_T towards the plate center would be advantageous and lead to sign $\sigma_L = -\text{sign}\sigma_T$ (Figure 7.23c). This may result in technological problems regarding the connection of such resistors, though (thermal mismatching of metal interconnects on the thin silicon pressure plate). Therefore, resistor arrangements are often located close to the edges of the pressure plates (Figure 7.23a).

Now, Equation (7.13) results in:

$$\frac{V_{out}}{V_0} = \frac{1}{2}(\pi_L - \pi_T)(\sigma_L - \sigma_T) \approx \pi_{44}(1 - v) \cdot \sigma_L \tag{7.124}$$

Applying Equation (7.118), it results for $x_1 \to b/2$, i.e.

$$\sigma_L = -\frac{1}{2}\left(\frac{b}{h}\right)^2 p \frac{V_{out}}{V_0} = -\frac{1}{2}\pi_{44}(1 - v)\left(\frac{b}{h}\right)^2 p. \tag{7.125}$$

Here, the negative sign lacks importance because the bridge's initial voltage can be easily reversely poled. With numeric values, Equation (7.124) becomes

$$V_{out} = -\frac{1}{2} \cdot 5 \cdot 10^{-10} \frac{\mathrm{m}^2}{\mathrm{N}} \cdot (1 - 0.063) \cdot 40^2 \cdot 10^5 \frac{\mathrm{N}}{\mathrm{m}^2} \cdot 5\mathrm{V} = 186.4\,m\mathrm{V}.$$

At full-scale pressure (air pressure), output voltage will become 186.4 mV.

Example 7.9 K-factors of piezoresistive sensors

We want to determine the K_L- and K_T-factors for the piezoresistive pressure sensor in Figure 7.23, using piezoresistive coefficients π_L and π_T.

Applying Equations (7.105) and (7.118), for a resistor on the very long plate:

$$R_L = \pi_L\sigma_L + \pi_T\sigma_T = K_L\varepsilon_L + K_T\varepsilon_T$$

$$R_T = \pi_T\sigma_L + \pi_L\sigma_T = K_T\varepsilon_L + K_L\varepsilon_T \tag{7.126}$$

where σ_L and σ_T are stresses and ε_L and ε_T the corresponding strains. Stresses and strains are related to each other via Equation (3.6). For the condition of a long plate, $\sigma_3 = 0$; $\sigma_T = 0$, it results that

$$\varepsilon_L = \frac{1}{E}\sigma_L - \frac{v}{E}\sigma_T, \; \varepsilon_T = \frac{1}{E}\sigma_L - \frac{\gamma}{E}\sigma_T = 0 \tag{7.127}$$

Thus

$$\sigma_Q = v\sigma_L, \tag{7.128}$$

which was already applied in Example 7.8, and

$$\varepsilon_L = \frac{1 - v}{E}\sigma_L. \tag{7.129}$$

Now, it results from Equation (7.125):

$$\pi_L\sigma_L + \pi_T\sigma_T = (\pi_L + \pi_T v) \cdot \sigma_L = K_L\varepsilon_L = K_L\frac{1 - v}{E}\sigma_L$$

and

$$K_L = \frac{E}{1 - v}(\pi_L + \pi_T v). \tag{7.130}$$

At the same time, it results from Equation (7.125):

$$\pi_{\mathrm{T}}\sigma_{\mathrm{L}} + \pi_{\mathrm{L}}\sigma_{\mathrm{T}} = (\pi_{\mathrm{T}} + \pi_{\mathrm{L}}v) \cdot \sigma_{\mathrm{L}} = K_{\mathrm{T}}\varepsilon_{\mathrm{L}} = K_{\mathrm{T}} \cdot \frac{1-v}{E}\sigma_{\mathrm{L}}$$

and

$$K_{\mathrm{T}} = \frac{E}{1-v}(\pi_{\mathrm{T}} + \pi_{\mathrm{L}}v). \tag{7.131}$$

With $E = 1.69 \cdot 10^{11}$ N $\cdot$m^2, $v = 0.063$, $\pi_{\mathrm{L}} = -\pi_{\mathrm{T}} = 5 \cdot 10^{-10}$ m^{2}N^{-1}, it results for (1 0 0) <1 1 0>-silicon that

$$K_{\mathrm{L}} = -K_{\mathrm{T}} = 84.5.$$

In comparison to metallic strain gauges ($K = 2$), (1 0 0) <1 1 1>-oriented monocrystalline silicon has a 40 times larger K-factor.

EXERCISES

1. Transform the lumped equivalent circuit in Figure 7.3a into that in Figure 7.3b.
2. Derive the transducer equations $\underline{I}$, $\underline{F} = f(\underline{V}, \underline{v})$ in analogy to Equation (7.20) for an electrostatic transducer whose movable electrode does not move parallel, as described in Figure 7.4, but vertically to the fixed bottom electrode.
3. Hall elements (Figure 7.24) are resistance elements for which a current I_x and, under the effect of magnetic flux density B_Z, a HALL voltage can be measured. Due to the coupling of HALL voltage $V_{\mathrm{H}} = V_y$ in direction y and current I_x in direction x, it can be assumed that this is a reversible transducer.
   (a) What kind of transducer is it?
   (b) Formulate the transducer equations.
   (c) What type of transducer represents the HALL sensor, for which it is commonly assumed that $I_y = 0$ and $I_x =$ const.?

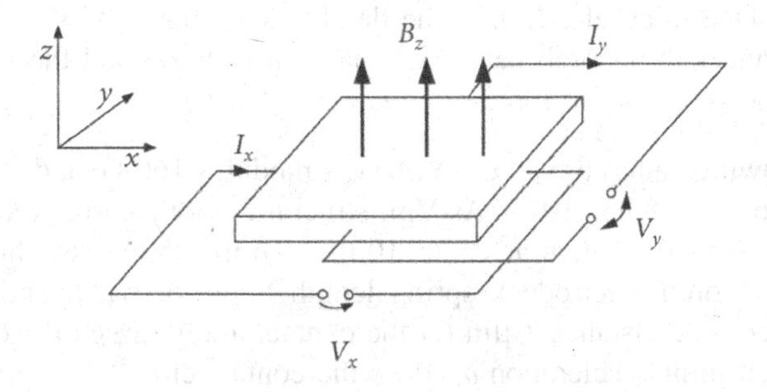

Figure 7.24 HALL element

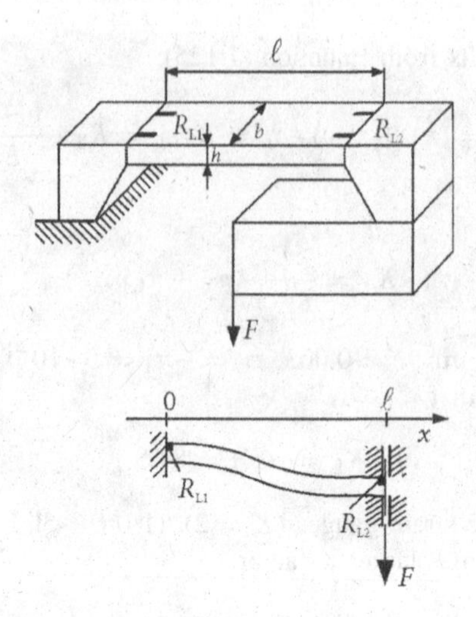

Figure 7.25 Piezoresistive force sensor

4. Determine the output voltage of the WHEATSTONE full-bridge circuit for the piezoresistive force sensor in Figure 7.25. Due to the length of the cantilever beam, it applies that $\varepsilon_T = 0$. Resistors R_L and R_T are located directly on the edges of the beam ($x = 0$ or, respectively, $x = \ell$). Measuring force $F = 10^{-3}$ N (corresponds to the weight of a mass of 1 g), supply voltage $V_0 = 10$ V, cantilever dimensions $h = 10$ μm, $b = 1$ mm, $\ell = 500$ μm. We use (1 0 0) <1 1 1> oriented silicon. Note that a full-bridge circuit does not use longitudinal and transversal resistors, but two different kinds of longitudinal resistors R_{L1} and R_{L2}.

5. Calculate in analogy to Example 7.9 the K-factors K_L and K_T for piezoresistors that are integrated into a long and narrow bending strip. We assume the cantilever to be transversally stress-free due to its small width ($\sigma_T = \sigma_3 = 0$). How do the values in Example 7.9 – which were calculated for wide deformation bodies – differ?

6. Why do we apply preferably comb cells for electrostatic transducers?

7. Figure 7.26 shows a threshold acceleration sensor manufactured using silicon micromachining. Due to acceleration a, the flexibly suspended mass m is displaced. The mass is connected to comb cells that enable a self-test and the calibration of the sensors using electrostatic forces.

   The following values are given: YOUNG's modulus 160 GPa, density 2300 kg/m^3, dielectric constant $8.85 \cdot 10^{-12}$ As/Vm, structural width 2 μm (spring width, width of comb elements), structural depth 10 μm (spring thickness, thickness of mass, thickness of comb electrodes), spring length 800 μm, width and length of mass 200 μm, electrode distance 1 μm (at the contact and between the comb elements).
   (a) At which limit acceleration a_G does the contact close?
   (b) How large does contact force F_K become for this threshold acceleration?
   (c) What effect does the mounting position of the sensor have?
   (d) What acceleration is required to produce a contact force of $F_K \geq 1$ μN?

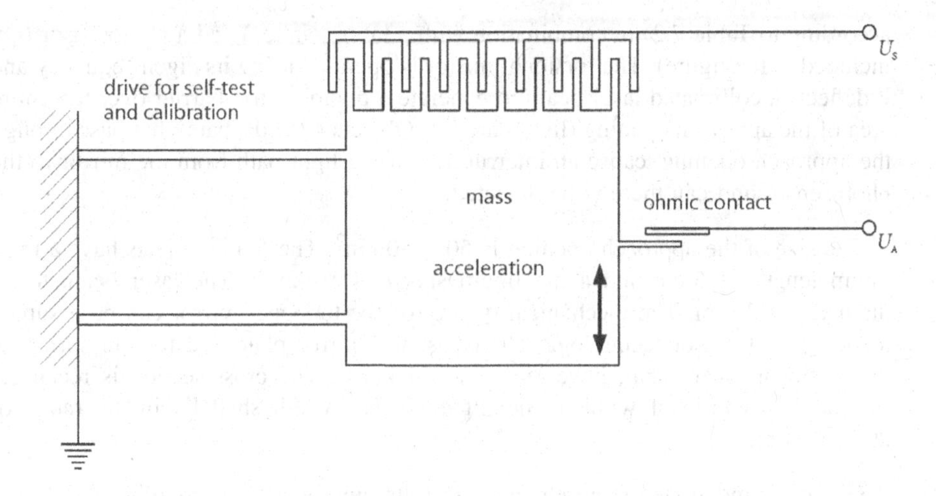

Figure 7.26 Threshold acceleration sensor

(e) What voltage V_S is required to carry out a self-test of the sensor?
(f) What voltage V_S is required to generate a contact force $F_K \geq 1$ μN during the self-test?
(g) How could the layout be changed in order to prevent a creeping contact making? (Note: use the pull-in effect.)

8. What is the relation of displacement angle φ and voltage V between the movable electrode and the base electrode for a micromachined one-dimensional torsional mirror (electrostatic transducer with distance variation according to Table 7.2)?

9. Figure 7.27 shows a surveillance device for approach opening 1. It consist of a micromechanic torsional mirror 2 (electrostatic transducer with distance variation

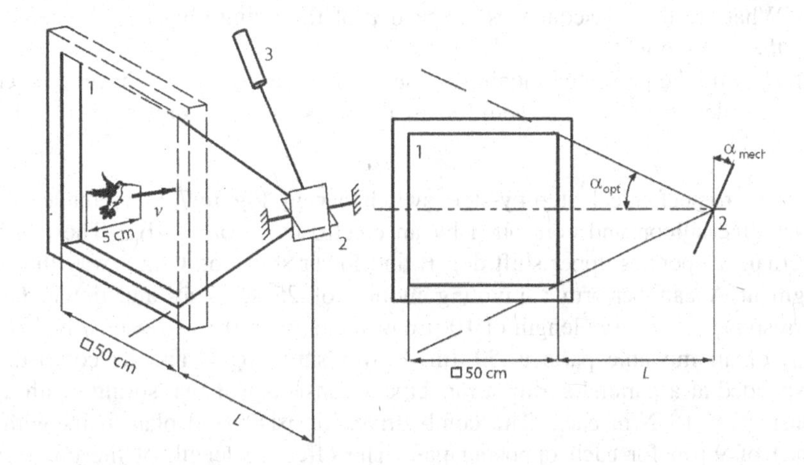

Figure 7.27 Surveillance device for an approach opening: 1 approach opening, 2 torsional mirror, 3 laser source, 4 flying object

according to Table 7.2) in combination with a laser source 3 and a photosensor (not included in the figure). The torsional mirror is operated using its eigenfrequency and it deflects a collimated laser beam to generate a photoelectric barrier over the entire area of the approach opening (light curtain). Objects 4 (birds, bats) that pass through the approach opening, cause an interruption of the light path from the mirror to the photosensor and can thereby be detected.

The size of the approach opening is $50 \times 50\ \text{cm}^2$. The flying objects have a minimum length of 5 cm and a maximum speed of 50 km/h. The laser beam has a diameter of 2 mm. The mechanical quality of the torsional mirror can be assumed to be $Q = 100$. For technological reasons, the mirror plate and torsion spring of the torsional mirror shall have the same thickness. The cross-section is rectangular and has a ratio of width to height of 2. The width shall lie in the range of $20 \ldots 60\ \mu\text{m}$.

Given the mentioned requirements, the following questions regarding the design of the torsional mirror are to be answered:

(a) How large is the sampling frequency and the required eigenfrequency of the micromechanical mirror?

(b) What distance L between mirror and approach opening has to be selected in order for the mirror to reach a mechanical displacement angle of $\alpha_{\text{mech}} = 12.5^\circ$ during resonance operation?

(c) What is the required edge length for a square mirror?

(d) What is the minimum thickness of a wafer that is used for manufacturing the mirror?

(e) What are the free design parameters of the electrostatic transducer regarding its parameters eigenfrequency and mechanical displacement angle? What influence does the mechanical quality have? What has to be taken into account when selecting the electrodes' basic distance? What values have to be chosen for the design parameters if the requirements regarding the parameters eigenfrequency and displacement angle have to be fulfilled?

(f) What are the consequences if speed v of the flying objects is higher or lower than 50 km/h?

(g) How do the presented monitoring device and the parameters have to be changed in order to detect – in addition to the passing of an object – its flight direction?

10. The micromechanical step-by-step switchgear in Figure 7.28 is made of single crystalline silicon and controlled by an electrostatic comb drive. The comb drive alternately operates upper shift dog B and lower shift dog C in a way that toothed segment A can perform a pivoting motion of 25 steps up and down. One step corresponds to a curve length of $10\ \mu\text{m}$ or an angle of 0.3°, respectively. The thickness of all movable parts is $50\ \mu\text{m}$. Lower shift dog C and its comb drive are suspended at a parallel spring arrangement consisting of two springs with a spring constant of 18 N/m each. The comb drive consists of 40 plate pairs with a distance of $9\ \mu\text{m}$ for each opposing pair. The effective length of the comb arms is $200\ \mu\text{m}$. The lower shift dog moves $7\ \mu\text{m}$ towards the teeth. The plate distance of the comb drive decreases accordingly. The comb drives are operated at a voltage

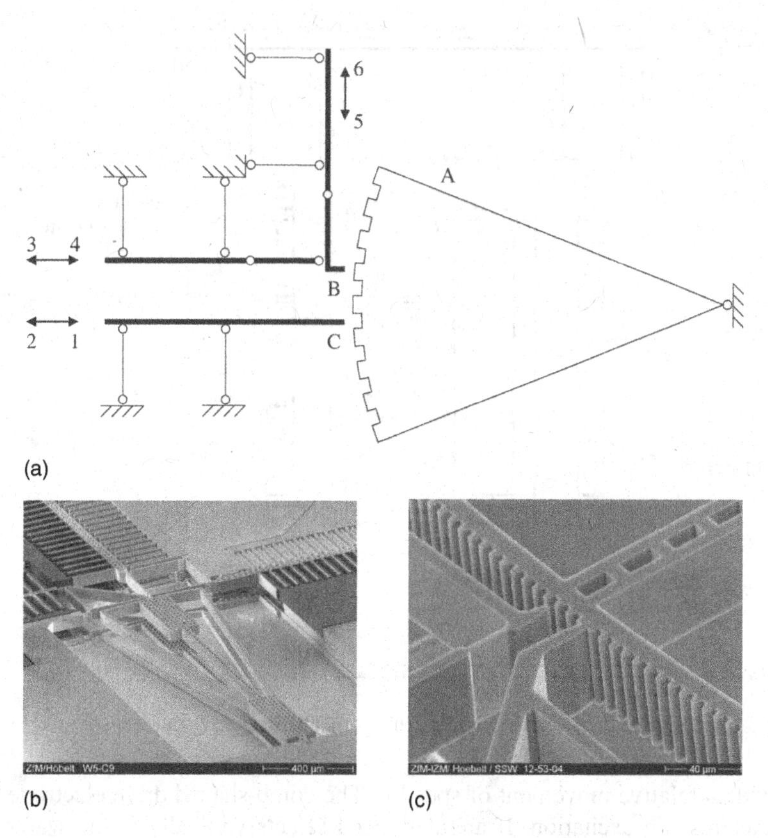

Figure 7.28 Micromechanical step-by-step switchgear: (a) working principle; (b) microscopic image from toothed segment A and comb drives for the shift dogs B and C; (c) detail: toothed segment A, feeding shift dog B, holding shift dog C. 1–2, 3–4 and 5–6 are the moving directions of the feeding shift dog B. Reproduced by permission from the Center for Microtechnologies, Chemnitz University of Technolog, Germany

of 50 V. What is the retaining force F that presses lower shift dog C into toothed segment A?

11. In order to apply micromechanical yaw rate sensors for detecting small yaw rates (typically smaller than 1000 °/s) in space, we can measure the CORIOLIS acceleration

$$\vec{a}_c = -2\vec{\Omega} \times \vec{v} \tag{7.132}$$

(see Section 1.4). Here, v represents the relative speed of a mass point in the reference coordinate system that rotates with angular speed Ω. If the two vectors are situated vertically to each other, there occurs a CORIOLIS acceleration a_c in the reference coordinate system. Figure 7.29 shows a suitable micromechanical sensor structure. The structure is produced applying SOI technology. External springs are used to suspend it in order for the entire structure to be able to carry out oscillating motions in direction x. This movement is called primary movement and serves to

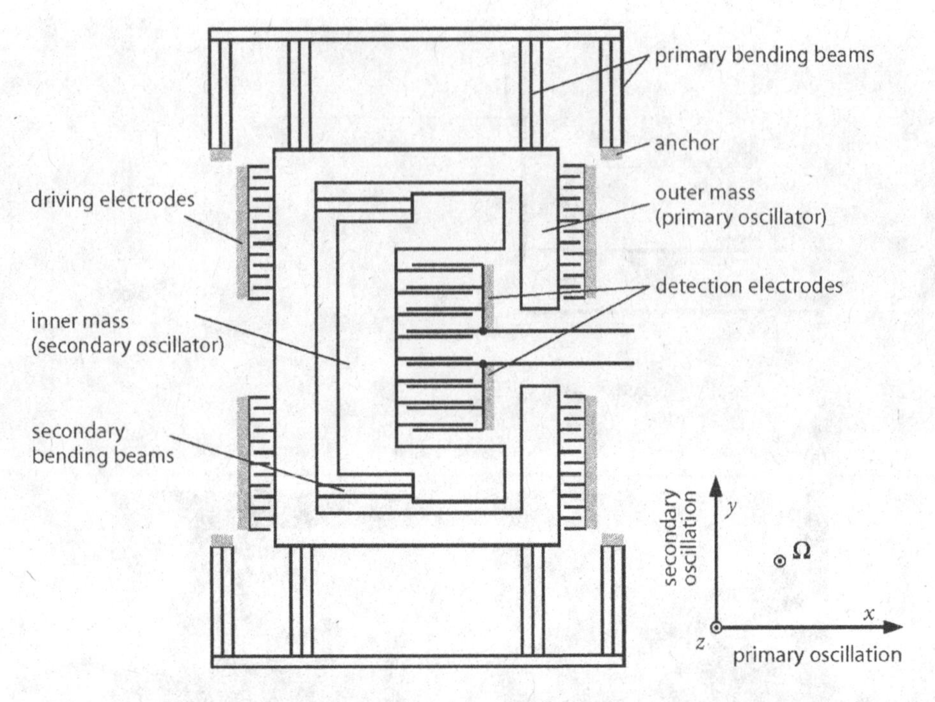

Figure 7.29 Structure of a micromechanical yaw rate sensor

provide a relative movement of speed v. The comb-shaped drive electrodes produce an electrostatic excitation. If angular speed Ω acts vertically to the wafer plane on the system (z-direction), it results in a CORIOLIS acceleration in the wafer plane, vertically to the primary movement (y-direction). The internal oscillating system consisting of internal mass and secondary cantilever beam is used to detect the CORIOLIS acceleration.

The sensitivity of the sensor increases with a larger amplitude of the primary motion. Therefore the structure should be dimensioned in order for it to operate in resonance and have a high resonance quality. The sensitivity of the system can be further increased if the eigenfrequency of the secondary oscillating system is tuned to that of the primary resonator. This way, it can also utilize the resonance quality factor.

The sensor structure consists of single crystalline silicon (E = 169 GPa). The mechanically active layer has a thickness of 50 µm. Due to the applied dry-etching technique, all cross-sections are near to rectangular. The external and internal mass have been determined to be $m_1 = 57$ µg and $m_2 = 35$ µg. The cantilever springs are assumed to have no mass.

(a) What dimensions does the primary cantilever beam have in order for the eigenfrequency of the primary movement to reach $f_{01} = 10$ kHz and assuming a spring width b_f of 5 µm?

(b) The primary motion is excited, as presented in Figure 7.29, by comb-shaped electrodes. The individual comb electrodes overlapp over a width of $b = 20$ µm, the electrode gap is $d = 2.5$ µm and the maximum available voltage is $V = 5$ V. For the electrostatic excitation, each drive comb will have $n = 30$ comb arms at the seismic mass. What is the maximum static displacement q_x that can be reached?

(c) How is the oscillation amplitude affected if the structure is excited in resonance? What parameter is affected by changing gas pressure in the sensor housing?

(d) At resonance, the primary oscillation is controlled at $f_{01} = 10$ kHz to have a constant amplitude of 10 µm. The internal oscillating system is dimensioned for its eigenfrequency to be $f_{02} = f_{01} = 10$ kHz, too. The resonance quality factor amounts to 1000. At what amplitude does the secondary movement oscillates for a yaw rate of 200°/s?

(e) The amplitude of the secondary oscillation is proportional to the CORIOLIS acceleration. We have a differential capacitor arrangement for detecting it. The electrode pairs consist of transversally moving comb electrodes with 20 tines each. Electrode distance is 2.5 µm and overlapping width 200 µm. What capacitance change can be reached per °/s?

(f) Due to technological tolerances and environmental influences during sensor operation, the eigenfrequencies of primary and secondary oscillation are not exactly identical. How can a continuous adjustment of the frequencies be carried out?

12. The electrodynamic transducer in Figures 7.10a and 7.11a is supposed to have an actuating drive and to displace a movable silicon element with a torque of $M = 5.6 \cdot 10^{-6}$ Nm. The magnetic flux density is assumed to be constant at $B = 0.25$ T. The 16 mm long and 6 mm wide movable silicon element is assumed to have a 2 mm wide area on its outside circumference to be used to include a planar coil. At the corners of the movable element the otherwise straight conductor is bent by 90°. The planar coil is contacted together with the aluminum coil conductors on the surface of the torsion spring elements (width 50 µm).

(a) What values can be chosen for current I and circuit loop number n of the planar coil in order to reach the required torque? The distance between the circuit loops is negligible as it is very small compared to its width. The current density in the interconnects shall not exceed that of the planar coil (the 90°-corners are neglected). For simplifying the calculation of the torque, we assume that all n circuit loops are situated in the middle of the area given for the planar coil.

(b) The thickness of the planar coil's conducting path is 2 µm and the resistivity of aluminum is $\rho_{Al} = 0.028$ Ωmm²/m. How large is the electric power consumed by the transducer for the given torque? How much is the temperature of the planar coil and the silicon element expected to increase during continuous operation of the transducer, if the heat is exclusively dissipated by convection and we assume a convection coefficient of $\alpha_K = 10$ W/m²K?

(c) Can the temperature be changed by varying values selected in a) and is there a minimum temperature?

What would the statements in (b) and (c) look like, assuming that the minimum distance required between the conductors of the planar coil is 5 μm?

13. For a given actuating application, we consider an electrostatic and an electrodynamic transducer as solution options. The actuator is to be operated in a control circuit. What are the differences between the transducer types regarding the control technology; which of the transducer types is preferable in this context?

REFERENCES

[FRÜHAUF05] Frühauf, J. (2005) *Werkstoffe der Mikrotechnik* (Materials in Microtechnology; in German). Fachbuchverlag Leipzig.

[KHAZAN94] Khazan, A. (1994) *Transducers and Their Elements*. Englewood Chiffs: PTR, Prentice Hall.

[KITTEL01] Kittel, Ch., Kroemer, H. (1980) *Thermal Physics*. 2nd edn. New York: Freeman.

[KOVACS98] Kovacs, G. T. A. (1998) *Micromachined Transducers Sourcebook*. Boston *et al.*: CB/McGraw-Hill.

[LEE98] K. B. Lee, Y.-H. Cho (1998) A triangular electrostatic comb array for micromechanical resonant frequency tuning. *Sensors and Actuators* A70, pp. 112–17.

[LENK01] Lenk, A, Pfeifer, G., Werthschützky, R. (2001) *Elektromechanische Systeme* (Electromechanical Systems; in German). Berlin: Springer.

[LENK75] Lenk, A. (1975) *Elektromechanische Systeme* (Electromechanical Systems). Vol. 3: *Systeme mit Hilfsenergie* (Systems with Energy Supply). Berlin: Verlag Technik.

[LIN94] Lin, L., Pisano, A. (1994) Thermal bubble powered microactuators. *Microsystem Technologies*, pp. 51–7.

[MEHNER94] Mehner, J. (1994) *Mechanische Beanspruchungsanalyse von Siliziumsensoren und -aktoren unter dem Einfluss von elektrostatischen und Temperaturfeldern* (Mechanical Load Analysis of Silicon Sensors and Actuators Influenced by Electrostatic an Temperature Fields; in German). Dissertation, TU Chemnitz.

[MESCHEDER04] Mescheder, U. (2004) *Mikrosystemtechnik* (Microsystem Technology; in German). 2nd edn. Stuttgart, Leipzig: B.G. Teubner.

[MIDDELHOEK98] Middelhoek: The sensor cube revisited. *Sensors and Materials* 10 (1998) 7, pp. 397–404.

[NGUYEN02] Nguyen, N. T, Wereley, S. T. (2002) *Fundamentals and Application of Microfluidics*. Boston, London: Artech House Publishers.

[ONSAGER31] Onsager, L. (1931) Reciprocal relations in irreversible processes. *Physical Review* 37, pp. 405–26 and 38, pp. 2265–79.

[KWAAITAAL93] Kwaaitaal, T. (1993) The fundamentals of sensors. *Sensors and Actuators* A39, pp. 103–10.

8 Design of Microsystems

The performance and also the cost efficiency of microsystems are mainly determined during the design phase. As a prototype production with a subsequent redesign is extremely costly, it is necessary to determine and optimize component behavior already during the design process. The results of the optimized design of a microsystem is a simulation model that both fulfils all system requirements under typical operational conditions as well as ensuring the reliability during critical load situations and the testability of characteristic parameters. In addition, economical aspects such as the required chip area, the number of process steps as well as the expected yield play an increasingly important role. In the following, we will describe the available design methods and tools and their current applications and limitations.

8.1 DESIGN METHODS AND TOOLS

It is specific to the design of microsystems that the individual functional components belong to different physical domains that are connected to each other. Sensor and actuator functions utilize interactions between mechanical, electrostatic, thermal and fluidic fields (Figure 8.1). However, certain interactions also cause undesirable side effects, such as cross-sensitivity. Subject of microsystem design is to define the essential component characteristics in computer models as well as the synthesis and optimization of the total system.

A special challenge of component and system simulation is to combine the heterogeneous behavioral models of sensor and actuator components with electronic control circuits and to analyze their interaction for different input signals and environments. To achieve that, the number of parameters of component models are reduced and transposed into a uniform modeling language such as VHDL-AMS[1] or Verilog-AMS[2].

Most properties of microtechnical devices can be calculated with a high degree of accuracy. However, the parameters of individual elements are subjected to variations that are mainly caused by manufacturing or assembly tolerances as well as by variances

[1] www.eda.org/vhdl-ams/
[2] www.eda.org/verilog-ams/

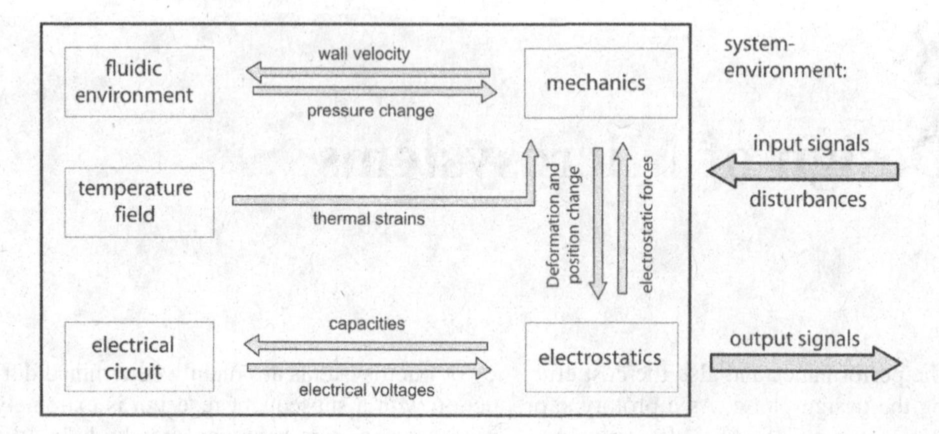

Figure 8.1 Interaction of physical domains in a microsystem

of material properties due to the specific process. Deviations of the characteristics of individual system components typically lie in the percentage range. However, calibration can be used to improve this situation for applications with high precision requirements.

The design process in microsystem technology covers the entire range from defining a requirement profile to a virtual computer model that represents the entire system behavior. Similar to the designing of machines, the design process can be divided into a conceptual and a realization phase. Both phases require engineers with creative thinking and design experience. It is helpful to use model libraries containing similar, already realized microsystems [MEHNER00].

The design process is divided into the following phases (Figure 8.2):

- During the *conceptual phase*, the designer evaluates the specifications in regards to whether they are complete and can be realized. Then, it follows the selection of a

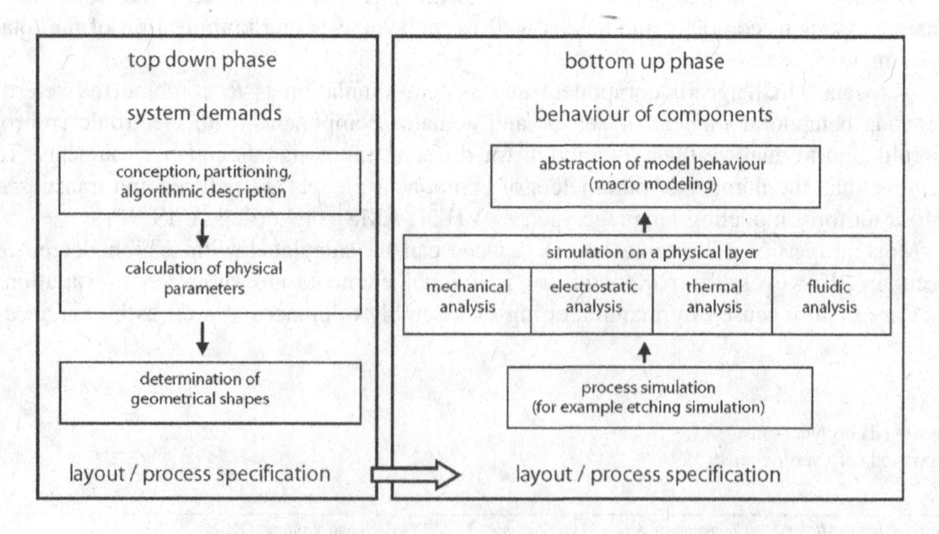

Figure 8.2 Important elements of the design process in microsystem technology

suitable operational principle, possible form and function elements and an appropriate manufacturing technology. For this decision, we also have to take into consideration the technological equipment available at the involved production sites.

- The technical design itself as well as modeling and simulations are carried out during the subsequent *realization phase*. Here, the total system is initially subdivided into separate function units, the so-called components. Typical components are electromechanical transducers (MEMS), analog and digital electronics as well as biochemical or microoptical components. Electromechanical components, in turn, are subdivided into function and form elements, e.g. into mechanical spring-mass-systems, capacitive position detectors, electrostatic drives and fluidic dampers.

The first modeling step describes the function elements and their interaction using simple analytical relations or graphic representations. Such low-level behavioral models are usually analyzed in mathematical calculation programs such as Mathcad[3] or Matlab/Simulink[4], but also by using network simulators such as PSPICE[5]. The goal of the low-level behavioral simulations is an approximate calculation of physical parameters and geometric dimensions that will fulfill component and system functions. Typical parameters are spring stiffness and mass parameters for mechanical function elements, capacitance-displacement characteristic for electrostatic transducers and damping constants or flow resistance for fluidic form elements.

During the next – and often difficult – step, the form elements that have become first choice during the concept phase are dimensioned and optimized in order to compare simulation results to the requirements. As a tool, we can use model libraries with commonly used form elements. To a certain degree, they are integrated into commercial design tools such as CoventorWare[6], IntelliSuite[7] and MemsPro[8]. However, design companies or production facilities often have libraries that were specifically developed for MEMS and that are integrated into the low-level behavioral models as parametric black-box elements. Numeric simulations with varied design variables are used during an iterative process to define suitable form elements and geometrical dimensions that correspond to the required physical parameters such as stiffness, mass and eigenfrequency. In general, model libraries are based on simplified analytical relations, which were presented in Chapter 6 for mechanical, fluidic and thermal function elements.

Figure 8.3 shows a low-level behavioral model in PSPICE, using the example of a torsional micro-mirror. The model represents the mirror plate as a rigid body with two degrees of freedom of motion in vertical direction u and a rotational movement around the spring bands ϕ that is used for light deflection in optical systems. PSPICE is a particularly suitable simulation environment if complex electronic circuits are coupled with electromechanical components. Via analogy relations, non-electric components are transformed into electrical elements and controlled sources or, as shown in Figure 8.3, are represented as signal flow charts (ABM Analog Behavioral Models). The model uses both blocks with

[3]www.mathcad.com
[4]www.mathworks.com
[5]www.pscice.com, www.orcad.com
[6]www.coventor.com
[7]www.intellisensesoftware.com
[8]www.memscap.com

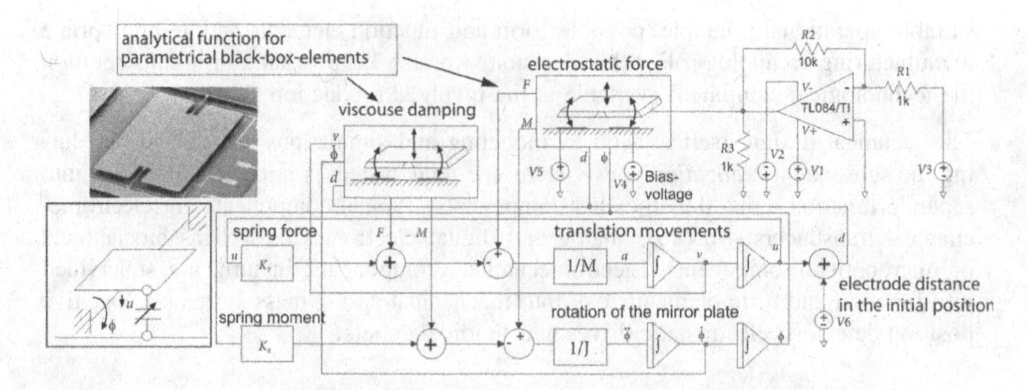

Figure 8.3 Low-level behavioral model of a torsional micro-mirror in PSPICE. Reproduced by permission from the Center for Microtechnologies, Chemnitz University of Technology, Germany

constant amplification factors (spring, mass) and parametric black-box elements with mathematical functions for describing the fluidic damping and the electrostatic force effect.

For a given set of physical parameters, there are often several different form elements that completely fulfill all requirements. Even the dimensioning of spring-mass systems produces ambiguous solutions. This is the reason why further information has be included into the selection and dimensioning of form elements. Such additional information often includes restrictions or limitations due to the manufacturing process. They are defined as design rules for the corresponding technology. Typical specifications in design-rule catalogues are minimum bridge width for spring elements, maximum structural dimensions of springs and mass bodies as well as realizable trench width and etching depth. The required chip area is another important criterion.

Mask layout is often automatically defined based on the model libraries of the form elements. Certain microtechnologies require manual corrections or additional specifications, though. The type of edge compensation, for instance, for wet-chemically etched bulk microstructures or the etch edge angle and width of mask undercutting for deep etching (DRIE) processes has to be interactively set. Now, the rough design, also called top-down phase, is completed and the developed component corresponds to the requirement profile. Due to the model simplifications and the corresponding risk of design errors, a second phase becomes necessary during which component behavior is tested. This phase is called the refined design or bottom-up phase (Figure 8.2).

The second phase starts with numerical simulations of technological process steps during which mask layout and process description are used to deduce the exact form of the microstructure. Etching simulations in SIMODE[9] or ANISE[10], for instance, can be used to exactly calculate anisotropic wet-chemical etching processes in bulk micromachining (Figure 8.4). Dry-etching processes with their specific characteristics such as undercutting of the mask or inclined etching edges are supported by CoventorWare and MemsPro. Process simulations provide exact three-dimensional volume models of the microstructure which can be automatically integrated via the corresponding interfaces (IGES, SAT, I-DEAS) into field calculation programs in order to numerically analyze the physical

[9]www.tu-chemnitz.de/etit/wetel/ausruestung
[10]www.intellisensesoftware.com/Anise.html

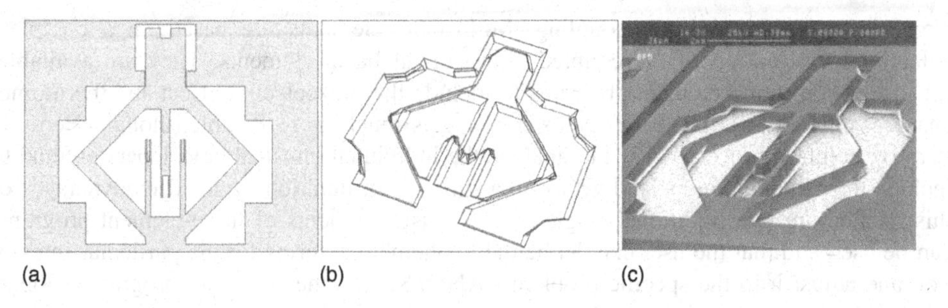

Figure 8.4 Simulation of anisotropic wet-chemical etching of silicon in SIMODE: (a) etch mask; (b) etching simulation; (c) etched structure. Reproduced by permission from the Center for Microtechnologies, Chemnitz University of Technology, Germany

behavior of the components. In addition, process simulation is used to verify internal stress in layer stacks or also the degree of coverage of metal electrodes at critical places, such as corners and edges.

The mechanical, electrostatic, thermal and fluidic behavior of the components is simulated using numeric programs based on finite element (FEM) or boundary element methods (BEM). The basic idea of both methods is that the volume (FEM) or the area (BEM) of the model to be calculated can be split into basic elements that can be simply described (hexahedron, tetrahedron or quadrangular elements). These basic elements are assigned material properties, boundary conditions and loads in order to be able to represent their behavior in mathematical equations. Similar to a network simulator, these equations are assembled and combined into a total system description. Solving the system equations results in an approximate solution for the physical parameters of the total system which – with an increasing number of basic elements – converges towards the exact value [BATHE02]. FEM-based field calculation programs such as ANSYS[11], ABAQUS[12] or NASTRAN[13] are standard tools of the computer-assisted design of technical components and are successfully used in a wide range of engineering disciplines [MÜLLER99].

Field calculation programs are preferably used for single-domain problems. Currently, it is thus possible to extract certain parameters of low-level behavioral models with a higher precision than can be achieved using analytical considerations. Force-displacement functions determined with FEM even take into account the compliance of the suspension at the anchor points, for instance, or nonlinearities during large displacements. In electrostatics, capacitance-displacement functions can thus be represented including the stray field, and for fluidic gap flows, it is possible to determine the damping or spring effect of the surrounding air. This way, the original low-level behavioral models are gradually refined and integrated into the macro-model for system simulation. The automated support for creating high-precision macro-models from FEM data is the subject of current research activities and is – to a certain basic degree – included in ANSYS and CoventorWare. The here applied methodology is called order reduction or macro-modeling (see Section 8.3.3).

Modern field calculation programs are also capable of analyzing coupled-domain problems. The interaction of individual physical fields are thereby included either as multi-field

[11]www.ansys.com
[12]www.hks.com
[13]www.mscsoftware.com

elements or by load vector coupling. Multi-field elements are based on a complete behavioral description of the coupled issue for all basic elements. They are available, for instance, in finite-element programs to include the piezoelectric effect and thermomechanical interactions. The load-vector coupling is much more flexible, though. Here, an iterative solution process is used to analyze the individual physical field separately und to include their interaction as load vector in a subsequent iteration cycle. The advantages of this method are that traditional single-domain basic elements of finite-element programs can be used and that the user can define the mathematical terms for the particular interaction and adjust it to the specific problems. ANSYS and other software programs support this procedure by a programmable multi-field solving tool.

Coupled field calculations are indispensable for designing sensors and actuators. Due to the currently required computing time, it can only be used for specific analyses of critical behavioral states, though. Coupled field calculations in the time and frequency range require several hours to days of computing time, equation systems with 10 000 to 100 000 degrees of freedom have to be solved for each iteration. For a system simulation that couples micromechanical component models and electric circuit models, such large computing times are not acceptable. In principle, field calculation programs can be coupled with electronics simulations [MEHNER00]. However, refining low-level models by extracted finite-element parameters and functions into macro-models is much more efficient and, for component and system simulations, sufficiently precise (Figure 8.5).

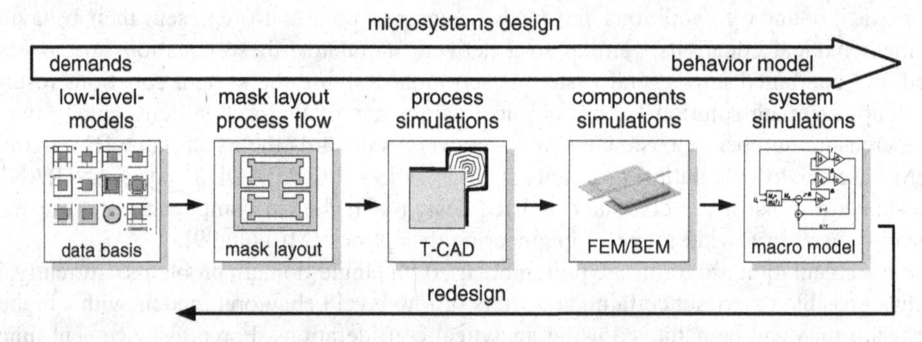

Figure 8.5 Design process in microsystem technology

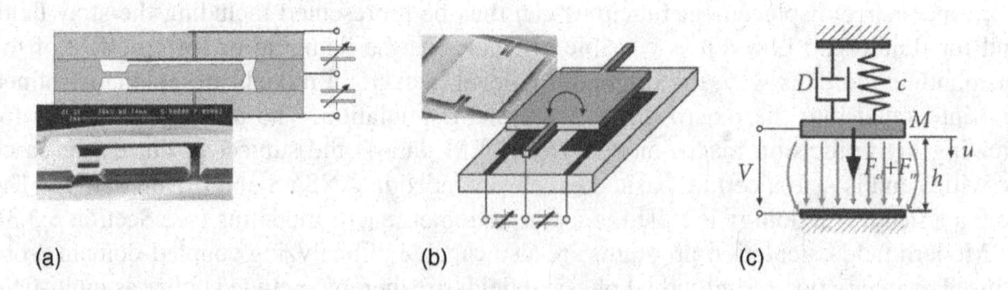

Figure 8.6 Sensors and actuators as systems with one degree of freedom: (a) capacitive sensor; (b) electrostatic actuator; (c) behavioral model of a capacitively coupled spring-mass oscillator. Reproduced by permission from the Center for Microtechnologies, Chemnitz University of Technology, Germany

8.2 SYSTEMS WITH LUMPED PARAMETERS

8.2.1 Behavioral Description of Electromechanical Systems

In microsystem technology, most electromechanical transducers can be approximately represented by a spring-mass-damper system with one degree of freedom of movement and one or several adjacent field spaces. The degree of freedom of movement describes the displacement of a reference point (e.g. the gravity center of a mass) in a load situation. The displacement is usually translatory or rotational; it can, however, also be curvilinear. In general, we therefore use the term generalized coordinate for displacement u. Figure 8.6 shows typical microstructures and the corresponding behavioral models.

The behavior of the electromechanical systems in Figure 8.6c is unambiguously defined by the force-displacement relation of the degree of freedom u and by the current-voltage relation at the electrodes.

The relation between force and displacement is

$$M\ddot{u} + D\dot{u} + cu = F_{\mathrm{el}} + F_{\mathrm{m}} \text{ with } F_{\mathrm{el}} = \sum_k \frac{V_{ij}^2}{2} \frac{\partial C_{ij}(u)}{\partial u}, \tag{8.1}$$

where M is the mass, D the damping constant, c the stiffness[14], F_{el} the acting force of the electrostatic fields and F_{m} the external mechanical forces. The difference of electrode voltages U_{ij} and their position-related capacitance functions C_{ij} specify the electrostatic forces.

The relation of current and voltage for the i-th electrode is defined as follows:

$$I_i = \frac{dQ_i}{dt} = \sum_r \dot{C}_{ij}(u)V_{ij} + C_{ij}(u)\dot{V}_{ij} \text{ with } \dot{C}_{ij}(u) = \frac{\partial C_{ij}(u)}{\partial u}\frac{\partial u}{\partial t}. \tag{8.2}$$

As opposed to the force calculation in Equation (8.1), Equation (8.2) does not take into account all capacitances, but only those that are directly related to the i-th electrode. The actual coupling of mechanical and electrostatic subsystems is described by capacitance-displacement functions $C_{ij}(u)$. These functions are determined analytically (e.g. by conformal mapping) or based on the results of finite element simulations.

8.2.2 Analysis of the Static Behavior of Electromechanical Systems

When simulating electromechanical behavior, three simulation types can be distinguished:

- static analysis for calculating the operating point,
- harmonic analysis for calculating the transfer functions,
- transient analysis for representing variations in time.

[14]In mechanics, spring stiffness is denominated with K. In order to conform with the other sections in this book, we use the symbol c.

In addition, the modal analysis for calculating eigenfrequencies and mode shapes has a certain special position. It is, however, not supported by electronics and system simulators and can therefore only be applied with specific mathematical tools as well as with finite element programs.

Static simulations are required for calculating the displacement state of microactuators when voltages are applied, for instance. In general, the displacement-dependent electrostatic forces require nonlinear equations that have to be solved iteratively. As a start, we want to describe this procedure using the model of a plate capacitor.

The equilibrium relation between mechanical and electrostatic forces is defined as

$$cu = F_{\mathrm{el}} = \frac{V^2}{2}\frac{\partial C(u)}{\partial u} = \frac{V^2 \varepsilon A}{2(h-u)^2}, \tag{8.3}$$

where ε is the permittivity, A the capacitor's plate area and h the electrode distance in the initial position. Formulating function $u = f(V)$ is not trivial, as there can be up to three states of equilibrium for a given voltage V. However, conversely it is possible to unambiguously determine the voltage required for a given displacement:

$$V = \sqrt{\frac{cu2(h-u)^2}{\varepsilon A}}. \tag{8.4}$$

Component behavior can be illustrated and interpreted by the force-displacement characteristic. Figure 8.7a shows electrostatic force functions for three different electrode voltages. For the lowest voltage (continuous line), there are three states of equilibrium in relation to the restoring force of the spring. There is a cross-section behind the electrode which – from a designing perspective – cannot be reached. A second point describes an unsteady state that is directly left if there are small disturbances (e.g. due to vibrations). The actual (stable) operating point is situated in the left part of the curve and describes the actual displacement of the actuator.

As was to be expected, a higher electrode voltage results in a larger displacement. Here, the stable operating point moves to the right and the unstable state to the left. For

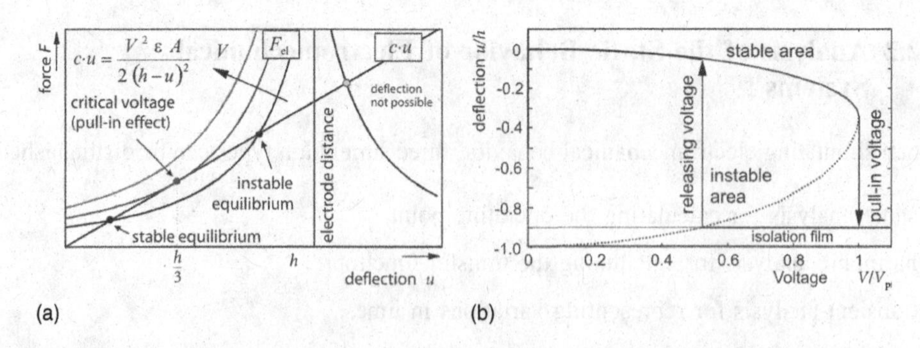

Figure 8.7 Static equilibrium of forces for electrostatic actuators: (a) force-displacement and (b) displacement-voltage relation

critical voltage V_{PI}, both operating points merge and the system becomes unstable. A further voltage increase results in an abrupt impact of the mass on the insulation layer or on a spacer (pull-in effect). In addition to the pull-in voltage, the release voltage is an important parameter. It describes the value to which the voltage has to decrease in order for the mass to become detached from the bottom electrode and to return to the stable operating range. Both parameters are particularly important for characterizing microrelais and microswitches.

The maximum displacement of a plate capacitor in the electrostatic field is limited to one-third of electrode distance h. This fact can be analytically verified by computing the zeros of the first derivation of the voltage-displacement function in Equation (8.4). It applies, however, only for translatory electrode movements in vertical direction and for constant stiffness.

For the numerical calculation of the static operating point, we commonly use the relaxation and the NEWTON-RAPHSON method. The relaxation procedure initially determines the electrostatic force for the current position. Based on this force and the mechanical stiffness, we can determine the displacement for the next iteration. The NEWTON-RAPHSON method computes the current error of the equilibrium of forces for each iteration and corrects it using a TAYLOR series expansion. The error is the difference between external (electrostatic forces) and internal forces (spring reaction) of the system. The relaxation procedure always calculates the actual (stable) operating point of the system. As opposed to this, the NEWTON-RAPHSON method converges towards the state of equilibrium that is closest to the initial value.

In order to linearize the voltage-displacement function in Figure 8.8, a positive and a negative polarization voltage V_P was applied to the two fixed electrodes. The figure shows, that a linear behavior is only guaranteed within ca. $\pm 10\,\%$ of the electrode distance.

Due to external loads (acceleration, pressure), for micromechanical sensors the static displacement of the mass is often determined by measuring the recharging currents. This requires sinusoidal polarization voltages for a permanent charge reversal of the capacitors. For small movements, the difference of the charging currents is proportional to the displacement. Bridge circuits or operational amplifiers are used to transform the difference in current into voltage signals. Figure 8.9 presents the corresponding behavioral model and the applied analytical relations.

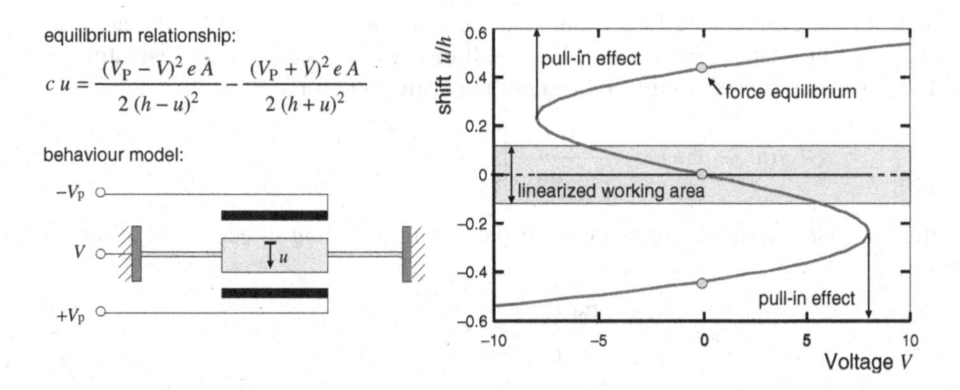

Figure 8.8 Voltage-displacement function of a microactuator

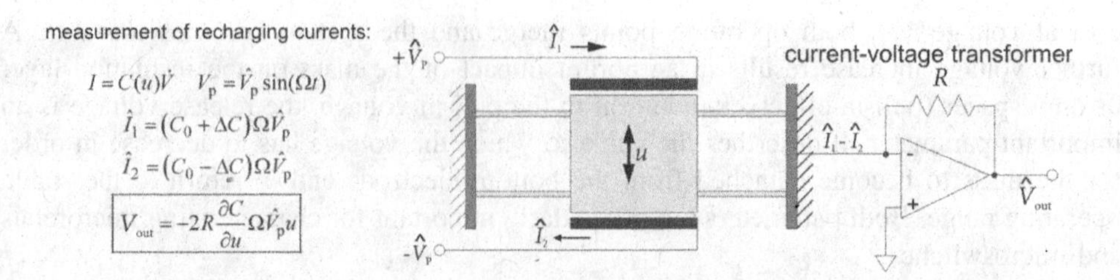

Figure 8.9 Low-level behavioral model of a capacitive sensor

8.2.3 Analysis of Electromechanical Systems for Harmonic Loads

Harmonic analyses are particularly important for sensors and actuators as, on the one hand, many microstructures are driven by sinusoidal loads and, on the other hand, amplitude and phase variations can be used to illustrate system behavior. By analogy to the AC analysis, complex equations are used to formulate the equilibrium conditions of electromechanical systems. For excitation frequency Ω, the balance of forces of the mechanical subsystems is given as

$$\left[-\Omega^2 M + j\Omega D + (c + c_{el})\right] \vec{u} = \vec{F}, \tag{8.5}$$

with $\vec{u}$ and $\vec{F}$ being complex numbers, that are used to describe the amplitude and phase of the displacement and the external force. In addition, we introduce a new term, electrostatic stiffness c_{el}, for harmonic analyses.

Electrostatic fields change both stiffness and resonance frequency of the system. Physically, the change in stiffness is caused by displacement-related electrostatic forces that – similar to displacement-related spring force – have to be assigned to the stiffness term in Equation (8.5). From a numeric perspective, an explicit formulation of c_{el} is not necessarily required for static and transient calculations, as it is possible to determine the equilibrium of forces for both analysis types by iteration. Neglecting c_{el} only affects the converging velocity, but not precision. For harmonic analyses, the result will be incorrect if this term is missing. That is why several software programs do not support this type of analysis (see Section 8.3.2).

The value of c_{el} is calculated at the operating point of the electromechanical system. Similar to the electrostatic force, this additional stiffness is affected by the applied DC voltage and the static displacement. For small displacements around the operating point, a TAYLOR series can be used to represent the electrostatic force:

$$F_{el}(u_= + \Delta u) = \frac{V_=^2}{2} \frac{\partial C(u)}{\partial u}\bigg|_{u_=} + \frac{V_=^2}{2} \frac{\partial^2 C(u)}{\partial u^2}\bigg|_{u_=} \Delta u + \cdots. \tag{8.6}$$

[15]

Here, the first derivation corresponds to the spring softening effect of the electrostatic field:

$$c_{el} = \frac{V_=^2}{2} \frac{\partial^2 C(u)}{\partial u^2}\bigg|_{u_=}. \tag{8.7}$$

[15] The index = refers to static quantities (DC or bias) at the operating point. The index ~ refers to sinusoidal quantities (AC, amplitude and phase) which are superimposed to their static counterpart.

Other important terms for behavioral modeling are amplitude and phase of the electrostatic force of sinusoidal voltages. In general, the resulting force can be split into a constant force component, a sinusoidal force with excitation frequency and a force portion with double excitation frequency. Here, it applies that

$$F_{\mathrm{el}} = \frac{(V_= + V_\sim)^2}{2}\frac{\partial C(u)}{\partial u} = \left[\frac{V_=^2}{2} + V_=\hat{V}_\sim \sin(\Omega t) + \frac{\hat{V}_\sim^2}{4}(1 - \cos(2\Omega t))\right]\frac{\partial C(u)}{\partial u}. \quad (8.8)$$

Mathematical algorithms for harmonic analyses only take into account the constant component and the portion of the force that coincides with the excitation frequency. For this reason, a harmonic analysis is only feasible if the applied polarization voltage $V_=$ is substantially larger than the superposed alternating voltage $V_\sim$. Under this condition, the portion of the force with double the excitation frequency becomes negligible.

Harmonic analysis requires a two-step solution. During the first step, the operating point of the electromechanical system is determined for the applied polarization voltages. The results are static displacement $u_=$ and electrostatic stiffness c_{el}. During the second step, the complex motion equation

$$\left[-\Omega^2 M + j\Omega D + \left(c - \frac{V_=^2}{2}\frac{\partial^2 C(u)}{\partial u^2}\bigg|_{u_=}\right)\right]\vec{u} = V_=\vec{V}_\sim \frac{\partial C(u)}{\partial u} \quad (8.9)$$

is solved for different excitation frequencies Ω. The graphic representation of the transfer function usually requires a large number of calculation points for the subsequent interpolation of the solutions in a frequency range.

Polarization voltage $V_=$ affects the oscillation amplitudes of electromechanical systems in two ways. In each case, the electrostatic force amplitude is amplified in proportion to the applied DC voltage, according to the right-hand term in Equation (8.9). In the case that the polarization voltage has a similar magnitude as the pull-in voltage, the electrostatic stiffness becomes effective and further increases the compliance of the system. A typical indication is the shift of the resonance to the left (Figure 8.10a). This effect is often used for calibrating microsystems.

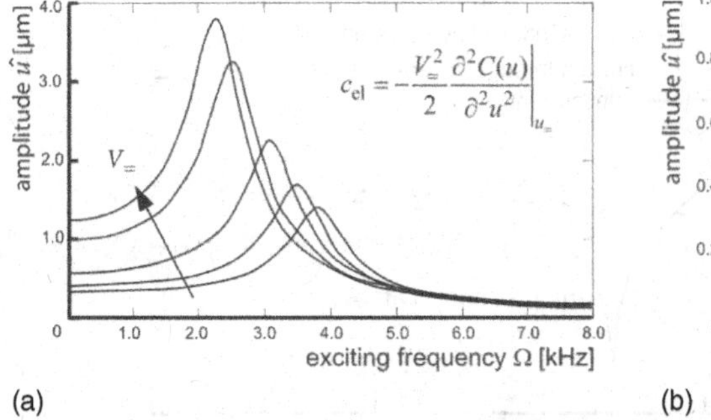

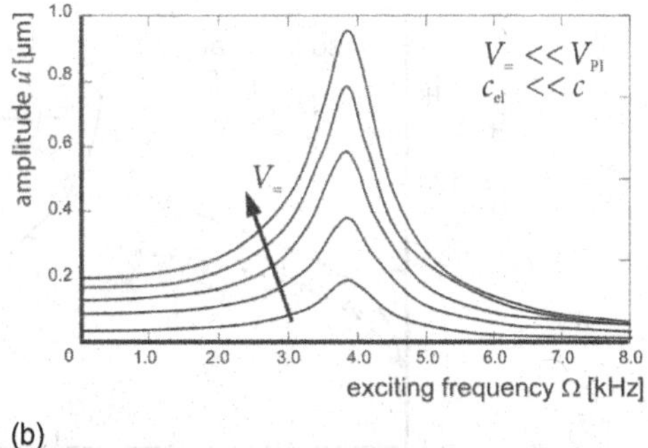

Figure 8.10 Transfer function of an actuator for different direct voltages: (a) influence on the electrostatic stiffness for a large polarization voltage; (b) the polarization voltage is substantially lower than the pull-in voltage

During a further computing step, the corresponding current can be determined based on the motion amplitude and the phase angle:

$$\vec{I} = \left.\frac{\partial C(u)}{\partial u}\right|_{u=} j\Omega\vec{u}V_= + C(u_=)j\Omega\vec{V}_\sim. \quad (8.10)$$

For an AC voltage amplitude of 1 V, Equation (8.10) directly results in the admittance of the electromechanical system.

Figure 8.11 shows a typical current function of a weakly damped electromechanical system. The curves can be used to read off two characteristic values, i.e. the eigenfrequency of the original mechanical system at 3.8 kHz and the resonance frequency shifted by the electrostatic field at 2.5 kHz. The absorption of the current in the eigenfrequency results from the extinction of the motion-induced current portion (first term in Equation (8.10)) by a portion with a phase shift of 180°, that is caused by the voltage change (second term in Equation (8.10)). This effect is, however, only available for an ideal plate capacitor with a vertical movement of the mass.

The sequential computing of the mechanical displacement and the electric current is only possible for voltage-controlled sensors and actuators. If the current is given or if the transducer is assigned as a component to an electric circuit, the voltage across the capacitor is initially not known. These systems require a simultaneous solution of both equations. The coupled formulation of electromechanical interactions results in

$$\begin{bmatrix} -\Omega^2 M + j\Omega D + c - \left.\dfrac{V_=^2}{2}\dfrac{\partial C(u)}{\partial u}\right|_{u=} & -V_= \left.\dfrac{\partial C(u)}{\partial_u}\right|_{u=} \\ j\Omega V_= \left.\dfrac{\partial C(u)}{\partial u}\right|_{u=} & j\Omega C(u_=) \end{bmatrix} \begin{bmatrix} \vec{u} \\ \vec{V}_\sim \end{bmatrix} = \begin{bmatrix} \vec{F}_\mathrm{m} \\ \vec{I} \end{bmatrix}, \quad (8.11)$$

where $\vec{F}_\mathrm{m}$ represents external mechanical forces and $\vec{I}$ the currents fed into the system. Equation (8.11) is compatible with KIRCHHOFF's description of electric circuits and is analogously used for the incorporation of electromechanical sensors and actuators into network models.

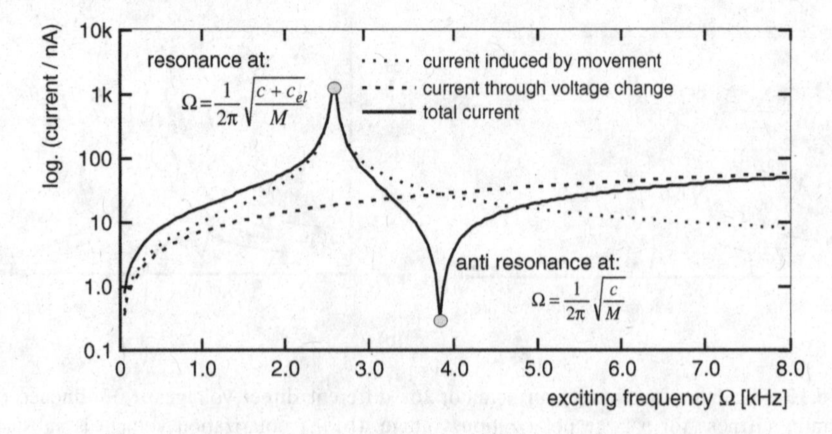

Figure 8.11 Frequency variation of the input current of an electrostatic actuator

8.2.4 Transient Analysis of Electromechanical Systems

Transient simulations of time behavior are mainly used for analyzing the settling time of sensors and actuators and for transducers with nonharmonic load functions. A typical example is the saw tooth deflection function of micromechanical torsional mirrors for image projection, where distortions and dynamic overshoot have to be reduced to a minimum.

Nonlinearity in electromechanical systems can be observed in the motion response at harmonic or discontinuous load functions. In Figure 8.12a, we have determined the displacement and the velocity of a mass body for a sinusoidal voltage excitation. As expected, the displacement of the structure occurs with exactly twice the frequency of the excitation signal. Note that the amplitudes in direction of the excitation electrode are larger than the amplitudes in opposite direction. The right-hand side of the figure shows the displacement function for a voltage pulse series. Here, an uncommon feature is the different cycle time of the oscillator for an applied voltage, in comparison to a zero voltage potential.

The equilibrium relation of the dynamic system consists of a tangential stiffness matrix (JAKOBIan matrix) and the residual part on the right-hand side

$$\begin{bmatrix} c - \dfrac{V_i^2}{2}\left.\dfrac{\partial^2 C(u)}{\partial u^2}\right|_{u_i} & -\left.\dfrac{\partial C(u)}{\partial u}\right|_{u_i} V_i \\ \left.\dfrac{\partial^2 C(u)}{\partial u^2}\right|_{u_i} \dot{u}_i V_i + \left.\dfrac{\partial C(u)}{\partial u}\right|_{u_i} \dot{V}_i & \left.\dfrac{\partial C(u)}{\partial u}\right|_{u_i} \dot{u}_i \end{bmatrix} \begin{bmatrix} \Delta u \\ \Delta V \end{bmatrix}$$

$$= \begin{bmatrix} F \\ I \end{bmatrix} - \begin{bmatrix} M\ddot{u}_i + D\dot{u}_i + cu_i - \dfrac{V_i^2}{2}\left.\dfrac{\partial C(u)}{\partial u}\right|_{u_i} \\ \left.\dfrac{\partial C(u)}{\partial u}\right|_{u_i} \dot{u}_i V_i + C(u_i)\dot{V}_i \end{bmatrix}, \quad (8.12)$$

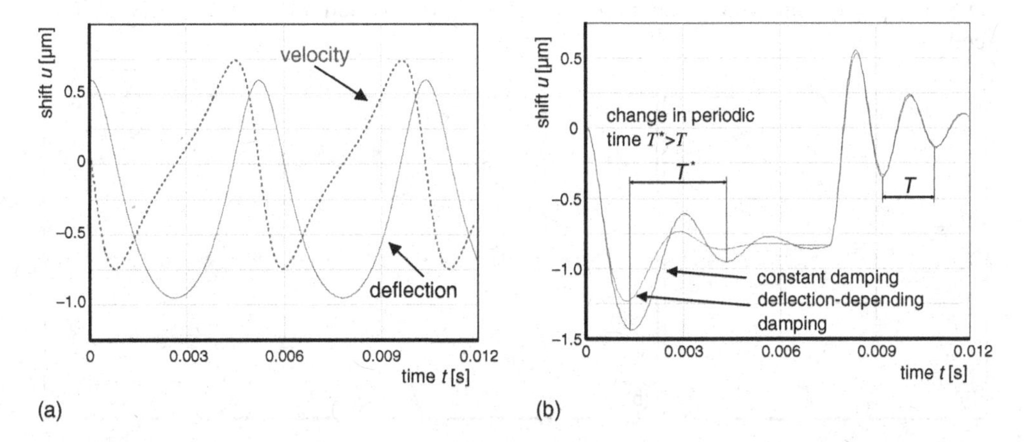

Figure 8.12 Transient motion behavior of electrostatic actuators: (a) sinusoidal voltage function; (b) rectangular voltage pulse series

that has been complemented, in relation to Section 8.2.1, by dynamic forces. For comprehensive explanations of the integration algorithms, see [BATHE02].

In addition to the direct implementation of the system equations and the solution procedures into mathematically oriented simulation programs, we often use commercial tools for model input and simulation of transient calculations. Behavioral models are then represented either by signal flow charts or by conservative KIRCHHOFF networks. The difference between the two models mainly consists in the way signals are exchanged between the individual model elements (blocks). Communication between blocks takes place through so-called terminals that are connected by lines or by a bus (combination of several signal lines). A characteristic feature of the signal flow system is that each line communicates only one signal, e.g. the displacement or the velocity of the mass body or the electric voltage, respectively, or the current of the electric subsystems. The signal is communicated in one direction (unidirectional). This direction is determined when terminals are assigned their function as input or output. Signals can be branched out from one input to several outputs. They can be only linked via special connecting elements, though. Signal flow charts are supported by Matlab/Simulink and some circuit and system simulators (SPECTRE, PSPICE).

Figure 8.13 shows a modeling example of electrostatic-mechanical transducers. In the presented case, the signal flow chart describes defined signal propagation from the input terminal (external mechanical force, voltage) to the output parameters (displacement, current) of the system. A change of the signal propagation, e.g. for current-driven actuators, requires a new conception of the model. Figure 8.13b illustrates the interaction of the signals and the processing in the functional blocks. In general, the applied blocks have three functions: They amplify the signal by a constant factor (mass, stiffness, damper), they process the input signals according to a mathematical function (capacity, derivation of the capacity, differentiation, integration) or they link several signals to an output signal (summation, product).

Figure 8.13a presents the higher-order black-box model for the behavioral analysis. In order to demonstrate the functionality, signal sources are connected to the inputs and the graphic monitors for the output signals. Electrical components that cannot, as usual, be directly connected between the terminals may constitute a problem as the signal flow charts do not apply current-voltage relations. Electric resistors therefore require an additional feedback branch.

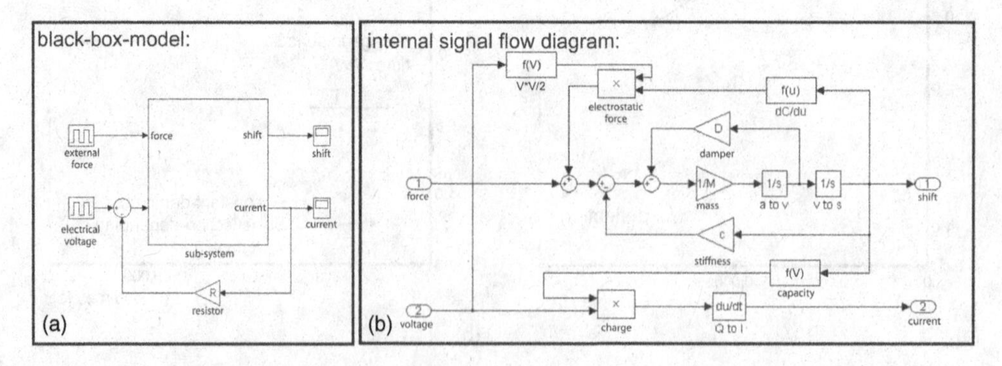

Figure 8.13 Signal flow chart of an electromechanical system in Matlab/Simulink: (a) system model of a capacitively coupled spring-mass oscillator; (b) internal signal flow chart

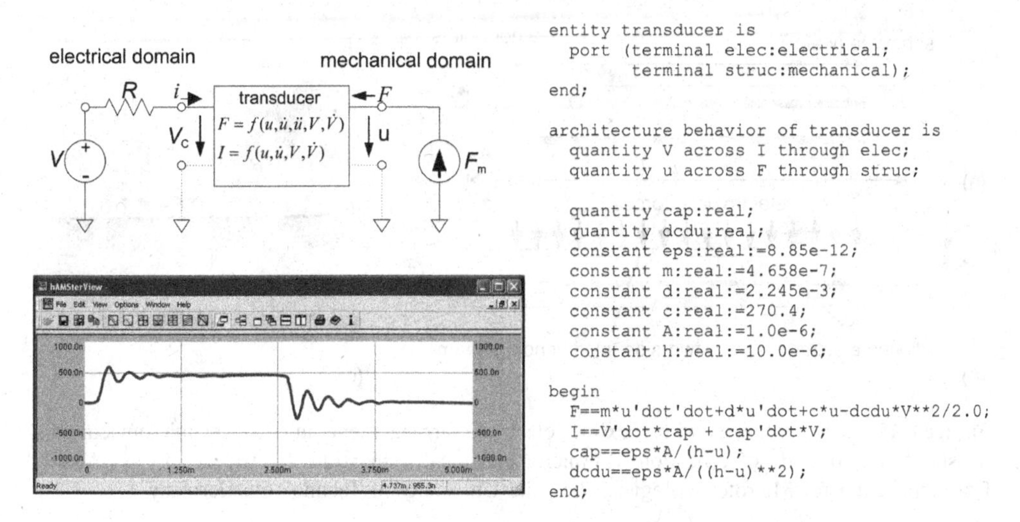

electrical domain
mechanical domain
transducer
R
i
F
V
V_C
u
F_m
F = f(u,u̇,ü,V,V̇)
I = f(u,u̇,V,V̇)
entity transducer is
port (terminal elec:electrical;
terminal struc:mechanical);
end;
architecture behavior of transducer is
quantity V across I through elec;
quantity u across F through struc;
quantity cap:real;
quantity dcdu:real;
constant eps:real:=8.85e-12;
constant m:real:=4.658e-7;
constant d:real:=2.245e-3;
constant c:real:=270.4;
constant A:real:=1.0e-6;
constant h:real:=10.0e-6;
begin
F==m*u'dot'dot+d*u'dot+c*u-dcdu*V**2/2.0;
I==V'dot*cap + cap'dot*V;
cap==eps*A/(h-u);
dcdu==eps*A/((h-u)**2);
end;

Figure 8.14 VHDL-AMS behavioral model of an electromechanical transducer

Alternatively, it is possible to model the electromechanical system using conservative KIRCHHOFF networks (Figure 8.14). Here, the transducer is defined by a block with four terminals that, on the left-hand side, describe the electrical parameters (current and voltage) and, on the right-hand side, the mechanical parameters (force and displacement). Similar to an electrical component, all lines transmit two signals, flow and effort. For electrical terminals, the flow variable is the current entering the system; for mechanical terminals, it is the external force in reference direction. The effort describes the difference in or decrease of the voltage between the terminals of electrical components. In mechanics, potential parameters are displacement and rotation. Both parameters can be set – similar to electric potentials with voltage sources – by external forces and torques via current sources. In a KIRCHHOFF network, the terminals of all blocks are always bi-directional.

Modern analog behavioral simulators support a large number of mathematical operations which can be used to represent specific properties such as nonlinearities and discontinuities. The command IF-USE, for instance, can be used to select model configurations or to change the properties of a model.

8.3 SYSTEMS WITH DISTRIBUTED PARAMETERS

8.3.1 Behavioral Description Based on Analytical Models

Numerous applications use electromechanical components with deformable form elements and electrode areas. For such microstructures, it is more complicated to select a characteristic degree of freedom of motion and to represent the system as a spring-mass-damper model than for rigid-body elements. The design approach is similar, though, and will be illustrated in the following.

Figure 8.15 shows a double suspended microbeam that is displaced towards the wafer surface by electrostatic forces. Depending on the applied voltage, the microbeam deforms into a bowl-shaped bending line with different amplitude. Even here, it occurs a pull-in

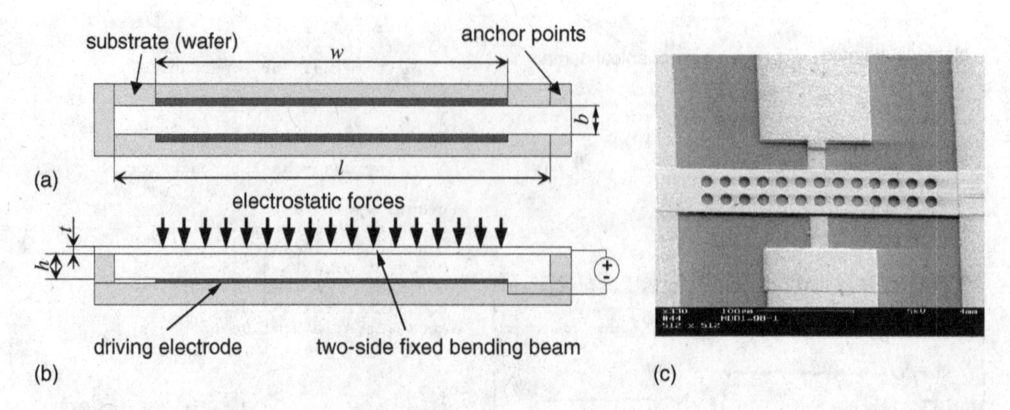

Figure 8.15 Microactuator with flexible electrode arrangement in surface micromachining: (a) structure; (b) mode of operation; (c) micromechanical realization. Reproduced by permission from the Center for Microtechnologies, Chemnitz University of Technology, Germany

effect after a critical voltage is passed, i.e. the flexible structure clings to the electrically insulated bottom electrode. Microactuators in similar structures are used as phase-shifters for high-frequency technology, as microswitches or for optical applications (adaptive diffraction grating).

For most flexible microstructures, the state of displacement for different load situations can be considered to be the scaling of one or several characteristic motion states (Figure 8.16). This property of mechanical form elements is used for the analytical and numeric behavioral description, i.e. the motion state is considered to be a superposition of so-called form functions. The scaling factors of the individual form function (which are called weights) are then used to unambiguously describe the actual displacement.

The following approach is used to define the approximate displacement u in each point of the structure

$$u(x, y, z, t) \approx \sum_{i=1}^{m} \phi_i(x, y, z)q_i(t) \tag{8.13}$$

where ϕ_i represents the form function and q_i its weights. In our example, one form function ϕ_i is adequate to describe the microsystem with sufficient accuracy. Suitable form functions are the lowest eigenvectors of the mechanical structure (see Section 8.3.3)

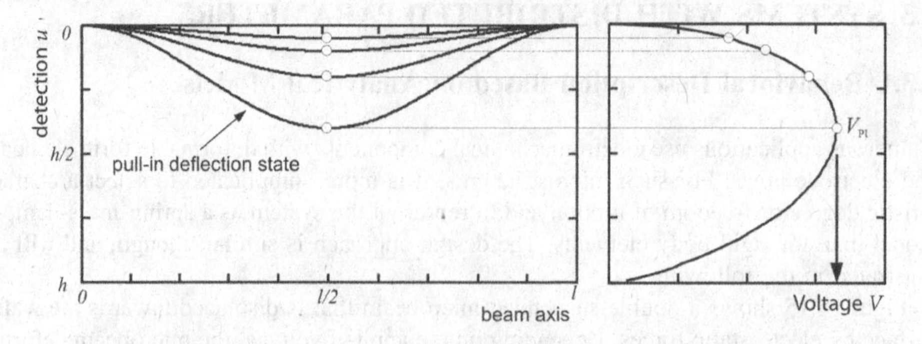

Figure 8.16 Bending lines of microbeams for different electrode voltages

or the static displacement for characteristic test loads. For this microbeam, an analytical solution of the bending line for both lumped and constant area load provides a form function that can be used for the behavioral description (Figure 8.17).

Similar to Section 8.2.1, energy and capacitance functions are necessary for an analytical description of electrostatic-mechanical systems.

For the mentioned cantilever beam, these functions can be analytically calculated (see Example 8.3). It applies that:

- potential energy of the beam (only bending):

$$E_\mathrm{B}(q_1) = \frac{EI}{2}\int_0^l u''(x)^2 dx = EI\int_0^{l/2}\left(q_1\frac{\partial^2\phi_1(x)}{\partial x^2}\right)^2 dx \tag{8.14}$$

- kinetic energy:

$$E_\mathrm{K}(\dot{q}_1) = \frac{\rho A}{2}\int_0^l \dot{u}(x)^2 dx = \rho A\int_0^{l/2}(\dot{q}_1\phi_1(x))^2 dx \tag{8.15}$$

- capacitance:

$$C(q_1) = \varepsilon b\int_0^l \frac{1}{h+u(x)}dx = 2\varepsilon b\int_0^{l/2}\frac{1}{h+q_1\phi_1(x)}dx, \tag{8.16}$$

where E is YOUNG's modulus, I the moment of inertia, ρ the density, A the cross-sectional area, ε the permittivity, b the width and l the length of the beam as well as h the electrode gap.

The corresponding generalized parameters such as mass, damping, stiffness and electrostatic force can be directly determined applying partial derivations of these functions according to the weight factors. In analogy to the already discussed discrete systems, we

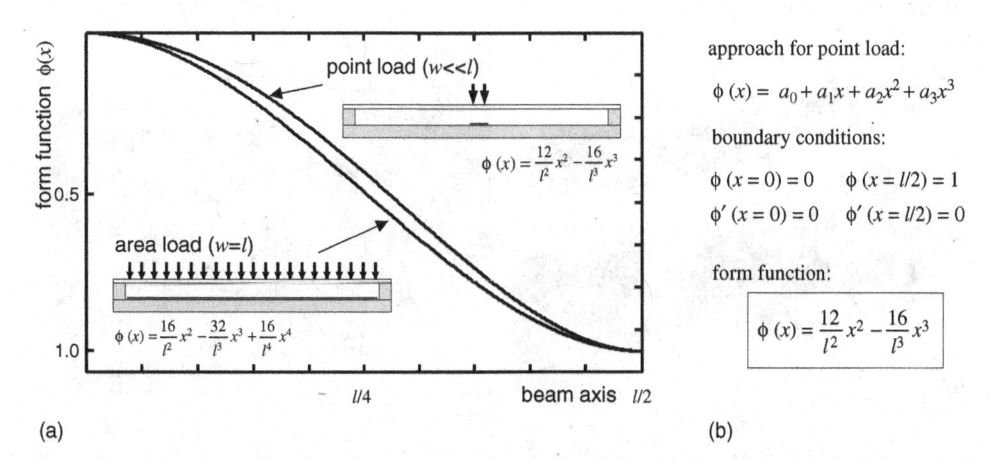

Figure 8.17 Form functions of a cantilever beam for different electrode lengths: (a) form functions Φ_i; (b) derivation of the form function for centrical lumped loads

Table 8.1 Calculation of characteristic parameters of flexible structures

Mechanical subsystem			
Stiffness	Mass	Damping	Electrostatic force
$c = \dfrac{\partial^2 E_{\mathrm{B}}(q_1)}{\partial q_1^2}$	$M = \dfrac{\partial^2 E_{\mathrm{K}}(\dot{q}_1)}{\partial \dot{q}_1^2}$	$D = 2\xi\sqrt{cM}$	$F_{\mathrm{el}}(q_1) = \dfrac{V^2}{2}\dfrac{\partial C(q_1)}{\partial q_1}$
(8.18)	(8.19)	(8.20)	(8.21)
$M\ddot{q}_1 + D\dot{q}_1 + cq_1 = F_{\mathrm{el}}(q_1)$ (8.22)			
Electrical subsystem			
$I = \dot{V}C(q_1) + \dfrac{\partial C(q_1)}{\partial q_1}\dot{q}_1\dot{V}$ (8.23)			

arrive at generalized coordinates of the relation force-displacement and current-voltage (Table 8.1).

The calculation procedures presented for rigid-body systems in Section 8.2.1 can also be used to determine the static and dynamic behavior. If we use several form functions or electrodes for representing the flexible electromechanical system, Equations (8.22) and (8.23) result in an ordinary differential equation system.

8.3.2 Numerical Methods Based on the Finite Element Method

Numerical methods are indispensable for calculating complicated form elements and load situations. In the past years, numerous commercial finite-element (FE) program systems have been introduced. In addition to mechanical field problems they can also analyze interactions with electrostatics and the surrounding fluidics. Figure 8.18c shows the detailed FE model of the microactuator in Section 8.3.1. As opposed to analytical solutions, FEM correctly includes the elasticity of the anchor points, the effect of perforation and the clinging to the bottom electrode.

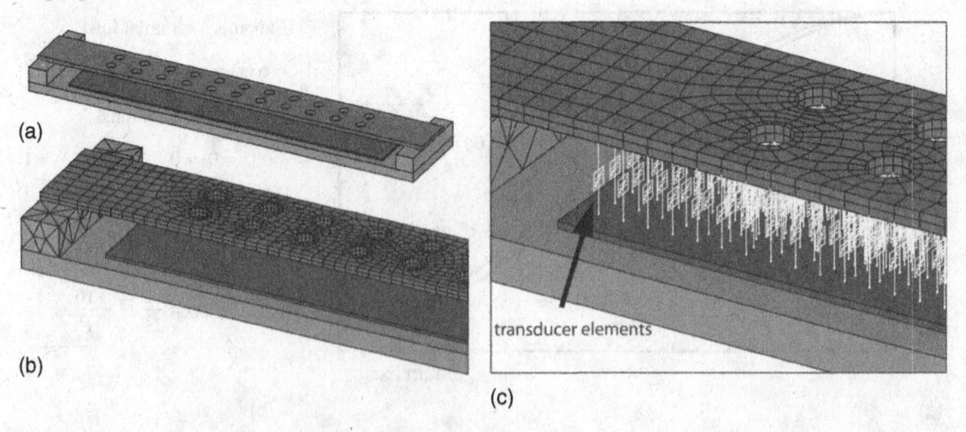

Figure 8.18 Finite element model of the microactuator in ANSYS/Multiphysics: (a) volume model; (b) finite element model from Figure 8.15c; (c) transducer element between the electrodes of the microsystem

FE systems either use simplifications and represent electromechanical interactions by quasi-analytical transducer elements or by a real three-dimensional cross-linking of the air space with electrostatic elements. Transducer elements [ANSYS04] are preferably used for modeling electrostatic fields in small gaps of flexible plates. The behavioral equations are formulated based on analytical relations that determine – from the local electrode distance and the electrode area assigned to the element – a capacitance function and calculate from that the electrostatic force, the change in stiffness and the electric current. As for each FE node, a transducer element is situated in the air gap of the mechanical model, it is possible to represent the varying electrostatic field distribution at tilted or flexible electrodes with high accuracy. In order to avoid singularities, the displacement of the corresponding nodes is restricted to a minimum gap. Due to this boundary condition, the elements can represent the actual clinging to a bottom electrode (Figure 8.19).

Transducer elements are available for all analysis types (static, harmonic, transient) and enable a high computing speed. 3D-elements constitute an alternative that can be used for modeling the electrostatic field. These elements have the advantage that the FE formulation of POISSON's equation allows an analysis of any field distribution, including stray fields. Disadvantages are the substantially larger computing effort due to the additional discretization of the air space (Figure 8.19) and the missing formulation of the electrostatic stiffness. Currently, many FE programs do not include a harmonic analysis for these elements [ANSYS04].

The electromechanical field problem is solved by a series of decoupled simulations during which the electrostatic and mechanical field problem is separately solved. The interactions are detected by so-called load vectors and are passed on to the other physical domain for subsequent equilibrium iterations (Figure 8.20).

8.3.3 Macro-modelling of Complex Systems by Order Reduction

Modern simulation methods use a combination of analytical and numerical procedures for describing the behavior of microsystems. The basic idea often is – similar to

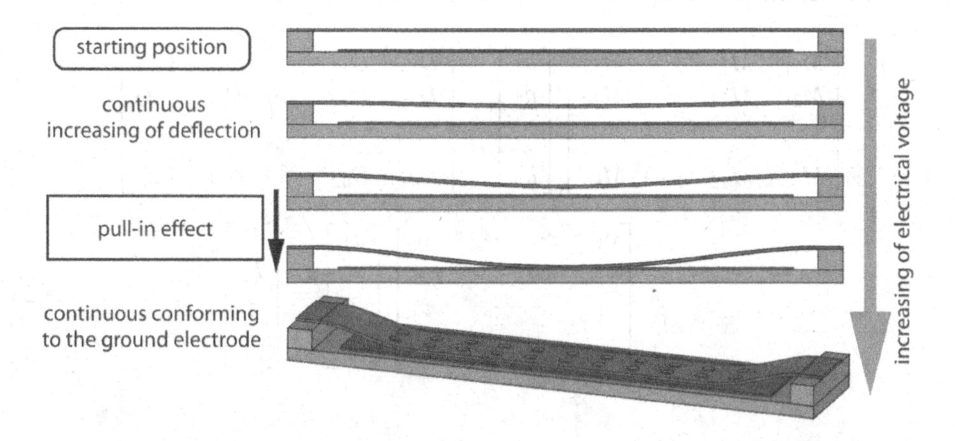

Figure 8.19 Displacement of microactuators for different electrode voltages

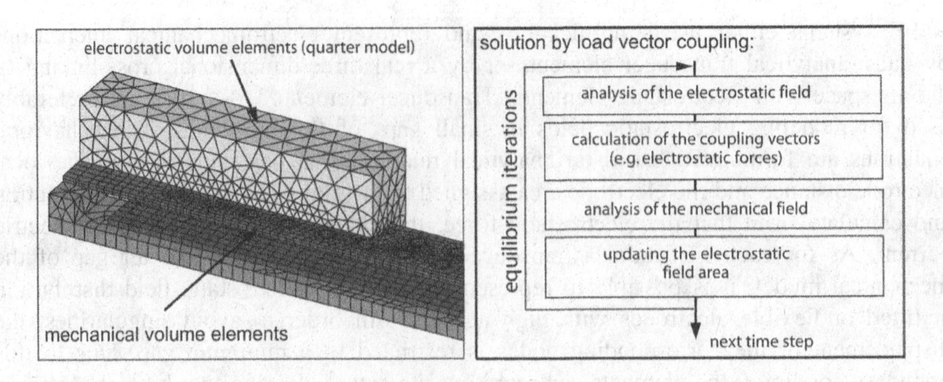

Figure 8.20 Simulation of electromechanical interactions using load vector coupling

Section 8.3.1 – to describe the deformation state by a superposition of form functions, with the parameters of the system equations being numerically extracted though, for instance by finite element methods. The representation of the total systems with only few form functions reduces computing time substantially in comparison to FE analyses. Its precision and applicability to complex and coupled systems are almost the same due to the combination of analytical solution methods with finite element calculations for data extraction.

A procedure that is similar to macro-modeling and has been successfully applied already for several years for solving mechanical field problems is called *method of modal superposition*. Modal superposition uses the natural oscillation modes of mechanical structures as a form function for representing the motion behavior during a harmonic or transient analysis [BATHE02]. The advantage of modal superposition consists in the substantial reduction of the number of degrees of freedom in comparison to a complete FE model. In comparison to the manual selection of the form functions in Section 8.3.1 another advantage is that the natural oscillation modes are automatically extracted from the geometric model of the form element. This is another essential requirement for automating the entire design process of microsystem technology.

A finite element set of equations with n degrees of freedom

$$\begin{bmatrix} M_{11} & M_{12} & \cdots & M_{1n} \\ M_{21} & M_{22} & \cdots & M_{2n} \\ \vdots & \vdots & \ddots & \vdots \\ M_{n1} & M_{n2} & \cdots & M_{nn} \end{bmatrix} \begin{bmatrix} \ddot{u}_1 \\ \ddot{u}_2 \\ \vdots \\ \ddot{u}_n \end{bmatrix} + \begin{bmatrix} D_{11} & D_{12} & \cdots & D_{1n} \\ D_{21} & D_{22} & \cdots & D_{2n} \\ \vdots & \vdots & \ddots & \vdots \\ D_{n1} & D_{n2} & \cdots & D_{nn} \end{bmatrix} \begin{bmatrix} \dot{u}_1 \\ \dot{u}_2 \\ \vdots \\ \dot{u}_n \end{bmatrix}$$

$$+ \begin{bmatrix} c_{11} & c_{12} & \cdots & c_{1n} \\ c_{21} & c_{22} & \cdots & c_{2n} \\ \vdots & \vdots & \ddots & \vdots \\ c_{n1} & c_{n2} & \cdots & c_{nn} \end{bmatrix} \begin{bmatrix} u_1 \\ u_2 \\ \vdots \\ u_n \end{bmatrix} = \begin{bmatrix} F_1 \\ F_2 \\ \vdots \\ F_n \end{bmatrix} \tag{8.24}$$

becomes a decoupled system with new load and motion variables:

$$\begin{bmatrix} 1 & 0 & \cdots & 0 \\ 0 & 1 & \cdots & 0 \\ \vdots & \vdots & \ddots & \vdots \\ 0 & 0 & \cdots & 1 \end{bmatrix} \begin{bmatrix} \ddot{q}_1 \\ \ddot{q}_2 \\ \vdots \\ \ddot{q}_m \end{bmatrix} + \begin{bmatrix} 2\xi_1\omega_1 & 0 & \cdots & 0 \\ 0 & 2\xi_2\omega_2 & \cdots & 0 \\ \vdots & \vdots & \ddots & \vdots \\ 0 & 0 & \cdots & 2\xi_m\omega_m \end{bmatrix} \begin{bmatrix} \dot{q}_1 \\ \dot{q}_2 \\ \vdots \\ \dot{q}_m \end{bmatrix}$$

$$+ \begin{bmatrix} \omega_1^2 & 0 & \cdots & 0 \\ 0 & \omega_2^2 & \cdots & 0 \\ \vdots & \vdots & \ddots & \vdots \\ 0 & 0 & \cdots & \omega_m^2 \end{bmatrix} \begin{bmatrix} q_1 \\ q_2 \\ \vdots \\ q_m \end{bmatrix} = \begin{bmatrix} f_1 \\ f_2 \\ \vdots \\ f_m \end{bmatrix} \tag{8.25}$$

Here, dimension m is substantially smaller than n. Ten natural oscillation modes are sufficient for most sensors and actuators in microsystem technology.

Multiplication of the transposed vector of the natural oscillation modes ϕ_i is used to project real loads F of the system onto modal loads f

$$f_i = \phi_i^T F. \tag{8.26}$$

The back transformation of modal variables into real displacement variables of the microstructure results from a superposition of m natural oscillation modes that have been previously weighted with the computed modal displacement:

$$u = \sum_{i=1}^{m} \phi_i q_i. \tag{8.27}$$

The method of modal superposition cannot only be applied to oscillation analysis, but to any analysis type and load situation.

In the following, we want to show how we can efficiently model the behavior of the microactuator in using modal superposition. Similar to analytical calculation procedures, capacitance and energy functions form a suitable approach even to the numerical macromodeling of coupled systems. The motion equation for m natural oscillation modes result in analogy to Table 8.1 as

$$\ddot{q}_i + 2\xi_i\omega_i\dot{q}_i + \frac{\partial W(q)}{\partial q_i} = \phi_i^T F_m + \sum_r \frac{V^2}{2}\frac{\partial C(q)}{\partial q_i}, \tag{8.28}$$

$$I = \sum_r \left(\dot{V} C(q) + \sum_{i=1}^{m} \frac{\partial C(q)}{\partial q_i} \dot{q}_i V \right), \tag{8.29}$$

where $W(q)$ represents the mechanical strain energy currently present in the system, ξ_i the damping ratio, ω_i the angular eigenfrequency and $C(q)$ the r capacitances between

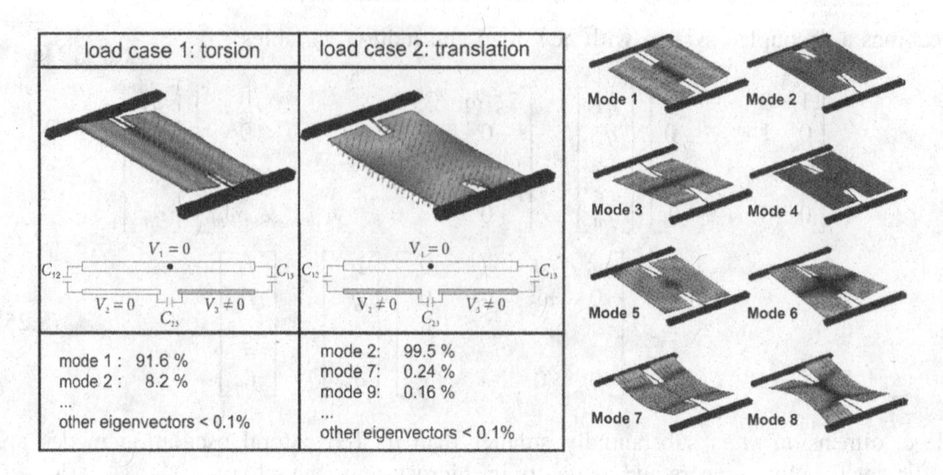

Figure 8.21 Automatic selection of relevant natural oscillation modes by test loads

the electrodes. Both strain energy and capacitances are multivariable functions that are extracted from the results of FE analyses using regression analysis. The total process is subdivided into three steps.

Step 1: *Determining relevant form functions for the reduced order model*: In the simplest case, it is possible to use the natural oscillation modes with the lowest frequencies for solving the coupled field problem. As Figure 8.21 shows, mostly, not all natural oscillation modes are actually excited. An experienced engineer will be able to manually select the modes that are relevant for the individual application. It is also possible though, to automatically determine the mode relevance through test loads and to activate only the form functions with the largest contribution to the deformation state.

Step 2: *Determining capacitance and strain energy functions*: Capacitance and strain energy functions are required for calculating the mechanical and electrostatic force on the natural oscillation modes as well as for the current-voltage relation. For simple form elements, the individual term is analytically available; for complex microsystems, the data have to be numerically determined using numerous finite element computations (parameter extraction). For this, the specific linear combination of the natural oscillation modes – for both the capacitances and the strain energy stored in the mechanical system – have to be calculated and filed in a value table (Figure 8.22). The required displacement is calculated by multiplying the natural oscillation modes with the modal displacement, which is then included as displacement constraints in the FE model. During a further step, the existing value tables are used to extract mathematical functions that can be applied to simply describe the parameters and their derivations (e.g. polynomials).

Step 3: *Export of the macro-model for component and system simulation*: With the mathematical functions for the capacitance and strain energy, the eigenfrequency and natural oscillation modes as well as with data regarding the damping of the oscillation modi, Equations (8.28) and (8.29) can be used to export the model of the electromechanical

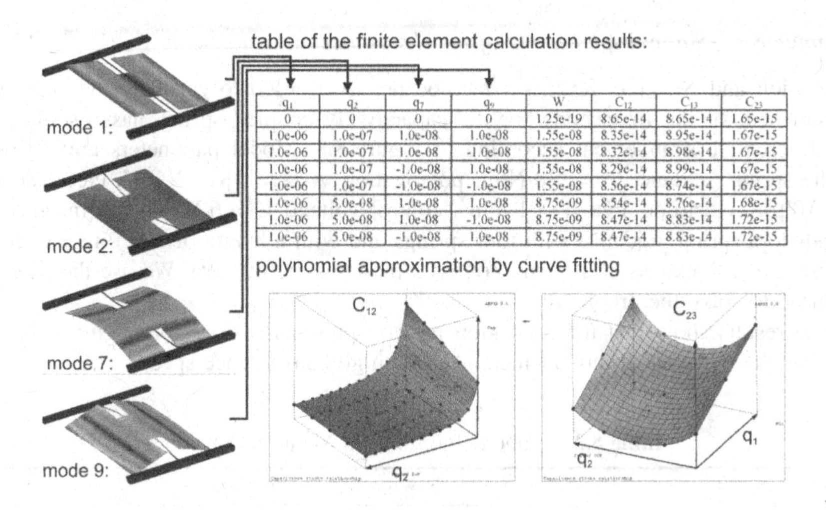

q_1	q_2	q_7	q_9	W	C_{12}	C_{13}	C_{23}
0	0	0	0	1.25e-19	8.65e-14	8.65e-14	1.65e-15
1.0e-06	1.0e-07	1.0e-08	1.0e-08	1.55e-08	8.35e-14	8.95e-14	1.67e-15
1.0e-06	1.0e-07	1.0e-08	-1.0e-08	1.55e-08	8.32e-14	8.98e-14	1.67e-15
1.0e-06	1.0e-07	-1.0e-08	1.0e-08	1.55e-08	8.29e-14	8.99e-14	1.67e-15
1.0e-06	1.0e-07	-1.0e-08	-1.0e-08	1.55e-08	8.56e-14	8.74e-14	1.67e-15
1.0e-06	5.0e-08	1-0e-08	1.0e-08	8.75e-09	8.54e-14	8.76e-14	1.68e-15
1.0e-06	5.0e-08	1.0e-08	-1.0e-08	8.75e-09	8.47e-14	8.83e-14	1.72e-15
1.0e-06	5.0e-08	-1.0e-08	1.0e-08	8.75e-09	8.47e-14	8.83e-14	1.72e-15

Figure 8.22 Determining capacity and strain energy function using FEM

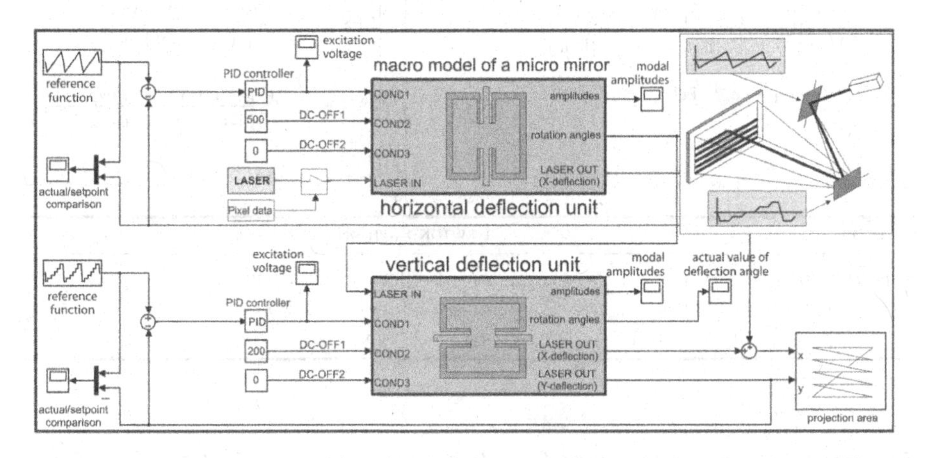

Figure 8.23 Application of macromodels for the system simulation in Matlab/Simulink

system into any simulation environment. As opposed to the low-level behavioral model in Figure 8.3, it does not only include the translatory displacement and rotation of the mirror plate, but also the plate warpage. Figure 8.23 shows the system model of an image projection system based on micromechanical torsional mirrors in Matlab/Simulink.

The goal of the system simulation in Figure 8.23 is an exact analysis of the interaction between the controller and the torsional micro mirror. In particular, we want to analyze the effect of the plate warpage and the real radiation variation on image quality and the maximum achievable resolution. Possible applications of laser projection systems are, for instance, head-up displays for vehicles, that project information of the navigating system or the instruments, respectively, as a virtual image in front of the vehicle.

Example 8.1 Static displacement of an electromechanical system

Relaxation and NEWTON-RAPHSON method are to be used to calculate the static displacement u of the mass body for the capacitively coupled spring-mass oscillator in the differential arrangement according to Figure 8.8. Model parameters are: stiffness of the spring guidance $c = 270$ N/m, polarization voltage $V_P = 100$ V, electrode gap $h = 10$ μm, electrode area $A = 1$ mm^2, control voltage $V = 6$ V. The stiffness corresponds to a spring guidance with four springs and with a length of 500 μm, a width of 50 μm and a thickness of 10 μm. YOUNG's modulus is 169 GPa. We use the program Mathcad to solve the problem.

The results show that the relaxation method presents a slow converging behavior, whereas the NEWTON-RAPHSON method has a high convergence speed.

Table 8.2 Solution algorithms in Mathcad notation

Relaxation method	NEWTON-RAPHSON method
$u(V) := \left\| \begin{array}{l} u_0 \leftarrow 0 \\ \text{for } i \in 0..20 \\ \quad \left\| \begin{array}{l} Fel \leftarrow \frac{(V_P-V)^2 \cdot \varepsilon \cdot A}{2 \cdot (h-u_i)^2} - \frac{(V_P+V)^2 \cdot \varepsilon \cdot A}{2 \cdot (h+u_i)^2} \\ u_{i+1} \leftarrow c^{-1} \cdot Fel \end{array} \right. \\ u \end{array} \right.$	$u(V) := \left\| \begin{array}{l} u_0 \leftarrow 0 \\ \text{for } i \in 0..anz \\ \quad \left\| \begin{array}{l} RES \leftarrow \frac{(V_P-V)^2 \cdot \varepsilon \cdot A}{2 \cdot (h-u_i)^2} - \frac{(V_P+V)^2 \cdot \varepsilon \cdot A}{2 \cdot (h+u_i)^2} - c \cdot u_i \\ JAC \leftarrow c - \frac{(V_P-V)^2 \cdot \varepsilon \cdot A}{(h-u_i)^3} - \frac{(V_P+V)^2 \cdot \varepsilon \cdot A}{(h+u_i)^2} \\ u_{i+1} \leftarrow u_i + JAC^{-1} \cdot RES \end{array} \right. \\ u \end{array} \right.$

Relaxation method	NEWTON-RAPHSON method
$u_1 = 0.393$ μm, $u_2 = 0.655$ μm, $u_3 = 0.833$ μm, $u_4 = 0.957$ μm, ..., $u_{10} = 1.236$ μm, ..., $u_{20} = 1.291$ μm	$u_1 = 1.150$ μm, $u_2 = 1.290$ μm, $u_3 = 1.295$ μm, $u_4 = 1.295$ μm, ..., $u_{10} = 1.295$ μm

Example 8.2 Transfer function of an electromechanical system

The goal is to calculate the transfer function for different polarization voltages V_P for the arrangement provided in Example 8.1. Damping ratio ξ is 0.1 and mass M of the oscillating system is 0.466 mg. This corresponds to a 200 μm thick plate with area A and a density of 2329 kg/m^3.

At first, we have to calculate the operating point: The static displacement $u_=$ is zero due to the selected symmetric polarization voltage. According to Equation (8.7), it applies to the electrostatic stiffness that:

$$c_{el}(V_=) = -2\frac{V_=^2 \varepsilon A}{h^3}.$$

For a polarization voltage of 100 V, the electrostatic softening becomes 177 N/m. Thus, total stiffness is reduced to 93 N/m. Due to the polarization voltage, the natural frequency for the undamped system shifts from originally 3832 Hz to 2250 Hz.

The program Mathcad is used to calculate the transfer function:

$$\mathrm{c\_d}(\Omega, \mathrm{Vp}) := -\Omega^2 \cdot \mathrm{M} + \mathrm{i} \cdot 2 \cdot \xi \cdot \sqrt{\mathrm{c} \cdot \mathrm{M}} + \left(\mathrm{c} - 2 \cdot \frac{\mathrm{Vp}^2 \cdot \varepsilon \cdot \mathrm{A}}{\mathrm{h}^3}\right) \quad \text{dynamical stiffness}$$

$$\mathrm{u}(\Omega, \mathrm{Vp}) := \left| \mathrm{c\_d}(\Omega, \mathrm{Vp})^{-1} \cdot \left(\mathrm{Vp} \cdot \mathrm{Vw} \cdot 2 \cdot \frac{\varepsilon \cdot \mathrm{A}}{\mathrm{h}^2}\right) \right| \quad \text{deflection}$$

$$\mathrm{f} := 1.100..6000 \quad \text{frequency range}$$

Figure 8.24 shows the motion amplitude $\hat{u}$ calculated in Mathcad for the capacitively coupled spring-mass system in Example 8.1 for an AC voltage amplitude of $V_{\mathrm{w}} = 1$ V and polarization voltages of between 40 V and 100 V.

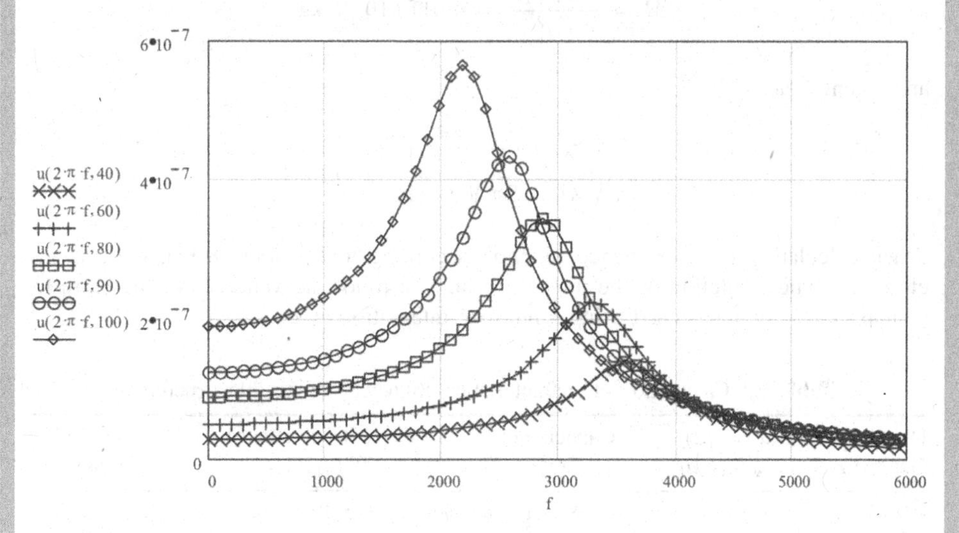

Figure 8.24 Motion amplitude of the capacitively coupled spring-mass oscillator

Example 8.3 Deformation behavior of flexible microstructures

The goal is to use form functions to calculate the mechanical displacement of a double suspended cantilever beam for electrostatic loads: beam length 200 μm, beam width 20 μm, beam thickness 1 μm, YOUNG's modulus 80 GPa, density 2700 kg/m^3, electrode gap 2 μm.

We use Equation (8.14) as form function of the beam displacement for a lumped load:

$$\phi(x) = 3 \cdot 10^8 x^2 - 2 \cdot 10^{12} x^3, \quad \frac{\partial^2 \phi(x)}{\partial x^2} = 6 \cdot 10^8 - 12 \cdot 10^{12} x$$

Equations (8.15) and (8.18) are used to arrive at potential energy and generalized stiffness:

$$E_{\mathrm{B}}(q_1) = -EIq_1^2 \frac{(6 \cdot 10^8 - 12 \cdot 10^{12} x)^3}{3 \cdot 12 \cdot 10^{12}} \Bigg|_{x=0}^{x=l/2} = 1.6 q_1^2$$

Thus, it results that

$$c = \frac{\partial^2 E_B(q_1)}{\partial q_1^2} = 3.2 \text{ N/m}$$

The kinetic energy and generalized energy results from Equations (8.16) and (8.19):

$$E_K(\dot{q}_1) = 3.714 \cdot 10^{-5} \rho A \dot{q}_1^2 = 2.006 \cdot 10^{-12} \dot{q}_1^2.$$

Thus, it follows that

$$M = \frac{\partial^2 E_K(\dot{q}_1)}{\partial \dot{q}_1^2} = 4.011 \cdot 10^{-12} \text{ kg}$$

and eigenfrequency

$$f_1 = \frac{1}{2\pi}\sqrt{\frac{c}{M}} = \frac{1}{2\pi}\sqrt{\frac{E_B(1)}{E_K(1)}} = 142 \text{ kHz}.$$

When calculating the capacitance function according to Equation (8.17), there is no closed analytical solution of the integral term. The following values have therefore to be approximately determined using a numeric integration procedure.

Table 8.3 Calculation of characteristic parameters for flexible structures

Displacement $u(x = l/2)$ Generalized coordinate q_1	Capacitance $C(q_1)$	Capacitance change $dC(q_1)/dq_1$	Required voltage V
−0.25 µm	18.9 fF	−5.40 fF/µm	17.2 V
−0.50 µm	20.4 fF	−6.78 fF/µm	21.7 V
−0.75 µm	22.4 fF	−4.42 fF/µm	23.2 V
−1.00 µm	25.0 fF	−8.90 fF/µm	22.7 V
−1.25 µm	28.8 fF	−18.9 fF/µm	20.4 V
−1.50 µm	35.2 fF	−34.5 fF/µm	16.7 V

Example 8.4 Numerical modeling of electromechanical systems using FEM

We want to use ANSYS to numerically analyze the static and harmonic behavior of the electromechanical system described in Examples 8.1 and 8.2. The electrostatic subsystem shall be represented by TRANS126 elements and the mechanical subsystem by COMBIN14 and MASS21:

```
! Static analysis of a capacitively coupled oscillator:
stiff=270          ! mechanical stiffness
dmp_rat=0.1        ! dampening measure
mass=0.466e-6      ! mass
area=1e-6          ! electrode area
```

```
eps=8.85e-12    ! permittivity of air
h=10e-6         ! electrode distance

/prep7
et,1,14,,1 ! spring-dampener element
et,2,21         ! lumped mass
et,3,126        ! converter element
! Assignment of physical properties
r,1,stiff,2*dmp_rat*sqrt(stiff*mass) ! c and D
r,2,mass        ! M
r,3,,,h         ! h
rmore,eps*area ! e *A

! Finite-element declaration:
n,1   $n,2,h $n,3,2*h
e,1,2
type,2 $real,2 $e,2
type,3 $real,3 $e,1,2 $e,2,3
fini

/solu
! Assignment of conditions and potentials:
d,1,ux,0        $d,3,ux,0 $d,1,volt,100
d,3,volt,-100  $d,2,volt,6
cnvtol,f,,1e-8
solve
fini

/post1
prdisp
fini

!Calculation of transfer function according to Example 8.2:
/solu
d,1,ux,0        $d,3,ux,0 $d,1,volt,100
d,3,volt,-100  $d,2,volt,0
cnvtol,f,,1e-8 $pstress,on
solve
fini

/solu
antype,harmonic
pstress,on $d,1,volt,0       $d,2,volt,1 $d,3,volt,0
hropt,full $harfrq,1,6000    $nsubst,100 $ kbc,1
solve
fini
```

```
/post26
nsol,2,2,ux    ! Output of transfer function
plvar,2        ! at node 2 in x-direction
```

The simulation results are identical to the analytical solutions in Examples 8.1 and 8.2.

REFERENCES

[ANSYS04] ANSYS Release 9.0 Software Documentation. ANSYS Inc. (2004).

[BATHE02] Bathe, K-J. (2002) *Finite-Elemente-Methoden* (Finite Element Methods; in German). Springer Verlag.

[BENNINI05] Bennini, F. (2005) *Ordnungsreduktion von elektrostatisch-mechanischen Finite Elemente Modellen auf der Basis der modalen Zerlegung* (Order Reduction of Electrostatic-mechanical Finite Element Models Based on Modal Decomposition). Dissertation, TU Chemnitz.

[MEHNER00] Mehner, J. (2000) *Entwurf in der Mikrosystemtechnik* (Design in Microsystem Technology; in German). Dresdner Beiträge zur Sensorik. Habilitationsschrift, TU Chemnitz.

[MÜLLER99] Müller, G., Groth, C. (1999) *FEM für Praktiker* (FEM for practitioners). Expert-Verlag.

9

Effect of Technological Processes on Microsystem Properties

The properties of microsystems depend on the geometric and material parameters of the function elements and components (see Section 7.1). The manufacturing can cause deviations of these parameters and subsequently deviations of the function parameters. Different manufacturing techniques will affect function parameters in widely different ways.

In practice, the goal is to keep function parameters of microsystems within certain tolerance limits. This means that the design parameters of microsystems have to be determined in order for the specifications to be fulfilled despite technology-related deviations. There may be critical parameters that affect the deviations of function parameters of the total systems beyond the tolerance limits. The microsystem design has to take into account such critical parameters in a specific way. In this context, we even use the term Design for Manufacturing.

In the following, we want to use examples to present the procedure for determining the function parameter deviations to be expected due to manufacturing. We will use the approach of [TRAH99] that is applied in commercial microsystem manufacturing.

9.1 PARAMETER-BASED MICROSYSTEM DESIGN

Microsystems are produced within a manufacturing process with a number m of process parameters P_i (Figure 9.1). Such process parameters are, for instance, etching time and temperature for wet chemical etching or implantation dose and energy for ion implantation as well as annealing time and temperature.

Process parameter P_i affects geometrical and material parameters of the corresponding function element. Process parameters of wet chemical etching determine, for instance, the thickness and lateral dimensions as geometric parameter G_i of a pressure plate for a pressure sensor. Implantation parameters influence the resistivity of silicon as material parameter M_i. The k geometric parameters G_i and the $(n - k)$ material parameters M_i are altogether n model parameters MP_j of the microsystem.

Each process parameter P_i affects a certain number of model parameters MP_j. In turn, each model parameter MP_j depends on a specific part of process parameter P_i.

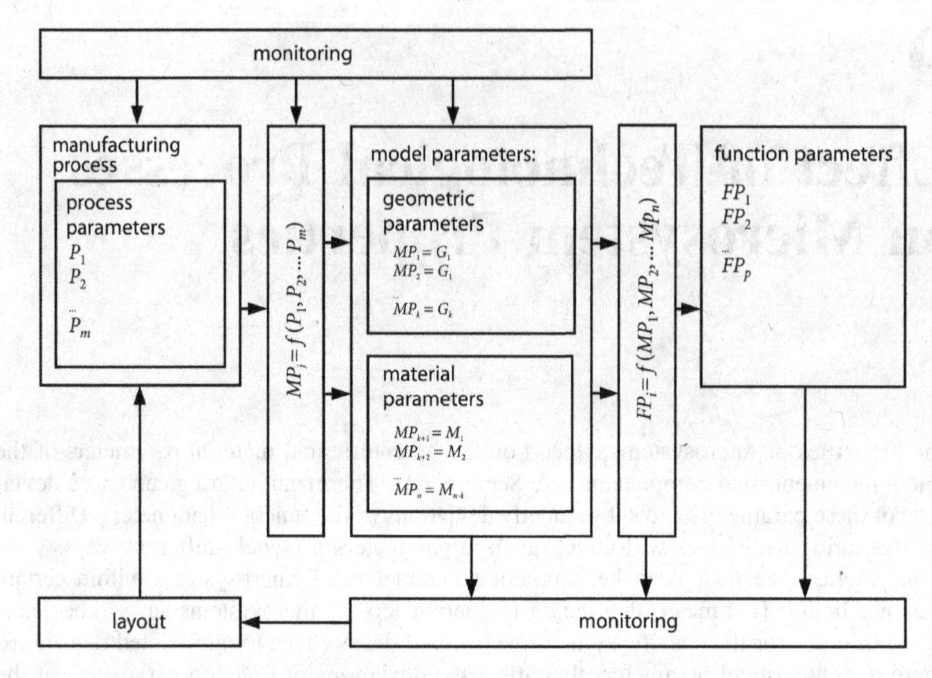

Figure 9.1 Model of a parameter-based microsystem design (acc. [TRAH99])

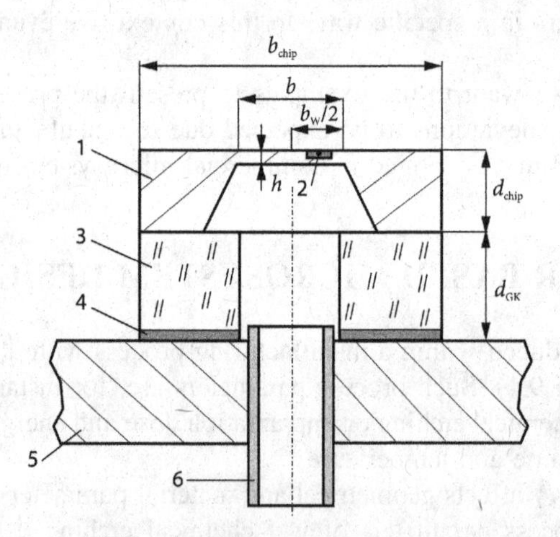

Figure 9.2 Piezoresistive pressure sensor with characteristic dimensions: 1 sensor chip, 2 piezoresistor bridge, 3 glass, 4 glue bond, 5 IC carrier, 6 pressure tube

The interrelation of model parameters MP_j and process parameters P_i is given by the relationship

$$MP_j = f(P_1, P_2, \ldots, P_m). \tag{9.1}$$

Process monitoring is used to monitor the individual distribution function of process parameters P_i and model parameters MP_j.

Function parameters FP_i of the microsystem depend on selected model parameters MP_j, i.e. the geometric and material parameters of the function elements:

$$FP_i = f(MP_1, MP_2, \ldots, MP_n). \tag{9.2}$$

Important function parameters are, for instance, sensitivity S, cross-sensitivities, temperature coefficients of the offset (TCO) or of the sensitivity (TCS) as well as nonlinearities of S, TCO or TCS. Other important parameters are reliability, yield and costs.

Example 9.1 Piezoresistive pressure sensor

We will use the piezoresistive pressure sensor in Figure 7.23 to determine model parameters MP_j that determine the function parameter sensitivity S. The idealized presentation in Figure 7.23 has to be transformed into a model that contains the relevant model parameters MP_j, that are affected by the corresponding process parameters during manufacturing (Figure 9.1).

According to Equation (2.1), sensitivity is defined as the ratio of output voltage V_{out} and measuring pressure p:

$$S = V_{out}(p)/p. \tag{9.3}$$

Table 9.1 Dependence of model parameters in Equation (9.5) from process parameters of piezoresistive pressure sensors according to Figure 9.2

Function parameter FP	Model parameter MP (Geometric parameter G, Material parameter M)	Process parameter P (selection)
FP_1 Sensitivity S	$MP_1 = G_1$: Plate thickness h	P_1 Etching time P_2 Concentration of etching solution P_3 Temperature of etching solution
	$MP_2 = G_2$: Plate width b	P_1, P_2, P_3 (see above); P_4 relevant dimensions of the lithography mask for etch resist
	$MP_3 = G_3$: Resistor position b_w	P_5 relevant dimensions of the lithography mask for piezoresistors
	$MP_4 = M_1$: Piezoresistive coefficient π_{44}	P_6 crystal orientation of (1 0 0)-Si wafer surface at wafer production P_7 orientation of <1 0 0> main flat at wafer production P_8 ion implantation dose P_9 ion implantation energy P_{10} annealing temperature P_{11} annealing time after ion implantation
	$MP_5 = M_2$: POISSON's ratio υ	P_6, P_7 (see above)
FP_2 Nonlinearity	$MP_1 \ldots MP_5$ and others	correspondingly
FP_i	...	...

Using Equations (7.124) and (7.119), for a pressure sensor with long rectangular plate, it results that

$$V_{out} = V_0\pi_{44} \cdot (1 - v) \cdot \sigma_L(p, b_w) = \frac{1}{4} V_0\pi_{44}(1 - v)\left(\frac{b}{h}\right)^2 p\left[1 - \left(\frac{b_w}{b}\right)^2\right] \quad (9.4)$$

or, respectively

$$S = \frac{1}{4} V_0\pi_{44}(1 - v)\frac{b^2 - 3b_w^2}{h^2}. \quad (9.5)$$

Equation (9.5) contains as model parameters the piezoresistive coefficient π_{44} and POISSON's ratio v (material parameters M_i) as well as thickness h and width b of the pressure plate and the position of the resistors b_w on the plate (geometric parameters G_i). Table 9.1 comprises a selection of possible process parameters P, that – according to Equation (9.1) – affect model parameters MP_1 to MP_5 and thus, according to general Equation (9.2) or special Equation (9.5), sensitivity S. Relationships $MP_j = f(P_i)$ have to be specifically stated for the applied manufacturing technique. Similarly, the correlation of further function parameters with relevant model parameters has to be determined in analogy to the derivation of Equation (9.5). For this, we can use the methods and procedures described in Chapter 8.

9.2 ROBUST MICROSYSTEM DESIGN

Deviations ΔP_i of process parameters P_i during the manufacturing process of microsystems cause, according to Equation (9.1) changes ΔM_j of material parameter M_i, which in turn, according to Equation (9.2) causes deviations ΔFP_K of function parameters FP_K.

In practice, technical specifications of microsystems often require the function parameters FP_K to lie within certain tolerance limits. For the pressure sensor in Example 9.1, this means for instance, that sensitivity S has to lie within the range $S_{targ} \pm \Delta S_{max}$. In order to achieve that, the corresponding model parameters MP_1 to MP_5 have to lie within specific limits. Equation (9.2) describes the relation between model and function parameters for target value FP_k, i.e. for the operating point OP_k. Deviation ΔFP_i due to deviation ΔMP_j of a model parameter results from the first term in a TAYLOR series expansion in

$$\Delta FP_{ij} = \left.\frac{\partial FP_i}{\partial MP_j}\right|_{OP} \cdot \Delta MP_j. \quad (9.6)$$

A precondition of this equation is that the correlation $FP_i = f(MP_j)$ is linear or that it can be linearized for operating point OP in case of small changes ΔMP_j. If this precondition is not fulfilled, Equation (9.6) has to be complemented by the corresponding higher-order terms for the derivation. Deviation ΔFP_{ij} can be related as relative deviation $\Delta FP_{ij,\text{rel}}$ to the maximum admissible deviation $\Delta FP_{i,\text{adm}}$ of function parameter FP_i:

$$\Delta FP_{ij,\text{rel}} = \frac{\text{deviation of } FP_i \text{ due to deviation of } MP_j}{\text{maximum admissible deviation of } FP_i}$$

$$= \frac{\Delta FP_{ij}}{\Delta FP_{i,\text{adm}}} = \frac{\left.\frac{\partial FP_i}{\partial MP_j}\right|_{\text{OP}} \cdot \Delta MP_j}{\Delta FP_{i,\text{adm}}} \tag{9.7}$$

Relative deviation contribution $\Delta FP_{ij,\text{rel}}$ is also called influence strength [TRAH99]. For an estimation of total deviation $\Delta F_{ij,\text{rel}}$, the effect of all deviation contributions of all model parameters $MP_j (j = 1 \ldots n)$ has to be taken into consideration. In analogy to the concept of determining the combined measuring uncertainty,[1] this can be achieved by considering deviations ΔMP_j to be the source of uncertainty contributions to the relative uncertainty of function parameter FP_i. Here, deviations ΔMP_j, $\Delta FP_{ij,\text{adm}}$ and $\Delta FP_{ij,\text{rel}}$ are considered to be absolute or relative parameter uncertainties $u(MP_j)$ and $u(FP_{ij,\text{rel}})$

$$\text{Var}(\Delta MP_j) \leftrightarrow u(MP_j), \tag{9.8}$$

$$\text{Var}(\Delta FP_{ij,\text{rel}}) \leftrightarrow u(FP_{ij,\text{rel}}), \tag{9.9}$$

The squares of the parameter uncertainties form parameter variances Var(MP_j) and Var$(FP_{ij,\text{rel}})$:

$$\text{Var}(MP_j) = u^2(MP_j), \tag{9.10}$$

$$\text{Var}(FP_{ij,\text{rel}}) = u^2(FP_{ij,\text{rel}}). \tag{9.11}$$

Using parameter variances Var is especially useful if the deviations of model parameters MP_j follow different distribution functions:

- If deviation ΔMP_j of MP_j follows a normal or GAUSSian distribution, parameter uncertainty $u(MP_j)$ results directly from standard deviation $s(MP_j)$, and it is valid for the variance:

$$\text{Var}(MP_j) = u^2(MP_j) = s^2(MP_j). \tag{9.12}$$

- In many cases, we only know the lower and upper tolerance limits $MP_{j,\text{l}}$ and $MP_{j,\text{u}}$. For a temperature two-step control during the manufacturing process, the temperature may, for instance, deviate from a certain operating point by a maximum of ± 3 K, without us knowing the specific temperature distribution. Due to a lack of

[1] German standard DIN 1319-4 (1999) Grundlagen der Messtechnik, Auswertung von Messungen, Meßunsicherhei; DIN (ed.) (1995) *Leitfaden für die Angabe der Messunsicherheit beim Messen*. Berlin: Beuth Verlag; Adunka, F. (1998) *Messunsicherheiten: Theorie und Praxis*. Essen: Vulkan-Verlag.

information, we have to assume an rectangular distribution with the corresponding variance:[2]

$$\mathrm{Var}(MP_j) = u^2(MP_j) = \frac{(MP_{j,\mathrm{u}} - MP_{j,\mathrm{l}})^2}{12} \tag{9.13}$$

According to Equations (9.12) or (9.13),[3] the variance contribution $\mathrm{Var}(MP_j)$ of the model parameter to the variance of the function parameter FP_i results from Equations (9.7) to (9.9) in

$$\mathrm{Var}(FP_{ij,\mathrm{rel}}) = \left(\frac{\partial FP_i}{\partial MP_j}\right)^2_{\mathrm{OP}} \cdot \frac{\mathrm{Var}(MP_j)}{(\Delta FP_{i,\mathrm{adm}})^2}. \tag{9.14}$$

The relative total variance $\mathrm{Var}(FP_{i,\mathrm{rel}})$ is the sum of the individual contributions of MP_j and results in

$$\mathrm{Var}(FP_{i,\mathrm{rel}}) = \sum_{j=1}^{n} \mathrm{Var}(FP_{ij,\mathrm{rel}}) = \frac{1}{(\Delta FP_{i,\mathrm{adm}})^2} \cdot \sum_{j=1}^{n} \left[\left(\frac{\partial FP_i}{\partial MP_j}\right)^2_{\mathrm{OP}} \cdot \mathrm{Var}(MP_j) \right]. \tag{9.15}$$

Using Equations (9.10) and (9.11), it is valid for parameter uncertainty $u(FP_{j,\mathrm{rel}})$ of function parameter FP_i that:

$$u(FP_{i,\mathrm{rel}}) = \sqrt{\sum_{j=1}^{n} u^2(FP_{ij,\mathrm{rel}})}$$

$$= \frac{1}{(\Delta FP_{i,\mathrm{adm}})} \cdot \sqrt{\sum_{j=1}^{n} \left[\left(\frac{\partial FP_i}{\partial MP_j}\right)_{\mathrm{OP}} \cdot u(MP_j) \right]^2} = \frac{u(FP_i)}{\Delta FP_{i,\mathrm{adm}}}. \tag{9.16}$$

Here, $u(FP_i)$ is the absolute variance of function parameter FP_i. Equations (9.15) and (9.16) are only valid in this form if the material parameters are independent of each other. Otherwise, the corresponding covariance terms have to be included into the equations, similar to the determination of measuring uncertainty[1], for instance.

Function parameter uncertainty $u(FP_i)$ corresponds to the 1σ-variance, if this function parameter is normalized. This means that 68.27 % of all systems lie in the 1σ-range, i.e. in the range $\pm\Delta FP_{ij,\mathrm{adm}}$. The smaller parameter variance $u(FP_{i,\mathrm{rel}})$ becomes, the larger the portion of microsystems for which function parameter FP_i falls within the admissible limits. Table 9.2 shows the portion of the elements that lie within the given limits $\pm\Delta FP_{i,\mathrm{adm}}$, depending on the correlation $u(FP_{i,\mathrm{rel}})$.

[2]Montgomery D.C.; Runger, G.C.: *Applied Statistics and Probability for Engineers*, 4th ed. Wiley 2006.
[3]For other distributions functions, other equations for $\mathrm{Var}(MP_j)$ are possible too.

Example 9.2 Safe system design

The goal is that a minimum of 99.73 % of all produced pressure sensors reach sensitivity S, that lies in the range of (100.0 ± 1.0) mV/kPa. We want to determine the value of the total admissible uncertainty $u(FP_i)$ of all model parameters $MP_1 \ldots MP_n$ due to deviation.

Table 9.2 Portion of microsystems for which a function parameter FP_i lies within the limits $\pm\Delta FP_i$

Normalization $\frac{1}{u(FP_{i,\mathrm{rel}})} = \frac{\Delta FP_{i,\mathrm{adm}}}{u(FP_i)}$	Portion	Chart
1	68.27 %	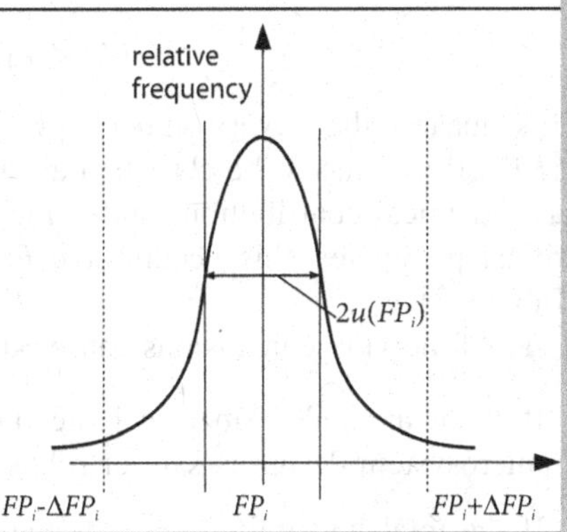
2	95.45 %	
3	99.73 %	
4	99.98 %	

According to Table 9.2, the corresponding required probability is

$$\frac{\Delta FP_{i,\mathrm{adm}}}{u(FP_i)} = \frac{1}{u(FP_{i,\mathrm{rel}})} \geq 3. \tag{9.17}$$

From that, it results that :

$$u(FP_i) \leq \frac{1}{3}\Delta FP_{i,\mathrm{adm}} = \frac{1}{3} \cdot 1.0\,\frac{\mathrm{mV}}{\mathrm{kPa}} = 0.33\,\frac{\mathrm{mV}}{\mathrm{kPa}}.$$

Total sensitivity uncertainty must not exceed 0.33 mV/kPa.

Range $\pm\Delta FP_{i,\mathrm{adm}}$ may be a tolerance band determined by specifications. In the case of adjustable parameters, $\Delta FP_{i,\mathrm{adm}}$ may also be the range that can be set by trimming.

According to Equation (9.16), all affecting model parameters MP_j contribute correspondingly to total uncertainty $u(FP_{i,\mathrm{adm}})$ of function parameter FP_i. The value of individual contributions $u(FP_{ij,\mathrm{rel}})$ can be used to estimate which parts of the total uncertainty are the most decisive. Table 9.3 shows how, according to [TRAH99], uncertainty portions $u(FP_{ij,\mathrm{rel}})$ can be classified based on their impact on total expected uncertainty.

Table 9.3 Classification of individual uncertainty contributions $u(FP_{ij,\text{rel}})$

Portion of uncertainty	Classification
$1 < u(\text{FP}_{ij,\text{rel}})$	hypercritical
$k < u(\text{FP}_{ij,\text{rel}}) < 1$	critical
$0 < u(\text{FP}_{ij,\text{rel}}) < k$	subcritical
$u(\text{FP}_{ij,\text{rel}}) = 0$	none

For microsystems, number n of model parameters that affect a function parameter can be larger than 100. Assuming that all model parameters MP_j have an identical contribution to total uncertainty $u(FP_{i,\text{rel}})$, it results from Equation (9.13):

$$u(FP_{i,\text{rel}}) = \sqrt{n} \cdot u(FP_{ij,\text{rel}}). \tag{9.18}$$

This means that every model parameter that becomes, applying Equation (9.17), $u(FP_{i,\text{rel}}) = 1$ has to be classified as critical. This is the case when all model parameters have identical contributions and if $u(FP_{ij,\text{rel}}) > 1/\sqrt{n}$ (see Table 9.3). For an individual critical parameter, this occurs at $u(FP_{ij,\text{rel}}) > 1$. The latter case is called hypercritical (Table 9.3).

The following conclusions can be drawn:

- If there are only subcritical, i.e. no critical or hypercritical model parameters, the microsystem design is safe or robust.
- If the total uncertainty exceeds the admissible value, we have to reduce those deviations $u(MP_j)$ of the model parameters that provide the largest contribution to total uncertainty.
- If required, uncertainty $u(MP_j)$ can be reached by using other manufacturing techniques with changed process parameters P_i.

Example 9.3 Robust design of a piezoresistive pressure sensor

We assume a piezoresistive silicon pressure sensor with a large rectangular plate according to Example 9.1. According to Equation (9.5), sensitivity S results from

$$S = \frac{1}{4} V_0 \pi_{44} (1 - \upsilon) \frac{b^2 - 3b_\text{w}^2}{h^2}. \tag{9.19}$$

Feeding voltage V_0 is 5 V. Material parameters are known with the following tolerance limits:

$$\pi_{44} = (5.00 \pm 0.05) \cdot 10^{-10}\ \text{m}^2/\text{N}$$
$$\nu = 0.063 \pm 0.001.$$

The pressure plate is manufactured by anisotropic KOH etching resulting in tolerances due to deviations of wafer thickness and etching time:

$$b = (400 \pm 4)\ \mu\text{m}$$
$$h = (10 \pm 5)\ \mu\text{m}.$$

Due to deviations during the lithographic process, the resistor bridge shows positional deviations:

$$b_W = (390 \pm 1)\ \mu\text{m}.$$

Maximum sensitivity deviation ΔS_{adm} must not exceed $\pm$ 10% of sensitivity. In this case, sensitivity S is the relevant function parameter FP_i. It depends on five model parameters $MP_j \in \{\pi_{44}; \nu; b; h; b_W\}$.

The most probable value of sensitivity S is the one resulting from the most probable (mean) values of the individual MP_j:

$$S = \frac{1}{4} V_0 \pi_{44} (1 - v) \frac{b^2 - 3b_{\text{w}}^2}{h^2}$$
$$= \frac{1}{4} \cdot 5\text{V} \cdot 5 \cdot 10^{-10}\ \text{Pa}^{-1} \cdot 0.932 \cdot \frac{400^2 - 3{\cdot}390^2}{100} = -1.73 \cdot 10^{-6}\ \text{V/Pa}. \tag{9.20}$$

The maximum admissible deviation of the sensitivity is

$$\Delta FP_{i,\text{adm}} = \pm \Delta S_{\text{adm}} = \pm 0.1 \cdot S = \pm 1.73 \cdot 10^{-7}\ \text{V/Pa}. \tag{9.21}$$

Table 9.4 shows the procedure of determining uncertainty $u(S_{\text{rel}})$ of sensitivity S that is related to this maximum admissible deviation and the effect of the individual model parameters. We will take the example of plate thickness h to illustrate the calculation of the influence. If $\Delta h = \pm 5\ \mu\text{m}$ is the upper and lower limit of plate thickness and we do not have any information regarding the distribution function, we have to assume an rectangular distribution of the thickness values in the range $h - \Delta h = 5\ \mu\text{m}$ to $h + \Delta h =\ 15\ \mu\text{m}$. According to Equation (9.13), the corresponding uncertainty is

$$u(h) = [(h + \Delta h) - (h - \Delta h)]/\sqrt{12} = \Delta h/\sqrt{3} = 2.89\ \mu\text{m}. \tag{9.22}$$

This deviation $\partial FP_i/\partial MP_j = \partial S/\partial h$ contributes as uncertainty contribution $(\partial S/\partial h) \cdot \Delta h$ to the total deviation of the sensitivity:

$$u(S_h) = u(FP_{ij}) = \partial S/\partial h \cdot u(h). \tag{9.23}$$

It results from Equation (9.18) that

$$\frac{\partial S}{\partial h} = \frac{1}{4} V_0 \pi_{44} (1 - v)(b^2 - 3b_{\text{w}}^2) \frac{d}{dh}\left(\frac{1}{h}\right) = -\frac{1}{2} V_0 \pi_{44} (1 - v) \frac{b^2 - 3b_{\text{w}}^2}{h^3} = -2 \cdot S \cdot \frac{1}{h} \tag{9.24}$$

and applying Equation (9.22)

$$u(S_h) = -S \cdot \frac{2 \cdot u(h)}{h} = 4.98 \cdot 10^{-7}\ \text{V/Pa}. \tag{9.25}$$

The maximum admissible uncertainty related to relative uncertainty $u(FP_{ij,\text{rel}}) = u(S_{h,\text{rel}})$ thus becomes $u(S_{h,\text{rel}}) = 2.89 > 1$.

Table 9.4 Determining critical design parameters for the function parameter sensitivity of piezoresistive pressure sensors in Example 9.1

MP_j	$u(MP_j)$ (acc. Equation(9.13))	$\frac{\partial FP_i}{\partial MP_j} \cdot u(MP_j) = u(FP_{ij})$ (acc. Equation(9.16))	$\lvert u(FP_{ij,\mathrm{rel}})\rvert = \left\lvert \frac{u(FP_{ij})}{\Delta FP_{i,\mathrm{adm}}} \right\rvert$	Classification $k = 1/\sqrt{5} = 0.45$ (acc. Tab. 9.3)
π_{44}	$\frac{\Delta\pi_{44}}{\sqrt{3}} = \frac{0.05 \cdot 10^{-10}\mathrm{m^2N^{-1}}}{\sqrt{3}}$	$S \cdot \frac{u(\pi_{44})}{\pi_{44}} = 9.97 \cdot 10^{-9}$ V/Pa	$5.78 \cdot 10^{-2}$	subcritical
ν	$\frac{\Delta\nu}{\sqrt{3}} = \frac{0.001}{\sqrt{3}}$	$S \cdot \frac{u(\nu)}{\nu} = 10.07 \cdot 10^{-9}$ V/Pa	$6.20 \cdot 10^{-3}$	subcritical
b	$\frac{\Delta b}{\sqrt{3}} = \frac{4\ \mu\mathrm{m}}{\sqrt{3}}$	$S \cdot \frac{2b \cdot u(b)}{b^2 - 3b_\mathrm{w}^2} = 8.48 \cdot 10^{-9}$ V/Pa	$4.91 \cdot 10^{10^{-2}}$	subcritical
b_w	$\frac{\Delta b_\mathrm{w}}{\sqrt{3}} = \frac{1\ \mu\mathrm{m}}{\sqrt{3}}$	$-S \cdot \frac{6b_\mathrm{w} \cdot u(b_\mathrm{w})}{b^2 - 3b_\mathrm{w}^2} = -1.27 \cdot 10^{-11}$ V/Pa	$7.36 \cdot 10^{-5}$	subcritical
h	$\frac{\Delta h}{\sqrt{3}} = \frac{5\ \mu\mathrm{m}}{\sqrt{3}}$	$-S \cdot \frac{2u(h)}{h} = 4.98 \cdot 10^{-7}$ V/Pa	2.89	hypercritical
Relative total variance $u(S_\mathrm{rel})$ (acc. Equation (9.13))			2.89	unsafe design
h[1]	$\frac{\Delta h}{\sqrt{3}} = \frac{0.5\ \mu\mathrm{m}}{\sqrt{3}}$	$-S \cdot \frac{2u(h)}{h} = 4.98 \cdot 10^{-8}$ V/Pa	0.29	subcritical
Relative total variance $u(S_\mathrm{rel})$ (acc. Equation (9.13))			0.29	safe design

[1] with electrochemical etch stop.

According to the classification in Table 9.3, the deviation in the plate thickness caused by wet chemical etching is hypercritical for keeping the specification regarding the maximum admissible deviation of the sensitivity. Table 9.4 shows that the plate thickness is the only model parameter that is critical for sensitivity. A robust or safe design requires measures to reduce the variations in plate thickness h during anisotropic etching. To achieve that, it is possible to apply electrochemical etch stop during KOH etching (see Section 4.6).

Assuming that the uncertainty then will decrease to $\Delta h = \pm 0.5\ \mu\mathrm{m}$, also the portion of deviation $u(S_{h,\mathrm{rel}})$ will be reduced to one-tenth:

$$u(S_{h,\mathrm{rel}}) = 0.29 < k = 1/\sqrt{5} = 0.45. \tag{9.26}$$

This reduces the influence of plate thickness on sensitivity to a subcritical level and the total design becomes safe. According to Equation (9.16), the deviation of the plate thickness still constitutes the main contribution to total deviation $u(FP_{i,\mathrm{rel}}) = u(S_\mathrm{rel})$.

In practice, we often use tolerance bands, such as $\pm n \cdot \sigma$-bands. Here we assume that, for $n = 3$ (3σ-band), 99.73 % of the components lie within the tolerance band, for instance. Uncertainty $u(FP_i)$ calculated according to Equation (9.16) constitutes the standard deviation of a normal distribution which means that we can use tolerance band $FP_i \pm n \cdot u(FP_i)$ in this case. Similarly, we can apply $n \cdot u(FP_{ij,\mathrm{rel}})$ instead of $u(FP_{ij,\mathrm{rel}})$ when classifying model parameters according to Table 9.3.

EXERCISES

9.1 Name a minimum of five material and ten geometric parameters for the yaw rate sensor in Figure 1.7 and assign a minimum of five process parameters to each of these model parameters.

9.2 Calculate for Example 9.4, how much the tolerance limits of Δh have to be reduced at least for plate thickness h not to be a critical parameter any longer.

9.3 It is required for Example 9.4 that maximum admissible sensitivity deviation ΔS is $\pm$ 1 %. Here, a special etch stop technique is used to achieve deviations Δh of a maximum of $\pm$ 0.1 μm for the plate thickness. Which model parameters are critical? Suggest technological measures to achieve a safe design.

REFERENCE

[TRAH99] Trah, H.-P., Franz, J., Marek, J. (1999) Physics of semiconductor sensors. In: Kramer, B. (ed.) *Advances in Solid State Physics*. Braunschweig, Wiesbaden: Vieweg. pp. 25–36.

10
The Future of Microsystems

As microsystem technology is able to combine electronic and non-electronic functions, it can promote the development of information and communication technology beyond the capabilities of microelectronics. With microsystems technology, it is possible to integrate information technology systems with sensors and actuators and thus directly analyze the state of objects and processes and affect them accordingly (see Figure 1.6). It can achieve completely new electrical, mechanical, optical, fluidic and other functionalities as well as miniaturization of known principles and applications (see Chapter 2).

During the more than 50 years following the discovery of the piezoresistive effect, a wide range of technologies for the fabrication of microsystems has been developed which, in turn, make it possible to open up new areas of application. It is commercially mainly of interest if microtechnical fabrication methods become feasible due to large production numbers (pressure sensors, acceleration sensors and yaw rate sensors in the automotive sector, printing heads for inkjet printers) or if miniaturization makes it possible to build components that could not be built earlier (micro-mirror arrays for digital light processors (DLP), catheter pressure sensors for biomedical technology, chemical and biological micro-analysis systems). Future developments in microsystem technology will continue to focus on these main issues.

10.1 STATUS AND TRENDS IN MICROSYSTEM TECHNOLOGY

In 2005, the market for silicon-based microsystem products amounted to approximately US$ 5.1 billion. By 2010, the market is expected to double this volume (US$ 9.7 billion) [ELOY06]. This corresponds to a growth rate of almost 14 % per year (see Table 10.1).

Currently, the microsystem industry shows the following trends [ELOY06], [YURISH05]:

- Microoptical components, and particularly DLP projection circuits, inkjet printing heads and micromechanical sensors for the automotive and processing industry, account for 90 % of the total world market and show a steady growth rate.
- Currently, it takes 3 ... 6 years to develop new microsystem products including new technological processes and market entrance. The adaptation of new products to existing technological procedures usually still takes about 2 ... 3 years.

Introduction to Microsystem Technology: A Guide for Students Gerald Gerlach and Wolfram Dötzel

Table 10.1 Development of the worldwide microsystem market 2005 to 2010 [ELOY06]

Area		2005 Turnover in US$ million	2005 Market share in %	2010[a] Turnover in US$ million	Total increase by %	Growth rate per annum in %
MOEMS[b]		1 292	25	3 154	144	15
Inkjet printing heads		1 532	30	2 015	31	5.6
Sensors	Pressure sensors	911	18	1 254	38	6.6
	Acceleration sensors	394	8	860	118	17
	Yaw rate sensors	398	8	801	101	15
	Silicon microphones	65	1	398	497	44
Microfluidic systems		404	8	849	110	16
RF microsystems		105	2	331	215	26
Micro fuel cells		0	0	65	–	–
Total		5 101	100	9 727	91	14

[a] Prognosis; [b] microoptoelectromechanical systems including digital light processors (DLP), barcode readers and infrared sensor arrays.

- The microtechnical fabrication of MEMS products is increasingly outsourced or carried out via foundries (semiconductor manufacturers that make their technological process available to third parties). Production outsourcing increases by 35 % per year. It promotes specialization among microsystem producers into component and system providers. The latter integrate microsystem elements and components (e.g. sensor circuits into complete modules and devices) and thus create added value.
- Increasingly, IC compatible processes are used in microsystem fabrication. CMOS or BiCMOS production lines can accommodate large production numbers and constitute the basis for mass market applications, e.g. in the automotive or information and communication industry. For large production numbers, monolithic integration provides cost and performance advantages.
- For markets that cannot absorb larger production numbers, hybrid integration continues to dominate microsystem technology. Here, the advantages are the higher yield (see Example 4.2 and Section 5.2.1 b), a more flexible manufacturing strategy and faster product development.
- As the larger part of microsystem technology works with markets that have small production numbers and as the industry is very heterogeneous, the standardization of both manufacturing processes and microsystem products is completely insufficient. Standardization efforts are mainly made by SEMI (Semiconductor Equipment and Materials International),[1] the trade organization of manufacturers of equipment and materials used in the fabrication of semiconductor devices. Some methods introduced in foundries have individually become quasi-standards (z. B. Bosch process, see Section 4.5.3, or LIGA process, see Section 4.12.2).[2]

[1] www.semi.org

[2] Wybransky, B. (2006) Standards and roadmaps in MST. *MST-News, International Newsletter on Micro-nano Integration* 2, pp. 45–6.

- Especially regarding consumer goods, the MEMS industry has an immense development potential. Automotive sensors and printing heads are examples of mass applications in everyday life. White goods, intelligent textiles as well as information and communication technology promote new microtechnical solutions. A vast market potential is also forecast for entertainment electronics[3] (e.g. Playstations, electronic toys) and for lifestyle products (sport equipment, intelligent pocket knives for outdoor activities).

For the areas presented in Table 10.1, the following trends can be discerned:

MOEMS: The area optical microsystems continues to be dominated by digital light processors (DLP) of the company Texas Instruments, which are particularly used for laser projectors. Infrared sensor arrays (IR-FPA Infrared Focal Plane Arrays) show a large growth rate due to new civil applications, such as security technology. Bar code readers and head-up displays will be shortly ready for market introduction.

Inkjet printing heads: The market development has been determined by the rapid spreading of cheap computer printers. The introduction of SPT (Scalable Printing Technology) by HP means that less disposable printing cartridges will be used. This will slow down market growth and, in the long run, even lead to market saturation. At the same time, inkjet printing heads are increasingly used for applications other than computer printers (see Section 10.6.2).

Sensors: The market share of microtechnical sensors of the total sensor market shows a steady growth rate and currently already exceeds 30 %. Here, automotive sensors are the fastest growing market with high production volumes. The sensors have to be extremely reliable and low cost. Automotive sensors are mainly used to increase car safety (preventing car crashes), to enhance passenger comfort, to reduce emissions as well as for the drive train (e.g. steer-by-wire) and for future car generations (hybrid and fuel cell cars).

Pressure sensors, acceleration sensors and yaw rate sensors: Pressure sensors for medical and automotive applications have a yearly growth rate of 12 % and are the fastest growing areas of pressure sensorics. Legal regulations in the US have promoted the market introduction of tire pressure sensors (especially by the companies Infineon and SensoNor) and created a large growth segment. Acceleration sensors have been used in large numbers for airbag and ABS systems. New applications comprise the prevention of read-head crashes in hard disks, GPS (Global Positioning System), human machine interfaces (e.g. 3D mouse). Yaw rate sensors are the core parts of ESP (electronic stability program) systems. GPS as well as further military and civil applications are expected to further promote growth rates.

Micromechanical microphones: Silicon-based microphones constitute an alternative to electrete microphones that currently dominate mobile phones and notebooks. Advantages of such micromechanical microphones are the CMOS compatibility of the production

[3]Fun Recreation with Microsystems. Special Issue of *MST-News, International Newsletter on Micro-nano Integration* (2006) 3.

(including the simplest SMD assembly possible), the integration of electronics as an on-chip solution, EMV resistance to jamming as well as the possibility to use microphone arrays for muting and direction detection.

Microfluidic systems: This area is mainly based on polymer substrates and not on silicon. Silicon components are especially used for pump elements and sensors. Due to the development of bio- and nanotechnology and the corresponding sensors, a growth similar to that of sensorics can be expected.

RF microsystems: Micromechanical switches and resonators use only little power and can be easily electronically configurated. Due to these properties, they are of interest to telecommunication and microwave technology, especially regarding wireless devices and satellites. FBAR (Film Bulk Acoustic Resonators) are already commercially available as substitutes for ceramic duplexer and quartz oscillators as well as RF switches. The growth is inhibited by the rapid price decrease in the telecom market.

Micro fuel cells: Several companies (e.g. Hitachi, NEC, Fujitsu, STM) have announced to market micro fuel cells based on methanol or hydrogen technology within the next few years. Starting from 2010, it is expected to have a large impact on the microsystem market.

Future developments of materials and technologies, packaging and reliability as well as microsystem design will substantially affect the trends presented for the different product areas in Table 10.1

Materials and technologies: Future microsystems will continue to use the established material and technological basis with silicon being the dominant material (see Chapters 3 and 4). The properties of silicon are well understood and it has further development potential. At the same time, there is extensive research going on in order to widen the range of available materials and technologies and thus open up new applications for microsystems. Examples comprise the development of functional polymers, nanocomposites and intelligent textiles. As opposed to inorganic materials, polymers can be easily processed. Large areas of stiff or flexible polymer substrates can be structured down to a resolution in the micrometer range, e.g. with the techniques described in Section 4.13, such as injection molding, hot embossing or nanoimprinting. Established bulk printing processes (silk screen, offset or inkjet printing) are used to apply polymer layers to these substrates and thus they can be used for sensor and actuator functions. In order to reach a certain functionality, it is possible to use polymer nanocomposites with specific conducting, semiconducting,[4] piezoelectric,[5,6], magnetic [7]

[4]Borchardt, J.K. (2004) Developments in organic displays. *Materials Today*, Sept. 42.

[5]Smay, J.E., Cesarano, J., Tuttle, B.A., Lewis, J.A. (2002) Piezoelectric properties of 3-X periodic $Pb(Zr_xTi_1\text{-}x)O_3$ composites. *Journal of Applied Physics* 92, pp. 6119–27.

[6]Sakamato, W.K., Marin-Franch, P., Tunnicliffe, D., Das-Gupta, D.K. (2001) Lead zirconate titanate/polyurethane (PZT/PU) composite for acoustic emission sensors. *2001 Annual Report Conference on Electrical Insulation and Dielectric Phenomena*, 10/14/2001–10/17/2001, Kitchener, Ont., Canada, pp. 20–3, ISBN: 0-7803-7053-8.

[7]Masala, O., Seshadri, R. (2005) Magnetic properties of capped, soluble $MnFe_2O_4$ nanoparticles. *Chemical Physics Letters* 402, pp. 160–4.

or optical[8,9] properties. In addition, it is possible to utilize the sensitivity of polymers to environmental factors, such as temperature, light and oxygen. Material and technology developments based on polymer nanocomposites are expected to produce new material characteristics due to the new properties of nanocomposites and their distribution in the polymer matrix. Large array arrangement in the areas microfluidics, Bio-MEMS, ultrasound transducer, RF-MEMS and tactile sensor arrays are considered to be most promising.
Another trend is the modification of surfaces in order to improve or conserve the functionality of microsystem components. Examples are wear-resistant layers for actuators, ultrathin layers for improving frictional properties or adsorption, capillary and adhesion behavior.
Regarding intelligent textiles, there are several new concepts. The microsystems are either 'hidden' in the textile and contacted via partially conducting structures or the sensor and actuator functions are directly integrated into the textile or metallized fibres of a textile.[10,11] Such material and technological developments can produce textiles with new functionalities that have a large commercial potential mainly in areas like communication, security and surveillance, medical and logistics.

Packaging and reliability: The widening commercial use of microsystem technology will increase the demands on packaging techniques regarding support and safety functions required in different applications. At the same time, they have to be cost-efficient and must not negatively affect the microsystem components. In the future, modeling and simulation will therefore to an even larger extent focus on the mechanical and thermal impact that packaging (see Chapter 5) has on microsystem components. Test chips are suitable to experimentally characterize such impacts.
Over the last years, reliability and lifetime concepts for microsystem technology have been developed. These are substantially different from reliability evaluations of macroscopic structures which are based on classical continuum theoretical models. These concepts integrate into the models local mechanical and thermal deformation processes as well as effects such as diffusion, delamination, migration; and they use modern experimental numerical methods for the analysis. These methods and measuring techniques are expected to become further refined and make it possible to provide more accurate forecasts of microsystems' lifetime and reliability.

Tendencies of microsystems design: Future microsystems will be characterized by a smart integration of advanced components which are manufactured by a variety of technologies and materials. Currently, CAD tools are specialized on individual components: Network simulators support the electronic design, finite element tools are utilized for MEMS, signal flow simulators for controller design and high level description languages for system evaluation and optimization. Computing power and missing interfaces among

[8]Belleville, P., Bonnin, C., Priotton, J.J. (2000) *Journal of Sol-Gel Science Technology*, 19, p. 223.
[9]X.-C. Yuan, *et al.*, Soft-lithography-enabled fabrication of large numerical aperture refractive microlens array in hybrid SiO_2-TiO_2 sol-gel glass. *Appl. Phys. Lett.* 86, (2005) pp. 86–9.
[10]Strese, H., John, L.G., Kaminorz, Y. (2005) Technologies for smart textiles. mst-news, *International Newsletter on Micro-nano Integration* 2, pp. 5–9.
[11]Gimpel, S., Möhring, U., Neudeck, A., Scheibner, W. (2005) Integration of microelectronic devices in textiles. *MST-News, International Newsletter on Micro-nano Integration* 2, pp. 14–15.

simulation tools are still a limiting factor for an ever increasing level of complexity. To overcome this bottleneck, new approaches of model order reduction and methodologies for linking heterogeneous models at different levels of abstraction have to be investigated.[12] The ultimate goal will be parametric models which cover not only the behavior of a single layout but also the effect of dimensional modifications and the change of physical quantities. Promising are black-box models where the governing equations have been automatically retrieved from specialized simulators (e. g. data sampling) and linked by simplified mathematical terms (e. g. fit functions) [GABBAY00].

10.2 MICROOPTICAL APPLICATIONS

10.2.1 Displays and Light Modulators

The field of optical displays has experienced rapid growth with the invention and commercialization of Digital Micromirror Devices (DMD) by Texas Instruments in the late 1990. It took about 15 years from the first patent on Mirror Light Modulators to create a first Digital Light Projector. The key idea of a DMD based display is to provide an array of bi-stable micro mirrors which individually control the light intensity of each pixel. The light modulation (gray scaling) concept is based on a permanent switching of the light direction, either pointing onto a screen or in a light trap where it is absorbed. Colored images are obtained by a superposition of green-red-blue light on the same spot (color-wheel technique, three-mirror arrays) [TEXINS].

Alternatively to digital mirror arrays, light projection displays can also be achieved by Resonant Micromirror Devices (RMD) which operate similar to a standard cathode ray tube. Advantage compared to DMD is that mirror arrays of several hundred thousand of cells can be replaced by simpler devices which just need a single bi-axial or two uni-axial mirror cells in order to realize light deflection in two spatial directions. Challenging are horizontal scan frequencies of several ten kilohertz which are required to obtain high-quality resolution [MOTAMEDI05]. Commercial light scanning units are manufactured by Microvison Inc. for head-up displays in automotive information systems, eye-glasses and cell phones.

Another promising concept for future display applications is known as 'Grating Light Valve Technology' (GLV) [TRISNADI04]. The modulation concept is different to reflective optics used in mirror devices. GLV controls the light intensity by means of diffraction from a miniaturized grating which consists of an array of beams able to move in vertical direction. In the 'off' state, the grating is inactive because all beams are in the same horizontal position. No light is transmitted to the screen. In the 'on' state, the grating is activated by electrostatic forces which drive the beams in alternating vertical positions. Usually, the screen is illuminated by the first diffractive order. GLVs can have many more pixels than micro mirror devices and modulate light several orders of magnitudes faster. The new technology mainly targets the high-end projection market as digital cinema projectors.

[12] EPOSS European Technology Platform on Smart Systems Integration: Towards a vision of innovative smart systems integration. http://www.smart-systems-integration.org

Future information technologies require faster communication networks, which may only be achieved by optical data transfer. Light modulators for fiber optic communication, adaptive optics and smart signal processing are key components for future telecommunication units. In the past, a series of MOEMS solutions as optic switches, fiber collimators, variable attenuators and tunable filters have been developed. Movable mirrors which block or modulate the light intensity play an important role in existing and future systems. Nevertheless, there is a strong demand for completely new functions and higher performance in adaptive and reconfigurable fiber optic networks.

Future optical devices require manufacturing technologies which integrate photonics, MEMS and electronic components with extremely small feature sizes. Optical displays and light modulators based on microstructures are limited in resolution and bandwidth because of the travel range, the chip area and the natural frequencies. New and upcoming principles based on interferometry and diffraction have sub-micron resolution and operate up to a MHz range what enables essential progress. However, the fusion of micro- and nanotechnologies in combination with new materials allow for structures with nano-scale operation and GHz response. A completely new spectrum of micro devices with astonishing features becomes visible.

10.2.2 Infrared Sensor Arrays [KIMATA98] [AKIN05]

According to PLANCK's radiation law, each object with a temperature above absolute zero emits electromagnetic radiation that depends on the temperature and emission properties of the radiating object. For technically relevant temperature ranges (ambient temperature up to a few 1000 K), the major part of this radiation lies in the infrared (IR) range. Measuring IR radiation makes it possible to detect objects even in the darkness and – if the objects' radiation emission properties are known – to determine their temperature without touching the object. IR sensors are therefore suitable for noncontact temperature measurement; the corresponding IR sensor arrays (FPA Focal Plane Arrays) can record thermal images. Measurements are preferably carried out in spectral ranges (3...5) μm or (8...13) μm as in these ranges, the atmosphere absorbs radiation only to a very low extent.

Thermal sensors are an important group of IR sensors [CANIOU99]. Here, the heat radiation Φ that strikes the sensor element is absorbed with absorption degree α resulting in an increase of the sensor temperature T_S in comparison to ambient temperature T_{amb} (Figure 10.1):

$$\Delta T = T_S - T_{\text{amb}}. \tag{10.1}$$

The sensor element itself now acts as a temperature sensor that is able to detect such very small changes in temperature.

If there is no radiation flow ($\Phi = 0$), temperature T_S of the sensor element is equal to the temperature of the heat sink of the sensor ($T_{\text{amb}} = T_S$).

If the incident radiation changes ($\Phi \neq 0$), part of the absorbed radiation heats the sensor element which has a heat capacity C_{th}; another part flows via heat conduction

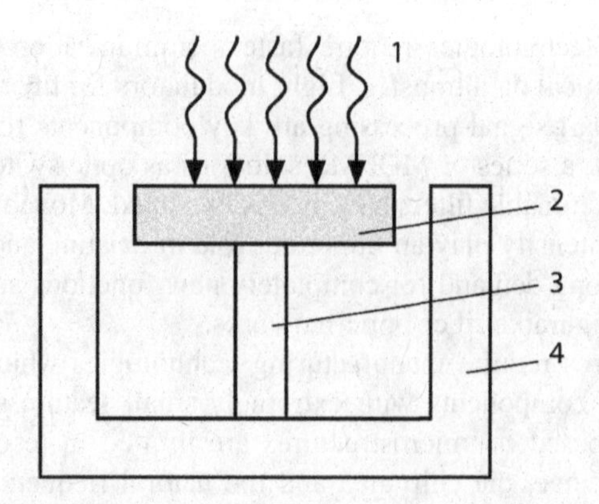

Figure 10.1 Operating principle of thermal infrared sensors: 1 incident radiation Φ, 2 sensor element (absorption coefficient α, heat capacity C_{th}, temperature T_S), 3 heat conduction with thermal conductance G_{th}, 4 heat sink (ambient temperature T_{amb})

from the sensor element to the environment:

$$\alpha\Phi(t) = C_{\mathrm{th}} \cdot \frac{\mathrm{d}[\Delta T(t)]}{\mathrm{d}t} + G_{\mathrm{th}} \cdot \Delta T(t) \tag{10.2}$$

If we look at sinusoidal changes of the incident radiation

$$\Phi(t) = \hat{\Phi} \cdot \sin \omega t, \tag{10.3}$$

($\hat{\Phi}$ amplitude of the incident radiation, ω angular frequency of the modulated frequency) it results for the amplitude of the temperature change

$$\Delta \hat{T} = \frac{\alpha\hat{\Phi}}{G_{\mathrm{th}}} \cdot \frac{1}{\sqrt{1 + \omega^2 \tau_{\mathrm{th}}^2}} \tag{10.4}$$

with $\tau_{\mathrm{th}} = \frac{C_{\mathrm{th}}}{G_{\mathrm{th}}}$ as the thermal time constant. Equation (10.4) shows that for a high sensitivity $S = \Delta\hat{T}/\Phi$ of thermal IR sensors, thermal time constant τ_{th} and thus also thermal capacity C_{th} have to be very small. For this reason, miniaturized solutions based on microtechnical fabrication are particularly suitable for IR sensors. According to Equation (10.4), the technical realization requires an extremely efficient thermal insulation between sensor element and heat sink. Thermal conductivity G_{th} is – in addition to convection via the surrounding gas – determined by the mechanical suspension of the sensor element. Infrared sensors should therefore be structured in a way that the sensor elements have a specific area (for 'catching' a specific heat radiation), but a low thermal capacity C_{th} (i.e. small thickness) which is suspended by long and thin connections (low thermal conductivity G_{th}) on a sensor chip. Typical representatives of such thermal sensors are bolometers and thermopiles.

(a) Bolometer arrays

Bolometers use the temperature dependence of an electrical resistor to transform temperature change ΔT between radiated sensor element and heat sink into an electrical signal. In order to comply with the mentioned structural requirements, the resistor material is deposited as a thin layer on the silicon surface. The bolometer resistance and the temperature coefficient have to be as large as possible to achieve high detectivity values (signal-to-noise ratio) [CANIOU99]. Therefore, in the past mainly semiconducting oxidic ceramics (vanadium oxide[13]) was used as sensor material. Recently, it has become possible to deposit amorphous silicon (a-Si) as a thin layer with a sufficient resistivity, which means that a-Si-bolometers with compatible properties are within reach.[14]

Surface micromachining[15] (see Section 4.10) is used for the fabrication of bolometers with the extracted areas of the sacrificial layers thermally insulating the sensor elements against the underlying silicon (Figure 10.2). The sensor elements are electrically – thus even thermally – connected to the underlying CMOS-ROIC (Read-out Integrated Circuit) only via thin metal columns. Figure 10.3 describes the manufacturing technology; Figure 10.4 shows scanning electron microscope (SEM) images of the sensor array.

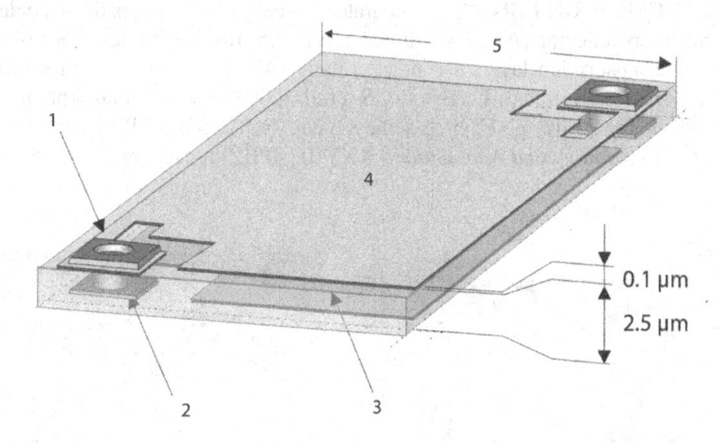

Figure 10.2 Schematic presentation of a micro bolometer pixel: 1 thermal insulation, 2 read-out circuit, 3 reflector, 4 bolometer element, 5 pixel pitch. Reproduced by permission from ULIS, France

[13] Wood, R.A. (2020) Uncooled thermal imaging with monolithic silicon focal plane. In: Andresen, B.F., Shepherd, F.D. (eds) Infrared Technology XIX. *Proceedings of SPIE*, Vol. 2020, pp. 322–9.

[14] Mottin, E., Bain, A., Martin, J.-L., Ouvrier-Buffet, J.-L., Bisotto, S. *et al.* (2002) Uncooled amorphous silicon technology enhancement for 25 μm pixel pitch achievement. In: Andresen, B., Fulop, G.F., Strojnik, M. (eds) *Infrared Technology and Applications XXVIII. SPIE Proceedings*, Vol. 4820, pp. 200–7.

[15] Hanson, C.M. (1996) *Thermal imaging system with a monolithic focal plane array and method.* US patent 5.512.748.

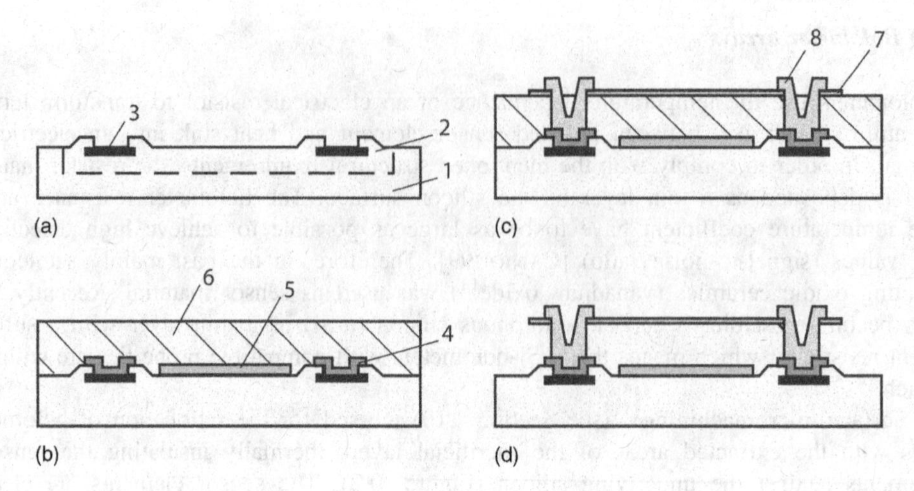

Figure 10.3 Manufacturing process for micro bolometers made of amorphous silicon: (a) CMOS base wafer with contact pads; (b) Coating and structuring of contact pads and radiation reflectors, spin-on and annealing of the polyimide sacrificial layer; (c) dry etching of contact holes, deposition and patterning of the a-Si bolometer layer including the radiation absorption layer, metal deposition and patterning in order to fill the contact holes; (d) removal of the polyimide sacrificial layer; 1 silicon wafer, 2 CMOS-RIOS (Read-out Integrated Circuit), 3 contact pads, 4 contact pad reinforcement, 5 radiation reflector (Al), 6 sacrificial layer (2.5 μm polyimide), 7 a-Si-layer (0.1 μm thick), covered with absorption layer, 8 contact column (Al). Reproduced from Mottin, E., Bain, A., Martin, J.-L., Ouvrier-Buffet, J.-L., Bisotto, S. *et al.* (2002) Uncooled amorphous silicon technology enhancement for 25 μm pixel pitch achievement. In: Andresen, B., Fulop, G. F., Strojnik, M. (eds): *Infrared Technology and Applications* XXVIII. SPIE Proceedings, Vol. 4820, pp. 200–7.

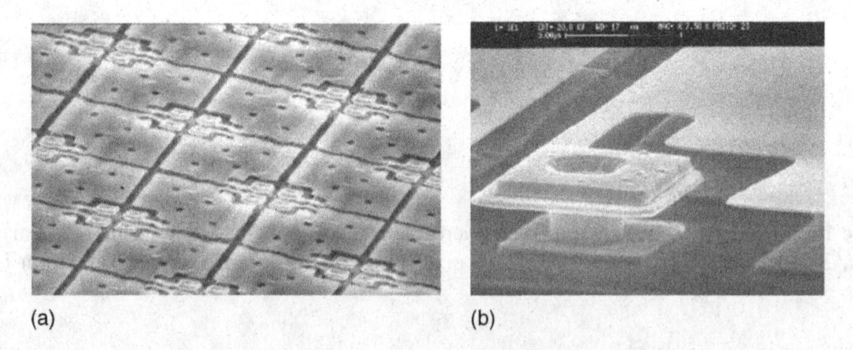

Figure 10.4 SEM images of an a-Si micro bolometer array: (a) pixel view; (b) suspension of a sensor element. Reproduced by permission of ULIS, France

(b) Thermopiles

Thermoelectric IR sensors utilize the thermoelectric effect (SEEBECK effect, see Example 7.4), where a voltage between two connection points of two different electric conductors occurs if – according to Figure 10.5a – there is a temperature difference

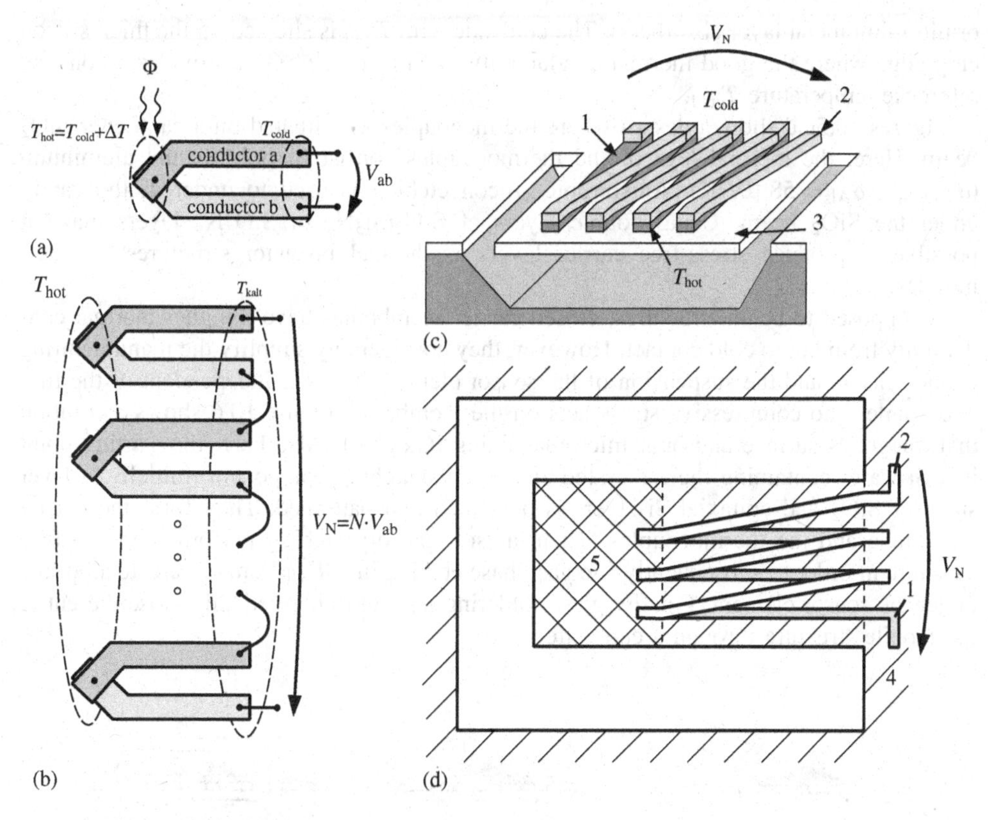

Figure 10.5 Thermoelectric radiation sensors: (a) thermocouple made of two different conductors a and b; (b) thermopile consisting of N thermocouples; (c) aluminum-polysilicon thermopile on a SiO_2 carrier (CVD oxide between poly-Si and Al is not included here); (d) top view of the structure in (c) 1 polysilicon, 2 aluminum, 3 oxide beam, silicon, 4 heat sink (Si) 5 absorber area, 6 etching cavity. Reproduced from Elbel, T., Lenggenhager, R., Baltes, H. (1992) Model of thermoelectric radiation sensors made by CMOS and micromachining. *Sensors and Actuators* A35, pp. 101–6.

between the ends of the two conductors:

$$V_{\mathrm{ab}} = (\alpha_{\mathrm{a}} - \alpha_{\mathrm{b}})\Delta T. \tag{10.5}$$

Here α_a and α_b are the SEEBECK coefficients of electrical conductors a and b. If N thermal elements are coupled in series, the output voltage increases by factor N, which means that output voltages are technically sufficient (Figure 10.5b):

$$V_{\mathrm{N}} = N \cdot V_{\mathrm{ab}} = N \cdot (\alpha_{\mathrm{a}} - \alpha_{\mathrm{b}}) \cdot \Delta T. \tag{10.6}$$

Thermopiles often have 40 to 100 thermocouples. The structural goal is to fit thermocouples which are made of different materials and have a numerically large SEEBECK coefficient with opposite signs onto a thermally insulated layer. The 'hot' contact with temperature T_{hot} is supposed to be heated by incident radiation Φ. This is usually achieved by using a heat absorption layer as cover (metallic black layers, low-reflection λ /4 layers or ultra-thin metal layer absorbers). The cold side with T_{cold} is situated on the thick silicon chip edge where the good thermal conductivity of Si (see Table 3.5) provides a constant reference temperature T_{cold}.

Figures 10.5c,d show a design where thermocouples are situated on a cantilever SiO_2 beam. Here, the thermal legs of the thermocouples consist of poly-Si and aluminium ($\alpha_{Poly\text{-}Si} - \alpha_{Al} = 58\ \mu V/K$). Anisotropic silicon etching is used to undercut the cavity under the SiO_2 beam. Using double layers of SiO_2/Si_3N_4 or SiO_xN_y layers make it possible to produce stress-free carrier layers as thermal insulator structures for thermopiles.

As opposed to beam structures, closed carrier membranes have a higher thermal conductivity from hot to cold contact. However, they considerably simplify the manufacturing of the sensors and the suspension of the sensor element is rather independent of the film stress unless no compressive stress acts on the membrane. Figure 10.6 shows a solution that integrates surface and bulk micromachining [GRAF06]. Also here, the starting point is a Si-wafer containing the integrated electronic switching component. Initially, a layer stack of SiO_2 and epitaxial Si-layers is deposited and patterned. They form the carrier membrane and the thermocouples. Then, anisotropic dry etching perforates the surface structure in order to subsequently use gas phase etching in ClF_3 plasma to create a hollow under the sensor element. Finally, glass soldering is used to hermetically close the entire thermopile structure through a cap wafer.

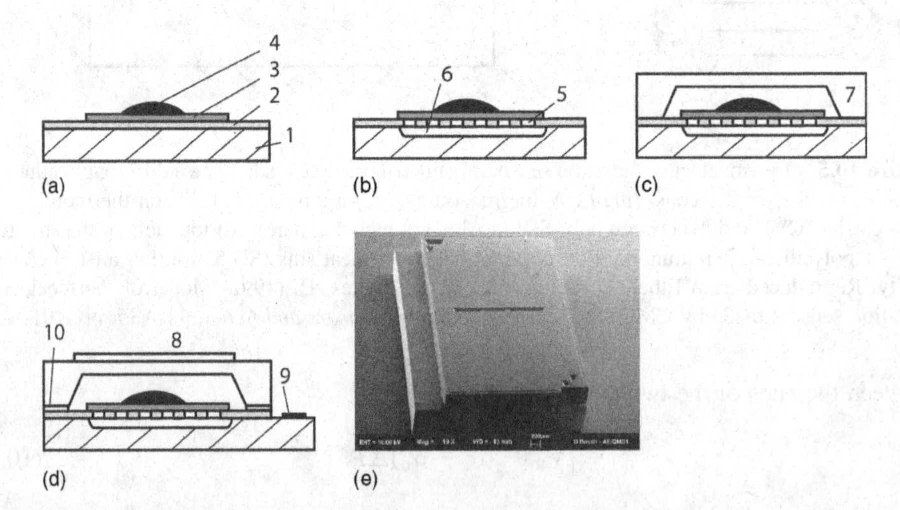

Figure 10.6 Fabrication of a thermopile using bulk/surface micromachining: (a) deposition of a SiO_2 carrier membrane, thermocouples and absorber layer; (b) perforation of SiO_2 carrier membrane; (c) bonding of a cap wafer in order to hermetically protect the IR sensor element; (d) cross-section of a finished sensor; (e) photograph; 1 Silicon wafer, 2 SiO_2 carrier membrane, 3 Thermocouple, 4 Radiation absorber, 5 Perforation holes, 6 Etching cavity, 7 Cap wafer, 8 IR filter, 9 Bonding pad, 10 Seal glass. (a)–(d) Reproduced from Arndt, M.; Sauer, M. (2004) Spectroscopic carbon dioxide sensor for automotive applications. In: Rocha, D., Sarro, P. M., Vellekoop, M. J. (eds): *Sensors 2004, The Third IEEE International Conference on Sensors*. Wien: IEEE. (e) Reproduced by permission from the Robert Bosch GmbH, Germany

Due to the large number of thermoelements in each thermopile and the required space, thermopiles are less suitable for sensor arrays with a large number of pixel. However, low-cost focal line and plane arrays with a low number of pixels can be advantageously used for more cost-efficient mass applications in the area of motion detection and surveillance.

10.2.3 Spectrometers

Optical spectrometry is widely used for material inspection, industrial process monitoring and chemical or gas analyses. Today's MEMS spectrometer products provide sufficient sensitivity and selectivity to detect very small amounts of gaseous species needed for security and biological applications. The primary goal of spectrometry is to evaluate the intensity of the incoming light at different wavelengths. It is well known that a prism or a periodic array of lines (optical gratings) diffract light into its colors. There are four methods to perform spectroscopy: (1) dispersive methods as optical multi-channel analyzers which measure the dispersed light from a grating by an detector array such that each cell senses a different wavelength; (2) a scanning slit spectrometer, in which the dispersed light is scanned by a single detector having a rotating mirror in the light path; (3) nondispersive methods as filter-wheel systems; and (4) digital-transform spectrometer which use FOURIER modulation techniques to observe the light intensity at different wavelength [DAY05]. Microtechnologies are mainly used to manufacture fixed and variable gratings and mirror systems to redirect the light path. A series of products are currently on the market.

Future spectrometer performance will be rated by obtained resolution, speed, sensitivity, size and manufacturing costs. New approaches will be promising such as programmable dark-field correlation spectrometer recently developed by Polychromix Inc. [POLY]. Correlation spectrometer distinguish between a series of gaseous species by comparing the spectrum of the incoming light with a reference pattern. Obviously, a reference spectrum can be obtained from a bank of tanks filled with gases and chemical substances to be analyzed. Polychromic Inc. replaced the tank by a MEMS chip what generates a synthetic spectrum according to the type of gas to be measured. Such an artificial reference spectrum can be produced by a reconfigurable grating consisting of thousands of beams which can individually be pulled down to the wafer surface. According to the vertical positions of the beams, a polychromatic light spectrum can be superimposed and will be directed on a detector cell. Characteristic spectrums are steadily programmed in the MEMS device and compared to the incoming light obtained from the environment. Gas concentration and contents of suspicious clouds can be measured from a large distance by a telescope or binocular having implemented such a microsystem.

10.3 PROBE TIPS

The development of ever smaller structures with dimensions in the nano- to micrometer range led to the necessity to develop the corresponding analytic methods with a lateral position resolution in the atomic range. The invention of the atomic force

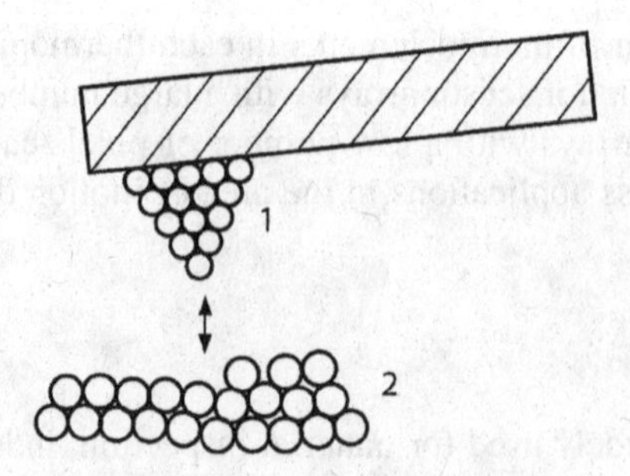

Figure 10.7 Principle of scanning probe microscopy. The thick arrow characterizes the interaction between tip and object surface; 1 Probe tip, 2 Object surface

microscopy (AFM) by BINNIG and ROHRER in 1982[16] (Nobel prize 1986) promoted a surprising development of different scanning probe techniques for analyzing surface and bulk properties on the nano-scale. The basic principle consists in that a 'nano finger' interacts with the surface of a specimen to be analyzed. If the probe has atomic dimensions and if the interaction length between tip and surface atom also lies in the atomic range, it is possible to achieve the desired atomic position resolution. The most important example of such scanning probe techniques is the Atomic Force Microscope[17] (AFM), where the tip of a force-measuring miniature bending beam scans the surface and the force between tip and surface is measured via the displacement of the bending beam. This force constitutes a measurement of the atomic surface contour of the object to be analyzed. Examples of new scanning probe methods that are currently being developed are:

- Piezoresponse microscopy[18,19] (PRM), which is used to analyze the piezoelectric properties;
- Scanning Spreading Resistance Microscopy[20] (SSRM), which can be used to determine the charge carrier distribution; and
- Scanning Near-field Optical Microscopy[21] (SNOM), where light interacts through a miniaturized aperture with the surface; using the optical near-field characteristics, there are no resolution restrictions due to wavelength.

[16]Binnig, G., Rohrer, H., Gerber, Ch., Weibel, E. (1982) Surface studies by scanning tunneling microscopy. *Physical Review Letters* 49(1), pp. 57–61.

[17]Binnig, G., Quate, C. F., Gerber, Ch. (1986) Atomic force microscope. *Physical Review Letters* 56(9), pp. 930–3.

[18]Abplanalp, M., Eng. L.M., Günter, P. (1988) Mapping the domain distribution at ferroelectric surfaces by scanning force microscopy. Applied Physics A 66 (1988), pp. 231–4.

[19]Abplanalp, M. (2001) *Piezoresponse Scanning Force Microscopy of Ferroelectric Domains*. Dissertation. ETH Zürich.

[20]Snauwaert, J., Blanc, N., De Wolf, P., Vandervorst, W., Hellemans, L. (1996) Minimizing the size of force-controlled point contacts on silicon for carrier profiling. *Journal of Vacuum Science and Technology* B 14(2), pp. 1513–17.

[21]Hecht, B., Sick, B., Wild, U.P., Deckert, V., Zenobi, R. *et al.* (2000) Scanning near-field optical microscopy with aperture probes: Fundamentals and applications. *Journal of Chemical Physics* 112 (18), pp. 7761–74.

For all these methods, resolution f is mainly determined by the effective tip radius r and distance d to the object (Figure 10.7). The interaction between tip and object decreases by exp($-d/\ell$), with ℓ being the effective decaying length. According to [BINNIG99], the spatial resolution can be approximated as

$$f = A \cdot \sqrt{(r+d) \cdot \ell}, \tag{10.7}$$

where the base geometry constant A for a spherical tip is A $\approx$ 3. For free electrons, decaying length ℓ for the scanning tunnel microscope is

$$\ell/\text{nm} \quad \approx \quad 0.1/\sqrt{\phi_{\text{eff}}/\text{eV}} \tag{10.8}$$

with effective tunnel barrier height ϕ_{eff} being the mean value of the work function of tip and sample material. For typical values of $\phi_{\text{eff}} \approx 4$ eV, there result decaying lengths of $\ell \approx 0.05$ nm. It follows that an atomic resolution requires tips with a tip diameter in the sub-nm range, which are positioned at an atomic distance to the object surface.

Figure 10.8 shows the principle manufacturing techniques that are used to produce tips [ALBRECHT90]. The tip is manufactured either by isotropic undercutting under a circular square etch mask (Figure 10.8a,b), by filling an anisotropically etched pyramid structure (Figure 10.8c) or by metal vaporization through an aperture (Figure 10.8d). These basic principles are then incorporated into the manufacturing technology for producing the corresponding scanning probe microscopes. Figure 10.9, for example, shows an atomically sharp Si-tip which is used for a scanning near-field microscope. The tip serves here as an optical wave conductor in order to focus the light wave that is emitted

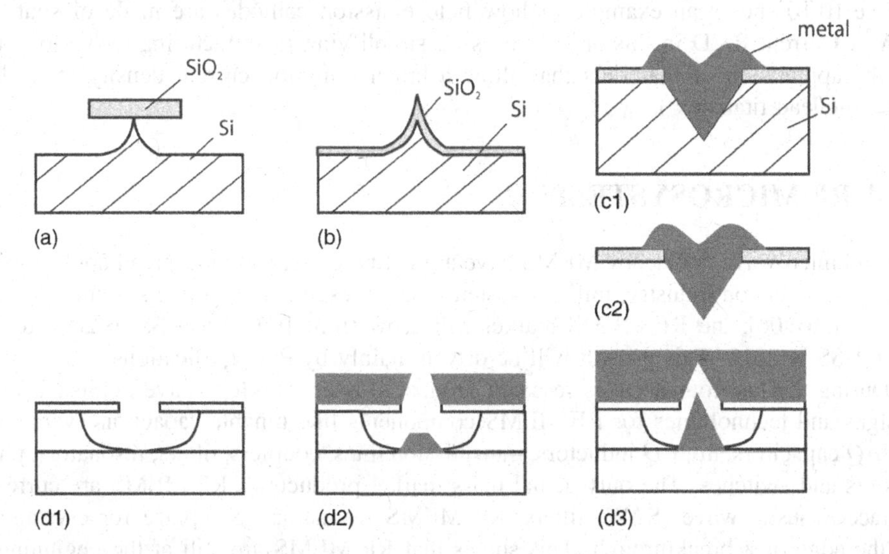

Figure 10.8 Principles for the manufacturing of tips on miniature bending beams for scanning probe microscopy [ALBRECHT90]: (a) Si-tip produced by isotropic undercutting of an etch mask; (b) SiO_2-tip produced by thermal oxidation of the Si-tip in (a); (c) metal tip, produced by metal vaporization in an anisotropically etched pyramid structure; (d) metal tip, produced by metal deposition through a mask aperture with subsequent lift-off of the mask structure

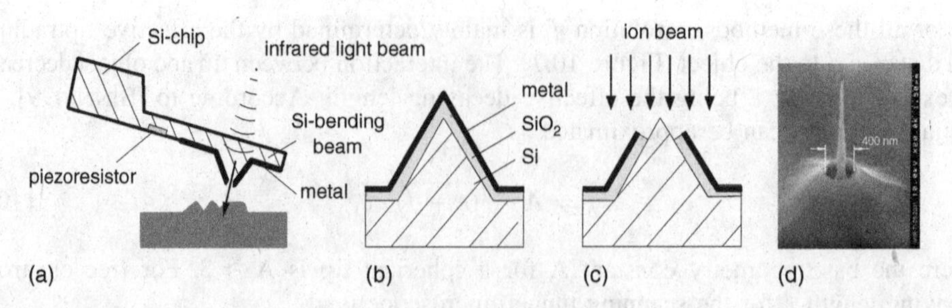

Figure 10.9 Piezoresistive scanning near-field microscope: (a) principle; (b, c) production of the SNOM tip by ion beam etching using Focused Ion Beam (FIB); (d) image of the SNOM tip. (a)–(c) Reproduced from [RANGELOW05] (d) Reproduced by permission from the Ilmenau University of Technology, Germany

through the metal aperture as an evanescent wave. The aperture diameter is (20...50) nm and is set by the ion beam process.

Current research focuses on further improving tip geometry and developing simple evaluation options for the displacement of the miniature beam that carries the tip. In the past, the standard solution was to direct a laser beam to the bending beam whose reflecting portion was evaluated using position-sensitive photodiodes. Piezoresistive [RANGELOW05] and capacitive variants have the potential to measure deformations in a simple and integrated way.

Today, the production of tips is a fairly established technology and can be applied to the manufacturing of field emitter cathodes. Even there the emission tips have to be atomically sharp as the electric field strength is inversely proportional to the tip radius. Figure 10.10 shows an example of how field emission cathodes are made of sputtered TiW[22]. Current R&D in this area focuses on simplifying manufacturing technology und on the application of materials that allow a larger emission current density (e. g. lead zirconate/lead titanate[23]).

10.4 RF MICROSYSTEMS

RF and microwave frequency MEMS have a great potential for widespread application in telecommunication industry, military systems, satellites and automotive systems. According to [ELOY06], the RF MEMS market will grow from 105 Mio. US$ in 2005 to 331 Mio. US$ in 2010. This growth will be driven mainly by new applications.

During the last four decades research groups all over the globe have published many designs and technologies for RF MEMS components like tunable capacitors (varactors), high-Q capacitors, high-Q inductors, transmission lines, couplers, filters, resonators, phase shifters and switches. The only actual mass market products of RF MEMS are currently surface acoustic wave (SAW) filters. RF MEMS resonators as replace for crystals are on the edge of a breakthrough. This shows that RF MEMS are still at the beginning of

[22]Kang, S.W., Lee, J.H., Yu, B.G., Cho, K.-I., Yoo, H.J. (1998) Novel structure of a silicon field emission cathode with a sputtered TiW gate electrode. *Journal of Vacuum Science and Technology* B 16(1), pp. 242–6.
[23]Rosenman, G., Shur, D., Krasnik, Ya, E., Dunaevsky, A. (2000) Electron emission from ferroelectrics. *Journal of Applied Physics* 88(11), pp. 6109–61.

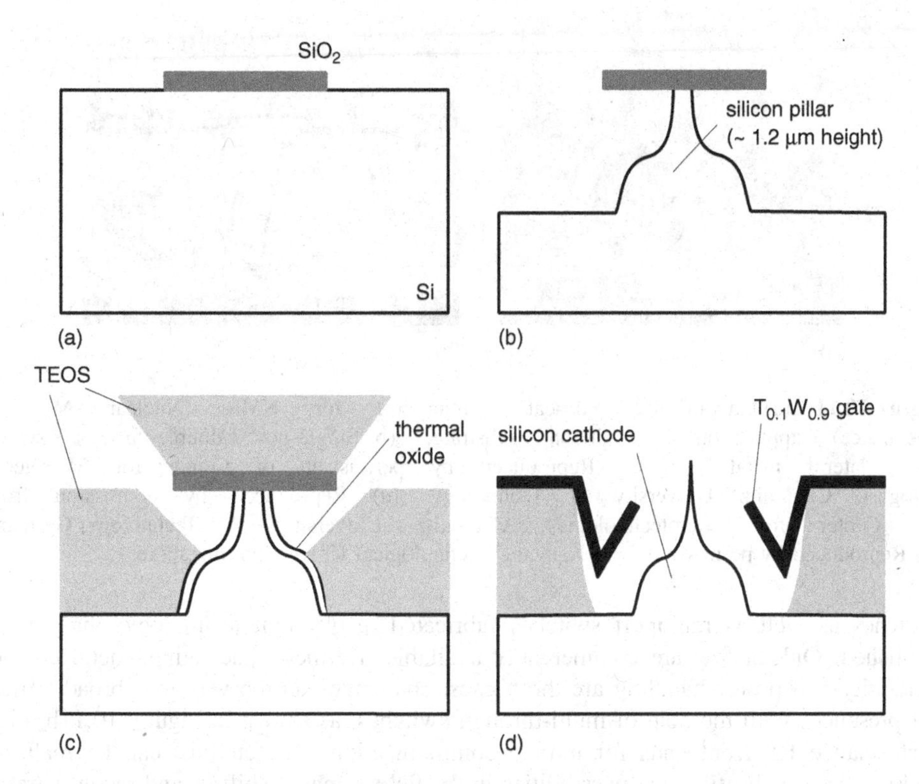

Figure 10.10 Manufacturing process of silicon field emitter cathodes using sputtered TiW[31]: (a) thermal oxidation to form an etch resist; (b) two-step dry etching of Si (isotropic, anisotropic); (c) thermal oxidation, subsequent TEOS oxide deposition; (d) masking, patterning and RIE etching of contact holes, gate-metal electrode deposition of TiW, isotropic sacrificial layer etching of SiO_2 with HF solution

their development. Many difficulties on the way to markets need to be resolved. Those are namely reliability, packaging, actuation voltage and cost. Especially in the field of actuators, RF MEMS show a great advantage over traditional devices. RF MEMS phase shifters using varactors or switches exhibit virtually zero power consumption compared to conventional semiconductor solutions. For passives like shown in Figure. 10.11a and transmission lines, the possibility to use air as insulator and dielectric MEMS technologies offer the potential for high-Q components.

The technologies used for RF-MEMS are traditional bulk micromachining, surface micromachining, wafer bonding technologies and LIGA. Microfabrication using polymer material is getting into focus.

Switches offer the biggest potential for mass applications of future RF MEMS. Possible fields of application are mobile telephony, WLAN, consumer and IT peripherals, base stations, automotive radar, RF test, satellites, microwave communications and military phased arrays. Besides thermal and piezoelectric actuation, magnetic and especially electrostatic actuation is believed to be the choice [VARADAN03] due to the low power and latching capability, respectively. Depending on the desired circuit layout, the required bandwidth, power handling capability, switching time, insertion loss and isolation, different types of switches are favorable. Capacitive and resistive switches, series and shunt

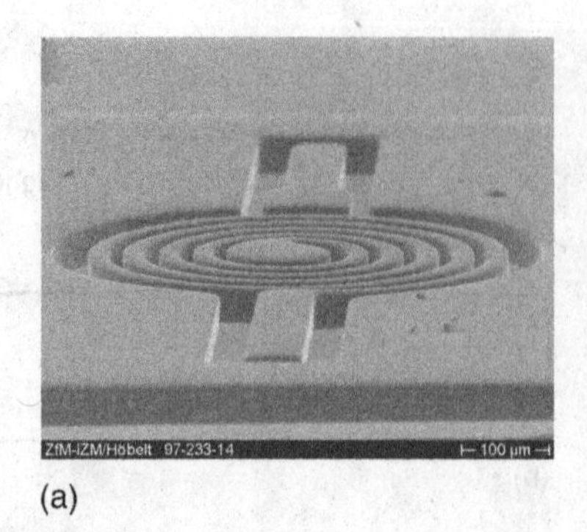

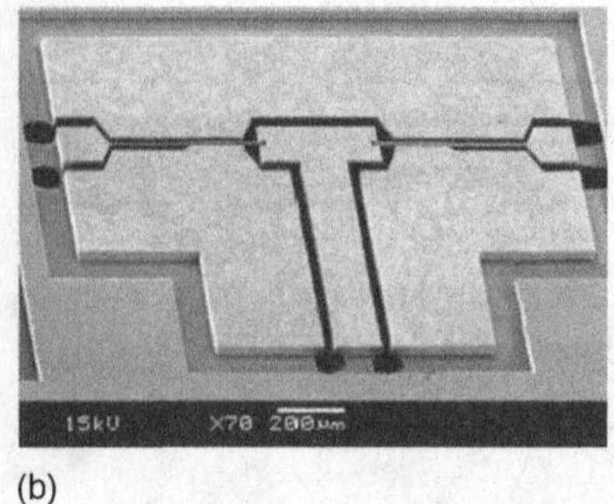

Figure 10.11 (a) Galvanically fabricated microcoil for NMR (Nuclear Magnetic Resonance) application on silicon substrate; (b) Single-port double-through switch using lateral metal' contacts. Reproduced by permission of Center for Microtechnologies, Chemnitz University of Technology. (a) Reproduced by permission from the Center for Microtechnologies, Chemnitz University of Technology, Germany (b) Reproduced by permission from Nanyang Technological University, Singapore

switches as well as multiport switches fabricated in different technologies have been published. Only a few are commercially available. Hermetic packaging, metal contact reliability and power handling are the biggest challenges on the way to a broader market presence. With the help of multi-through switches, as shown in Figure. 10.11b, high performance RF front ends for mobile communication and satellites can be realized. Switch arrays will offer new possibilities in the field of phase shifters and reconfigurable antennas [CETINER04].

Apart from SAW filters, the application of micromachined RF components for consumer electronics has been prevented so far by the not competitive fabrication cost for the particular application. Especially in the upcoming field of reconfigurable or smart antennas, there is a need for new technologies using low-cost substrates and materials. Within the classical microtechnology a higher integration level offers advantages with respect to cost. Nevertheless, particularly in consumer electronics, where short development cycles and cost are the main criteria, the integration level and device performance are only of secondary interest. In the case of reconfigurable antennas with large substrate sizes, the use and development of technologies for low-cost substrates like liquid crystal polymers (LCP) is essential.

10.5 ACTUATORS

For microactuators, the trends focus on improving performance, reliability and application potential. The dominant actuators will continue to be those based on electrostatic, electromagnetic, electrodynamic and thermomechanical principles as well as on transducer effects of shape memory alloys, piezoelectric and magnetostrictive material (see Table 7.1). Starting from the known transducer principles for microactuators, the goal is to improve them by optimizing models, designs and technologies as well as by developing materials with specific properties. Further trends comprise the integration of sensors and actuators in order to broaden the functionality of microsystems[24,25,26] as well as the combination of actuators with additional mechanisms such as toggle joints or folded structures.[27] For the individual applications, the choice of suitable actuator mechanisms depends on forces, strokes and response times, on the existing technological base and on the intended operating mode, e.g. quasi-static or resonant, controlled or uncontrolled.

Electrostatic transducers will – due to the advantages stated in Table 7.4 – even in the future be part of typical actuator mechanisms. The goal is to modify design and technology of basic types of electrode systems for distance and area variations (see Tables 7.2 and 7.3) in order to generate – for low operating voltages – large forces (driving torques), large actuating strokes or the desired capacity functions according to Equation (7.56). Electrode systems with parallel plates and distance variation have limitations due to the restricted displacement angle, the large operating voltages and the risk of pull-in effects. For specific applications, a design with coupled oscillators[28] can be used to avoid these disadvantages. Comb drives constitute an alternative. They prevent the pull-in effect and can be operated at lower operating voltages.

For *electromagnetic and electrodynamic microactuators*, the demand is expected to grow. Due to their substantially higher energy density, they are preferred to electrostatic systems, particularly in applications that require large strokes (up to several mm) and forces (up to several mN). It is possible to manufacture design variations for translatory and rotary movements. All required materials are available for microtechnology. Improvements can be achieved by adapting materials and technologies for hard and soft magnets, conductors, substrate materials and insulators to specific applications as well as by optimizing design and dimensioning. A special challenge constitutes the design of planar coils regarding the thermal rating of conductor and substrate material as well as realizing permanent magnets in order to generate a maximum direct magnetic flow. For this, bulk permanent magnets or thin-film hard magnets can be used. They can be structured using sputtering or spinning-on photoresists containing hard magnetic particles.

Miniaturized *piezoelectric actuators* are commercially available in several fields of application. Examples are positioning mechanisms in autofocus cameras and in robotics, printing heads in inkjet printers, drives in medical pumps, medicine dosing systems as well as endoscopes with movable mirrors. The future development of piezoelectric microactuators is driven by the application areas communication technology, medicine

[24]Forke, R., Mehner, J., Scheibner, D., Geßner, T., Dötzel, W. (2006) Micromachined force-coupled sensor-actuator system for frequency selective vibration monitoring. *Proc. of the 10th International Conference on New Actuators*. Bremen, pp. 928–31.

[25]Ballas, R.G., Greiner, P.F., Schlaak, H.F. (2005) Smart piezoelectric bending microactuator – an integrated inductive non-contact sensor to detect tip deflection. *Mikrosystemtechnik Kongress 2005*, VDE Verlag Berlin, Offenbach, pp. 309–12.

[26]Lotz, P., Bischof, V., Matysek, M., Schlaak, H.F. (2006) Integrated sensor-actuator-system based on dielectric polymer actuators for peristaltic pumps. *Proc. of the 10th International Conference on New Actuators*. Bremen, pp. 104–7.

[27]Carpi, F., De Rossi, D. (2006) A new contractile linear actuator made of dielectric elastomers with folded structure. Proc. of the 10th International Conference on New Actuators. Bremen, pp. 101–3.

[28]Kurth, S., Kaufmann, Ch., Hahn, R., Mehner, J., Dötzel, W. *et al.* (2005) A novel 24-kHz resonant scanner or high-resolution laser display, *Proceedings of the SPIE* 5721(1), pp. 23–33.

and environmental technology.[29] Sol-gel techniques or sputtering can be used to deposit piezoelectric thin films, e.g. made of lead zirconate titanate (PZT) on flexible silicon structures. Examples of piezoelectrically driven microscanners confirm that large displacements can be achieved at low voltages and low power.[30,31] Both deposition techniques are complex steps, though, and cannot be easily integrated into standard process technology for silicon MEMS. Challenges comprise manufacturing technology, cost reduction, minimization of dimensions and triggering voltages; for medical application, it even comprises the integration of a sensor for failure detection of the piezoelectric microactuator.

In addition to piezoelectric microactuators based on silicon, piezoelectric MEMS drives using piezo-polymer composite technology will become established. They are simple, cost-efficient and can be produced in a large variety of forms, e.g. by injection molding. Piezoelectric functionality can be achieved either by PZT powder in polymer materials or by structured PZT ceramics with the polymer material being cast around the ceramics in a casting mold.

Shape Memory Alloys (SMA) are already being commercially used for clams, switches and valves,[32] among others. The shape memory effect is based on a temperature-dependent martensite-austenite phase transition. Electric heating can be used to achieve this phase transition which results – at high output force – in a changed form of the material. There are several materials with shape memory effect. Currently, mainly Ni-Ti alloys are used in actual applications. Advantages include a very characteristic form change at phase transition, favorable mechanical properties, good processability and biocompatibility. Shape memory alloys are particularly advantageous for actuator applications that require small movements for large forces. Often, they form a bi-stable mechanism, either only through heating or cooling processes of the SMA actuator or through interaction of the SMA actuator with a loaded spring.

Sputtering can be used to deposit metal shape memory alloys as thin layer on the substrate which then can be patterned. For hybrid manufacturing technologies, laser cutting or wet chemical etching is used to structure cold-rolled SMA sheets into SMA actuators which then can be integrated with the substrate. SMA actuators can also be imbedded into a polymer substrate material, e.g. SU8.

Challenges for future SMA actuators in microsystem technology comprise the technological optimization of cost-efficient batch production as well as the search for suitable alloys with shape memory effect for specific applications. In biomedicine, Ni-Ti alloys are expected to be replaced by other alloys due to the risk of Ni-hypersensitivity.[33] Ni-Ti-based SMA neither are suitable for high-temperature applications, as transformation temperature for the phase transition lies at a maximum of ca. 90 °C. For the form

[29] Uchino, K. (2006) Piezoelectric actuators 2006 – expansion from IT/robotics to ecological/energy applications. *Proc. of the 10th International Conference on New Actuators*. Bremen, pp. 48–57.

[30] Smits, J.G. *et al.* (2005) Microelectromechanical flexure PZT actuated optical scanner – static and resonant behavior. *Journal of Micromachining and Microengineering* 15, pp. 1285–93.

[31] Kotera, H. *et al.* (2005) Deformable micro-mirror for adaptive optics composed of piezoelectric thin films. *IEEE-Intern. Conference on Optical MEMS,* Piscataway, pp. 31–2.

[32] Couto, D., Brailovski, V., Demers, V., Terriault, P. (2006) The design diagram for optimal shape memory alloys linear actuators. *Proc. of the 10th International Conference on New Actuators*. Bremen, pp. 924–27.

[33] Inamura, T., Kim, J.I., Kim, H.Y., Hosoda, H., Wakashima, K. *et al.* (2006) Shape memory properties of textured Ti-Nb-Al biomedical β-titanium-alloy. *Proc. of the 10th International Conference on New Actuators*. Bremen, pp. 904–7.

Table 10.2 Materials for electroactive polymers (EAP) [SOMMER06]

Ionic EAPs (ion conductivity)	Electric EAPs (electron conductivity)
Polyelectrolyte gels	Electrostrictive polymers
Ionic polymer-metal-composite	Dielectric elastomers
Conducting polymers	

transformation of the SMA actuators to function reliably, transformation temperature has to substantially exceed the ambient temperature at which the actuators are operated.

Actuators based on electroactive polymers (EAP) are forecast to have a high potential in the areas microfluidics, haptic displays and prothetic actuators [SOMMER06]. The main reasons are the low material and processing costs of polymers as well as their flexibility regarding size and design. Polymers are therefore a promising alternative to silicon-based components or typical actuator materials, such as piezoceramics, magnetostrictive materials and shape memory alloys. Most EAP types can be included in either of the categories given in Table 10.2. Electric EAPs show a faster response and are more energy-efficient than ionic ones. They also have advantages regarding their chemical and environmental stability; they require larger operating voltages, though.

Vibration measurement in order to monitor the state of machines and plants can be considered an example of possible improvements of the conductivity of sensor systems through the integration of an actuator (see Examples 6.3 and 7.5).

Currently, piezoelectric wideband sensors with sophisticated signal-analyzing electronics are used to monitor the spectrum. Because of high costs, permanent monitoring is only practicable at expensive equipment or in safety-related applications. Most information about wear states and process conditions can be found in the vibration spectrum between some Hertz and 10 kHz. Generally, it is sufficient to analyze only a few spectral lines in order to get the relevant information. Thus, micromechanical frequency-selective vibration sensors without a subsequent Fast Fourier Transformation (FFT) are powerful means for this type of application. The main drawback of resonant sensors is that monitoring of frequencies below 1 kHz requires excessively large chip area. To bypass this problem and to extend the frequency range down to a few Hertz, a sensor-actuator system can be used[24]. It consists of two separated but electrostatically coupled mechanical spring-mass-damping systems (Figure 10.12).

Following the superheterodyne principle, well known from radio receivers, the mechanical signal of low frequency is transformed to a higher intermediate frequency. The first oscillator operates as a wideband vibration sensor, it detects vibration signals up to 1 kHz proportional to the occurring acceleration. The second oscillator is a high-Q resonator with its eigenfrequency set to approximately one decade above the latter. In terms of the superheterodyne principle, this resonator can also be described as the amplifying filter. Due to mechanical vibrations, the wideband sensor couples electrostatic forces to the high-Q resonator, which is additionally stimulated by a carrier signal.

Amplitude modulation occurs in the same way as in the mixer stage of the superheterodyne receiver, with lower and upper sidebands which carry the information of the vibration signal. By variation of the carrier frequency it is possible to adjust the lower (or upper) sideband exactly to the resonance frequency of the second oscillator. Thus, the resonator is actuated at its resonance frequency. Thanks to the high Q-factor of the

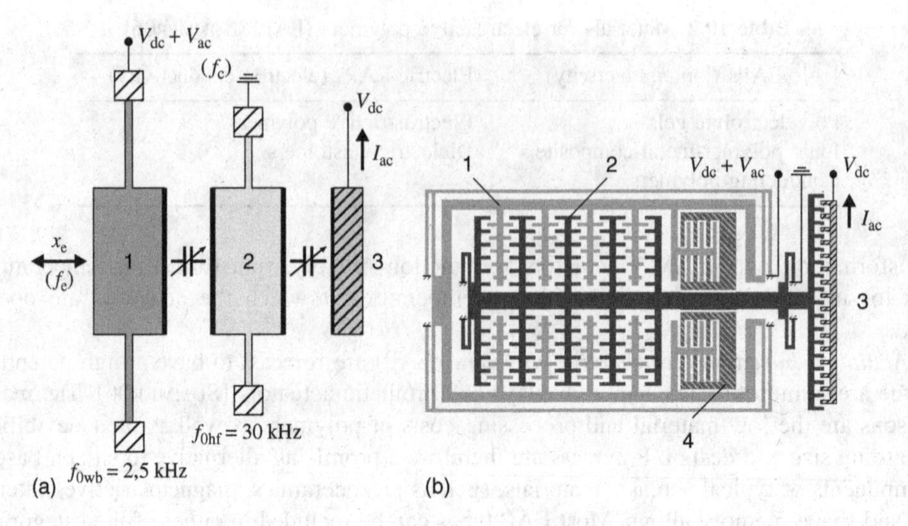

Figure 10.12 Sensor-actuator system for frequency selective vibration monitoring: (a) basic principle; (b) simplified layout; 1 mechanical receiver, 2 amplifying filter, 3 detection, 4 damping elements

second oscillator, the signal is amplified in a very small band with a high signal-to-noise ratio. Hence, the sense frequency can be tuned by variation of the carrier frequency. The motion signals are detected by capacitive pick-up.

10.6 MICROFLUIDIC SYSTEMS

Microfluidics is a growing and important field of microsystem technology. In the late 1980s and early 1990s, microfluidics research community mainly focused on the development of microflow sensors, micropumps, and microvalves. Since mid-1990, with chemists joining the field, microfluidics has gained a rapid development with numerous applications in analytical chemistry and biochemical analysis. Compared to other microsystems, microfluidic systems have many distinctive characteristics. Size and material are the two key characteristics, which differ microfluidic systems from other microsystems.

While most microsystems have an overall size in micrometer scale, microfluidic systems may have sizes on the order of centimeters. The most important characteristics of microfludic systems is the microscale that determines the flow behavior and new effects. The substrate material for microfluidic system does not need to be silicon as in conventional microsystems. Because of the detection limit of molecules, microfluidic systems require a minimum size and cannot be miniaturized without limit. This fact leads to much larger microfluidic devices and a small number of systems that can be fabricated in batch on a silicon wafer. Material cost, processing cost, and the yield rate make silicon-based microfluidic systems too expensive to be accepted by the commercial market. Since mid-1990 a number of microfluidic systems have been made of polymeric materials.

Compared to conventional microsystems, the design of microfluidic devices is relatively simple. Sometimes, a microfluidic device consists only of a passive microchannel system. Actuation and detection do not necessarily need to be integrated into the microfluidic systems. Because the overall system size is not critical in microfluidics, large-scale fabrication of polymeric microfluidic systems is possible with established replication and forming techniques. To maintain the precision of microstructures, the mold for replication can be fabricated in silicon with traditional micromachining technologies. For applications in chemical industry and in highly corrosive fluids, microfluidic systems may need to be fabricated in materials such as stainless steel or ceramics. Alternative technologies are therefore needed for such materials.

The potential of biotechnology promise microfluidic systems to be a huge commercial success. Microfluidic systems will have its impact in the current analytical laboratory instrumentation market of 10 billion US$. Furthermore, 'labs-on-a-chip' will help to accelerate drug discovery processes and to lower their costs. Conventional drug discovery processes are usually time-consuming and expensive because of the huge amount of reactions to be tested. Massively parallel analysis on the same microfluidic chip allows higher screening throughput. Therefore, microfluidic systems are able to screen combinatorial libraries with extremely high throughput, which previously was not imaginable with manual experiments. The overall cost can also be lowered by the smaller quantities of chemicals, which are usually expensive.

10.6.1 Micropumps and Microvalves

Micropumps and microvalves were one of the first microfluidic devices ever reported. For more details, readers can refer to the recent reviews on micropumps [NGUYEN02], [LINK04] and microvalves [OH06]. In the past, the need for flow control in microfluidic systems drove the development of these two microfluidic components. However, due to their complex structures and the corresponding high fabrication cost, most recent disposable lab-on-a-chip applications use external pumps and valves for fluid delivery.

Based on their working concepts, micropumps can be classified as mechanical and nonmechanical pumps [NGUYEN02] or displacement and dynamic pumps [LASER04]. In the first classification scheme, mechanical concepts are usually the miniaturized version of their macroscale counterparts such as check-valve pumps, peristaltic pumps, valveless rectification pumps, rotary pumps, and ultrasonic pumps. Mechanical pumps require an actuator. Therefore, they can also be further classified according the actuation concepts: pneumatic, thermopneumatic, thermomechanic, piezoelectric, electrostatic, electromagnetic, electrochemical or chemical. Mechanical pumps are suitable for device sizes on the order of millimeters and centimeters. The large viscous force associated with the higher surface to volume ratio makes mechanical pumps less attractive for the micrometer scale.

Microscale effects can be utilized for more efficient pumping. These effects are nonmechanical and are previously negligible in the macroscale. Non-mechanical pumping concepts in microscale are electrohydrodynamic, electrokinetic, surface tension driven, electrochemical and magnetohydrodynamic. Many of these concepts only require simple electrodes in their implementation. Thus, nonmechanical pumps can be implemented with a low cost in microfluidic systems.

Another way to look at micropumps is categorizing them into displacement pumps and dynamic pumps. Displacement pumps are mechanical pumps that exert pressure on a fluid through its moving boundaries. Dynamic pumps continuously add energy to the working fluid to increase or maintain its momentum. The added energy can be mechanical, electrical, magnetic or chemical. Most of the nonmechanical pumps belong to this category.

There are two types of microvalves: passive and active. Passive microvalves are used in mechanical micropumps as check valves. Active valves are further classified as mechanical and nonmechanical valves according to their concepts. Mechanical valves are devices fabricated with conventional micromachining technologies. The closing force in mechanical valves is generated with a number of available microactuators: thermopneumatic, thermomechanical, piezoelectric, electrostatic, and electromagnetic. Nonmechanical valves use novel materials and microscale effects to control the flow. Hydrogels and ferrofluids belong to this class of materials.

10.6.2 Inkjet Print Heads

Inkjet print heads belong to one of the most successful silicon-based microfluidic devices. The print head is the core of an inkjet printer. The print head can be micromachined in silicon using a batch process. Together with the ink cartridge, the print head is disposable. Due to the huge demand in personalized computing and office solutions, inkjet printers and inkjet print heads form the largest segment in the printer market, which is dominated by four main players: Canon, Hewlett-Packard, Epson, and Lexmark.

Inkjet print heads are categorized into three main concepts: thermal, piezoelectric, and continuous. Because of its simple implementation, thermal inkjet print heads can be found in most consumer inkjet printers. The print head consists of an array of ink chambers. Each chamber has a nozzle and at least one microheater (Figure 7.15). A current pulse through the microheater causes the ink to form a bubble in an explosive manner. The high pressure generated propels an ink droplet onto the paper. The bubble collapses, when the current is no longer applied to the heater. Surface tension force refills the ink into the chamber. The chamber, the microheater and the nozzle can be easily implemented with conventional microtechnology.

Another most common inkjet print head type uses a piezoelectric actuator in each nozzle to propel the ink droplets. The implementation of this concept is more complex and thus the fabrication cost is also higher. However, piezoelectric ink jet allows a wider variety of inks compared to other types. This unique characteristics and the low temperature involved allow piezoelectric inkjet heads to be used in the emerging field of inkjet-based material deposition. Different functional materials can be printed directly on a substrate, for instance in the case of polymeric microelectronics where circuits can be printed layer by layer directly on a flexible polymeric substrate.

Continuous inkjet is one of the oldest inkjet technologies. Nowadays, continuous inkjet print heads are mainly used for marking and coding of products. In this concept, liquid ink from a reservoir is supplied to a nozzle. At a high speed, a piezoelectric actuator breaks up the ink stream into a continuous stream of ink droplets. The droplets are charged at the nozzle and can be deflected by the electrostatic field between a pair of electrodes. By this way, controlled printing on a substrate can be achieved. Compared to the other two methods, the droplets in continuous inkjet have high kinetic energy and allow very long distance from the nozzle to the target substrate.

In the near future the development of inkjet print heads may go beyond conventional printing applications. One of the most exciting new applications for inkjet print head is the dispensing of chemical and samples on a screening assay. Such an array consists of thousands or even millions of tiny dots on a substrate. Each dot consists of a compound, which responses to a molecule to be screened. This array of dots can be simply deposited with the inkjet technology.

10.7 CHEMICAL, BIOLOGICAL AND MEDICAL SYSTEMS

10.7.1 Microreactors

It is apparent in the introduction of the previous section that microfluidics has the most impact in chemical, biological and medical applications. One of the unexplored areas is the chemical reaction in microscale. The unique microscale conditions allow the realization of reactions and products that were not possible in macroscale. The constraint of the small production volume can be overcome by using multiple microreactors running in parallel. Due to the small size of microreactors, a large production volume can be achieved by numbering-up instead of conventional scaling-up approach. Microsystem technologies offer all the key factors for miniaturization of chemical reactors. A number of components such as reaction chamber, catalyst, temperature control and flow control can be integrated in a single system. This batch process will enable a potentially low-cost mass production.

The unique reaction conditions achievable with microreactors are the fast thermal response as well as high thermal and concentration gradients. The fast thermal response makes temperature control in microreactors more precise. Further advantages for chemical reactions are a uniform temperature distribution in the reaction chamber, a short residence time, and a high surface-volume ratio. The latter is advantageous for catalyst-based reactions, where a large surface area means a better reaction condition. The higher heat and mass transfer rates in microscale allow reactions to be carried out under more aggressive conditions than previously in macroscale reactors. The shorter residence time in microreactors allows unstable intermediate products to be transferred quickly to the next process. Residence time on the order of microseconds opens up a new reaction pathway which is inaccessible in conventional reactors. Furthermore, the large surface-to-volume ratio allows effective suppression of homogeneous side reactions in heterogeneously catalyzed gas phase reactions.

The large ratio between surface heat losses and heat generation in microreactors makes chemical reactions become safe because of the suppression of flames and explosions. Furthermore, a small reaction volume leads to a small quantity of chemicals released accidentally. Therefore, large-scale hazards can easily be avoided. Since different types of sensors can be integrated in a microreactor, the operation condition can be monitored and controlled. Failed reactors can be detected, isolated, and replaced. With an array of microreactors working on redundancy principle, the replacement process can be carried out automatically by replacing the defect microreactor with another one in the same array.

The possible parallel process and high throughput make microreactors cost-effective, especially for chemical analysis. The required small volume of expensive reagents also minimizes the overall cost of the reaction. Because microreactors can be fabricated in batch cheaply, numbering-up production can be carried out by replication of reactor units. The numbering-up approach reduces the time of design-construction cycle, because lab-scale results are ready to use in the industrial plant. The number of microreactors can be tailored to the required production volume. This numbering-up approach is especially suitable for fine chemical and pharmaceutical industries, where only a small production capacity is needed.

10.7.2 Lab on a Chip

Labs on a chip refer to chemical and biochemical systems, where a whole lab-based analytical protocol can be integrated into a single chip. Analytical chemists, biochemists, and chemical engineers now take advantage of the new effects in microfluidics to design devices with better performance. As noted in the introduction to this section, a lab-on-chip platform does not need to be very small, because the microscopic length scales sometimes is not really beneficial, especially for sensing purpose. For instance, the sensitivity of a chemical sensor is limited by the analyte concentration in a sample. The relation between sample volume V in liter and analyte concentration c_i is given by [MANZ90]:

$$V = (\eta N_A c_i)^{-1}, \tag{10.9}$$

where η is the sensor efficiency ($0 < \eta < 1$), N_A the AVOGADRO number (mol^{-1}), and c_i the concentration of analyte in mol per liter. The above equation shows that the required sample volume or consequently the size of a lab-on-a-chip device depends on the concentration of the sample. If the sample volume is too small, the sample may not contain any target molecules. Thus, the sample is useless for detection. The required sample volume can be determined by the known required concentration. For instance, common human clinical chemistry assays require analyte concentrations between 10^{14} and 10^{21} copies per milliliter. The concentration range for a typical immunoassay is from 10^8 to 10^{18} copies per milliliter. Desoxyribonucleic acid (DNA) probe assays for genomic molecules, infective bacteria, or virus particles require a concentration range from 10^2 to 10^7 copies per milliliter. Immunoassays with their lower analyte concentrations require sample volumes on the order of nanoliters. Non-preconcentrated analysis of the DNA present in human blood requires a sample volume on the order of a milliliter. Some types of samples, like libraries for drug discovery, have relatively high concentrations.

10.8 ENERGY HARVESTING AND WIRELESS COMMUNICATIONS

In microsystem technology, there is an increasing number of applications that require autonomous power supply. Examples are autonomously operating microsystems or

Table 10.3 Energy sources in the environment for supplying energy for autonomous or distributed microsystems

Energy form	Effective principle	Estimated power density[1] in mW/cm^2 (solar and thermal energy) or mW/cm^3 (kinetic energy)	Characteristics
Solar energy	Photoelectric effect (photovoltaics)	Outdoors: 0.15...15 Indoors: < 0.01	Large areas necessary for indoor
Thermal energy	SEEBECK effect (see Section 7.1.3, Equation (7.40))	0.015 ($\Delta T = 10$ K)	Large temperature differences in narrow space necessary; very low efficiency
Kinetic energy[2]	Electrostatic effect (see Section 7.2.1)	0.05...0.1	Simple integration into microsystems; low efficieny
	Piezoelectric effect (see Section 7.2.2)	0.2	Required integration of piezoelectric transducer element (e.g. as thin layer)
	Electrodynamic transducer (see Section 7.2.3)	< 0.001	Integration of coils into microsystems is difficult

[1] Forres, E. O., Rincon-Mora, G. A.: Energy-harvesting chips: The quest for everlasting life. *Electronic Engineering Times* 1386, October 3, 2005, pp. 62
[2] also as vibration- or acoustic energy

distributed wireless self-organizing networks. Applications of such systems are automotive, planes, distributed machines or plants, facility technology, intelligent cloths or the military. Energy for such systems can be generated from wireless induction, such as in RFID (Radio Frequency Identification) systems, batteries or accumulators or by utilizing energy sources in the environment of such components or systems. The latter approach is also called 'energy harvesting' and has the advantages that microsystems can be continuously supplied with energy over the entire lifetime and that no spatial proximity to an energy sauce (which is the case for RFID systems) is necessary. Table 10.3 lists possible energy sources.

In this context, kinetic energy is of special interest as it has less restrictions than solar and thermal energy and a comparatively high efficiency in the range of several percent. A major focus of current R&D is on piezoelectric generators. As thin-film arrangements, they can be integrated into microsystems with an effort that is controllable. Figure 10.13 shows such a generator with a bending beam and a piezoelectric PZT layer that has been deposited on the beam as a mechanoelectric transducer (see Table 7.5). The interdigital electrodes deposited on it make it possible to utilize the d_{33}-mode (see Equation (7.65)), which allows a higher efficiency (due to the high coupling factor k^2, cp. Table 7.6) than other operating modes.

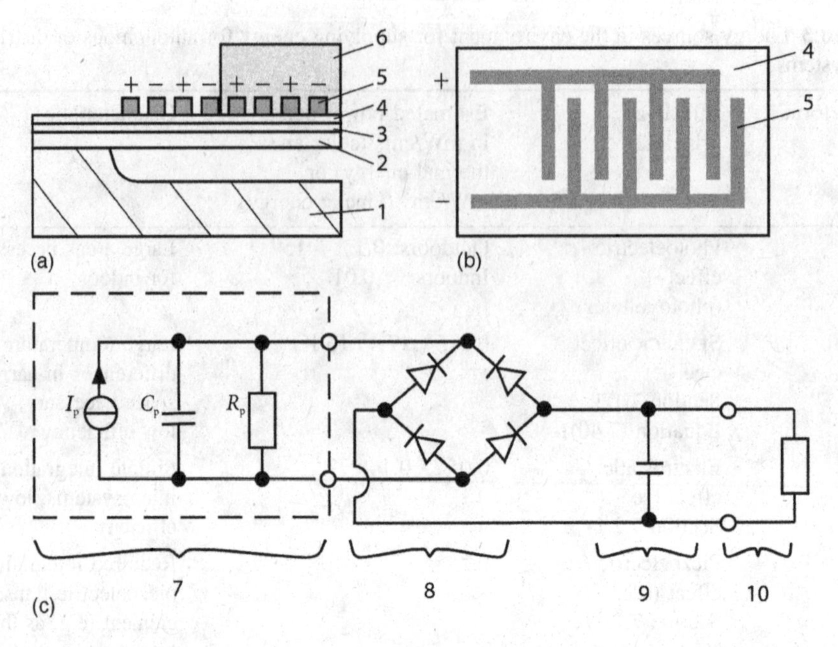

Figure 10.13 Schematic presentation of a piezoelectric energy generator: (a) cross-section, (b) top view; (c) electric equivalent circuit for read-out electronics; 1 Si chip, 2 carrier bending beam (SiO_2), 3 buffer layer, 4 piezoelectric layer (PZT), 5 interdigital electrodes (Ti, Pt), 6 inert mass (SU-8 photoresist) 7 piezoelectric energy generator, 8 rectifier, 9 energy storage, 10 load resistor. Reproduced from Jeon, Y.; Sood, R.; Steyn, L.; Kim, S.-G. (2003) Energy harvesting MEMS devices based on d_{33} mode piezoelelectric Pb(Zr,Ti)O_3 thin film cantilever. *CIRP Seminar on Micro and Nano Technology*, Copenhagen, Denmark, November 13–14

The manufacturing of the piezoelectric generator comprises the following steps[47]:

1. Deposition of the layer that subsequently forms the carrier bending beam (400 nm SiO_2, PECVD),
2. Deposition of a buffer layer (50 nm ZrO_2, sol-gel process or sputtering)
3. Deposition of the PZT layer (480 nm, sol-gel process or sputtering)
4. Patterning of PZT, ZrO_2 and SiO_2 layers through RIE (mask #1)
5. Deposition of interdigital electrode (20 nm Ti, 200 nm Pt, evaporation)
6. Patterning of electrode (lift-off, mask #2)
7. Generation of an inertial mass through spin-on (SU-8 photoresist) and photolithography (mask #3)
8. Baring of the movable structure through undercutting of Si below the SiO_2 bending beam using XeF_2 plasma etching. When etching Si, XeF_2 shows a high selectivity regarding SiO_2, ZrO_2, PZT and Ti/Pt. Therefore no further etch mask is required.

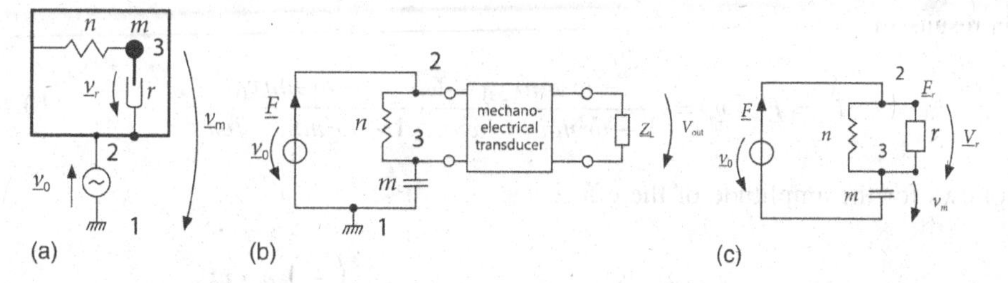

Figure 10.14 Simplified lumped model of a system for energy harvesting: (a) mechanical scheme; (b),(c) electric equivalent circuits, m mass, n compliance of the spring, r damping, Z_L load impedance, 1, 2, 3 system points

Figure 10.13c shows the electric circuit of such a piezoelectric generator. The generator itself can be considered to be current source I_p with an internal capacity C_p.[34] This results from transforming the equivalent circuit in Table 7.6, if we add one force source and then look at the electric contact behavior. R_p characterizes the dielectric losses. The mechanoelectric energy is then stored in a capacitor or an accumulator.

The simple equivalent circuit in Figure 10.14 will be used to present the requirements of an energy generating system. In principle, such arrangements consist of oscillating spring-mass systems that are excited by the oscillations of sinusoidal velocity $\underline{v}_0$. The movement of the mass is damped (friction $r = \underline{F}_r/\underline{v}_r$). Active power P, that is withdrawn from the system – i.e. that can be transformed into electric active power – is 'consumed' by damping r.

$$P = \underline{F}_r \cdot \underline{v}_r \tag{10.10}$$

In practice, the mechanoelectric transducer (e.g. an electrostatic, piezoelectric or electrodynamic transducer, see Sections 7.2.1 to 7.2.3) and load impedance $\underline{Z}_L$ (Figure 10.14b) form damping r. Mechanoelectric transducer and load impedance $\underline{Z}_L$ have to be adapted in order for them to form a damping element r, as only this way it is possible to draw active power (Figure 10.14c). Power P according to Equation (10.10) can be calculated using the following relations resulting from the lumped model in Figure 10.14c:

$$\frac{\underline{F}_r}{\underline{F}} = \frac{r}{r + \frac{1}{j\omega n}} = \frac{j\omega nr}{1 + j\omega nr} \tag{10.11}$$

$$\frac{\underline{v}_0}{\underline{F}} = \frac{1}{j\omega m} + \frac{1}{r + \frac{1}{j\omega n}} = \frac{1 - \omega^2 mn + j\omega nr}{j\omega m\,(1 + j\omega nr)} \tag{10.12}$$

$$\frac{\underline{v}_r}{\underline{v}_0} = \frac{j\omega m}{j\omega m + r + \frac{1}{j\omega n}} = \frac{-\omega^2 mn}{1 - \omega^2 mn + j\omega nr}. \tag{10.13}$$

[34] Khazan, A.D. (1994) *Transducers and Their Elements*. Englewood Cliffs: PTR Prentice Hall.

This results in

$$P = \underline{F}_r \cdot \underline{v}_r = \frac{-\omega^2 mnr\underline{v}_0}{1 - \omega^2 mn + j\omega nr} \cdot \frac{-\omega^2 mn\underline{v}_0}{1 - \omega^2 mn + j\omega nr}. \tag{10.14}$$

It follows for the amplitude of the effect

$$|P| = |\underline{F}_r \cdot \underline{v}_r| = \frac{(\omega^2 mn)^2 r \cdot v_0^2}{(1 - \omega^2 mn)^2 + (\omega nr)^2} = \frac{\left(\frac{\omega}{\omega_0}\right)^4 r \cdot v_0^2}{\left[1 - \left(\frac{\omega}{\omega_0}\right)^2\right]^2 + \left(\frac{\omega}{\omega_0}\frac{1}{Q}\right)^2} \tag{10.15}$$

with resonance frequency

$$\omega_0 = \frac{1}{\sqrt{mn}} \tag{10.16}$$

and quality factor

$$Q = \frac{1}{\omega_0 nr}. \tag{10.17}$$

v_0 is the amplitude of the sinusoidal excitation velocity.

Equation (10.15) applies in principle to all types of mechanoelectrical transducers. This means that for an ideal transducer, the choice of the transducing principle does not affect the energy volume that can be converted. Figure 10.15 shows the mean output power

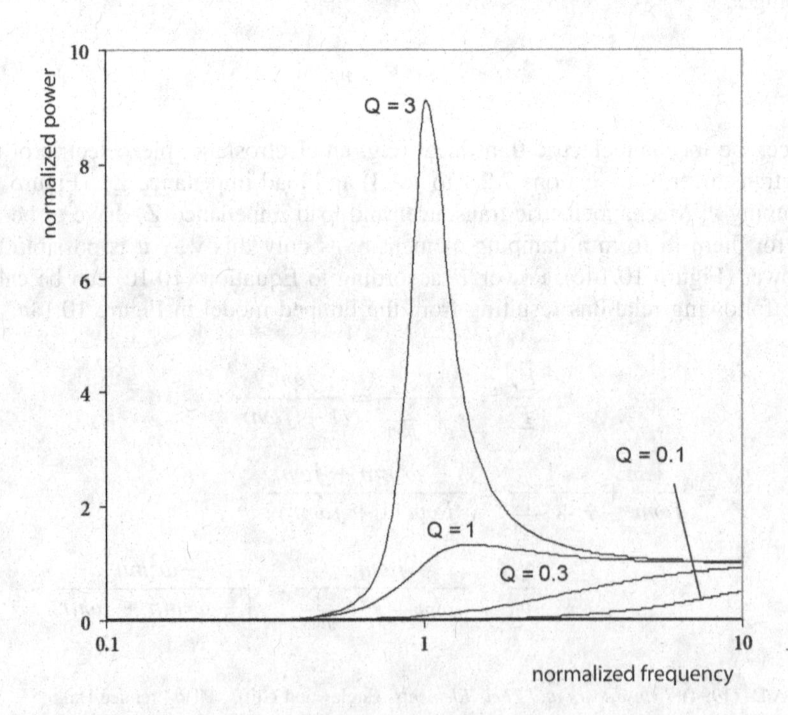

Figure 10.15 Piezoelectric generator: mean withdrawable power P in relation to quality factor Q

that is available in relation to quality factor Q. We can see that maximum power occurs at the resonance frequency. Quality factor Q determines the frequency selectivity of the energy generator. Large values of Q (low damping) lead to a power generation in only a small frequency band around resonance frequency ω_0, whereas a large damping (small values of Q) generates energy in a wide frequency band. The latter is of advantage if the external oscillation excitation for the energy generator occurs in a very wide band.

If the transducer characteristics are known, quality factor can be easily set via load impedance $\underline{Z}_L$. For an electrodynamic transducer according to Figure 7.9b, force F acting in magnetic field B on a coil (wire length ℓ) connected to load resistor R_L is:

$$F = \frac{(B\ell)^2}{R_\mathrm{L}} v = r\,v. \tag{10.18}$$

This results in quality factor Q

$$Q = \frac{1}{\omega_0 nr} = \frac{\omega_0}{\omega_0^2 nr} = \frac{\omega_0 mn}{nr} = \frac{\omega_0 m\,R_L}{(B\ell)^2}. \tag{10.19}$$

Here, R_L has to be selected according to optimum frequency (ω_0) and the spectrum of the oscillating movement

$$R_L = \frac{(B\ell)^2}{\omega_0\,m} \cdot Q. \tag{10.20}$$

In the resonance case $\omega = \omega_0$, it follows from Equation (10.15)

$$P = r \cdot v_0 \cdot Q^2 \tag{10.21}$$

or, correspondingly, Equation (10.17)

$$P = \frac{v_0\,m}{n \cdot r}. \tag{10.22}$$

Here, it follows that[35]

- the mass within the available volume of the energy generator should be as large as possible;
- the maximum velocity and thus the maximum displacement should also be as large as possible;
- compliance n of the spring should be dimensioned in order for the excitation frequency to correspond to the resonance frequency of the system; and
- electric load impedance $\underline{Z}_L$ has to be set in order for the quality factor Q to become large enough for a sufficient displacement of the mass.

[35] Williams, C.B., Yates, R.B. (1995) Analysis of a microelectric generator for microsystems. In: *Transducers '95, Eurosensors IX, Stockholm, Sweden, June 25–29, 1995*, pp. 369–72.

Recent research projects look at

- developing optimum structures for special applications that fulfill the named criteria; and
- developing manufacturing processes for the named transducer principles allowing a simple fabrication and a direct integration into the microsystem.

10.9 MICRO FUEL CELLS

Autonomous microsystems require an individual energy supply that operates, if possible, continuously. Partially, it is possible to obtain the energy from the environment (energy harvesting, Section 10.8). If that is not possible, an individual power supply has to be integrated into the microsystem. Typically, rechargeable batteries (accumulators) are used for this. They have the advantage that energy can be supplied without battery exchange; they have a lower energy density, though (Table 10.4). The working principle of a battery is based on the transport of ions, e.g. Li^+, from an anode through a membrane to a cathode. Therefore, only the ions close to the boundary area can participate in the chemical reaction. For this reason, a battery that is thicker by factor two will not provide twice the energy volume. This means that the scaling laws from Chapter 2 cannot be applied here. Fuel cells constitute an alternative with the energy density of the fuel being about one order of magnitude higher.

Figure 10.16 schematically shows the effective principle of fuel cells (using the example of a hydrogen fuel cell). The fuel (H_2) enters the cell at the anode and diffuses through a graphite layer that is electrically conducting in order to allow a current transport. At the catalysis layer (5 . . . 10 μm thick layer consisting of about 3 nm large Pt particles that have been deposited on a ca. 30 nm large graphite particle), H_2 dissociates into protons (H^+) and electrons. The proton exchange membrane allows the protons to pass, but not the electrons. Therefore, the electrons are forced through the outer circuit and flow through external load R_L. On the catalytic side of the cathode, H^+ ions are – under the effect of oxygen-converted into water.

Table 10.4 Comparison of fuel cells and rechargeable batteries

Principle	Type	Chemical reaction	Energy density in Wh/ℓ
Rechargeable batteries	Li polymer	$6C + 2LiCoO_2 \rightarrow LiC_6 + 2Li_{0.5}CoO_2$	360
	Ni-MeH[a)]	$NiOOH + MeH \rightarrow Ni\,(OH)_2 + Me$	205
	Ni-Cd	$2NiOOH + Cd + 2H_2O \rightarrow 2Ni(OH)_2 + Cd(OH)_2$	120
Fuel cells	H_2	$2H_2 + O_2 \rightarrow 2H_2O$	2300[b)]
	CH_3OH	$2CH_3OH + 3O_2 \rightarrow 2CO_2 + 4H_2O$	1 900

[a)] Metal hydride, [b)] By using liquid hydrogen and air.

Reproduced from Maynard, H.L., Meyers, J.P. (2002) Miniature fuel cells for portable power: Design considerations and challenges. *Journal of Vacuum Science and Technology* B 20(4), pp. 1287–97.

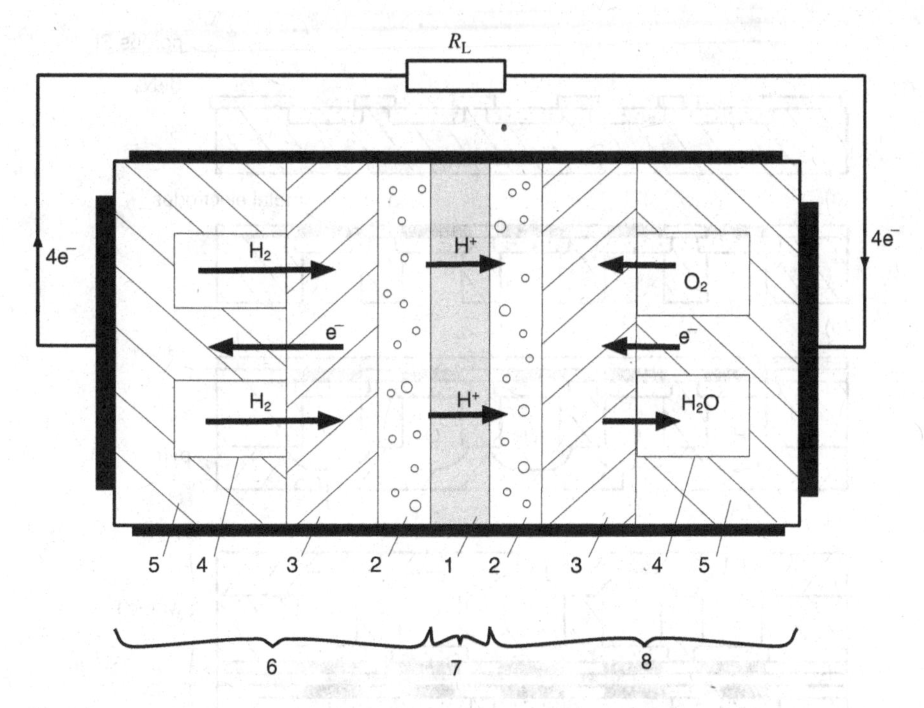

Figure 10.16 Principle of a hydrogen fuel cell. 1 proton exchange membrane (Nafion), 2 catalyst layer (platinum particles in electronically and ionally conducting matrix), 3 graphite diffusion layer, 4 channel for substance transport (H_2, O_2 or H_2O), 5 electron-conduction cover layer, R_L load resistor 6 anode, 7 proton exchange, 8 cathode

Methanol-based systems operate in a similar way with the two chemical reactions occurring as follows:

$$\begin{aligned} &\text{Anode:} \quad 2CH_3OH + 2H_2O \xrightarrow{\text{Pt - Ru}} 2CO_2 + 12H^+ + 12e^- \\ &\text{Kathode:} \quad 12H^+ + 12e^- + 3O_2 \xrightarrow{\text{Pt}} 6H_2O \end{aligned} \tag{10.23}$$

Cell voltage V_0 in a load-free state (kinetic regime) of the fuel cell is determined by the change of free energy

$$\Delta G_0 = -nFV_0$$

with n being the electron number per reaction and F the FARADAY constant. For hydrogen, $\Delta G_0 = -\ 237$ kJ/mol and thus $V_0 = 1.23\ V$. In the load case, the voltage typically decreases to $V = (0.6 \ldots 0.8)V$.

In a Pt catalyst, gaseous hydrogen outstandingly burns as fuel. To achieve acceptable energy densities, it has to be highly pressurized though. Currently, there are tests to store hydrogen, for example, also in graphite or other carbon forms (e.g. CNT Carbon

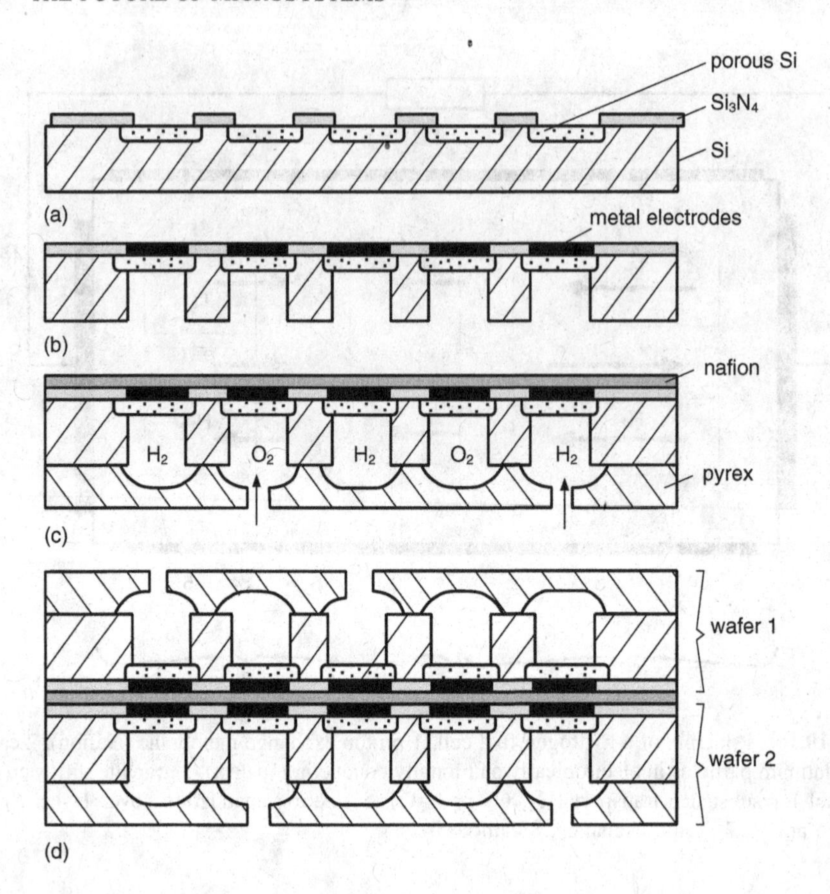

Figure 10.17 Manufacturing of a hydrogen micro fuel cell: (a) after anodizing the porous Si isles; (b) after DRIE and electrode deposition; (c) after bonding with Pyrex wafer and hot molding of the nafion ion exchange membrane, (d) after completion of the cell

Nanotubes). Regarding liquids, methanol is the fuel that catalytically dissociates most easily. Mainly nafion, a polymer electrolyte, is used as proton exchange membrane.[36,37]

Figure 10.17 shows an example of an MEMS-based solution for the manufacturing of hydrogen fuel cells. The diffusion layer consists of porous silicon that was produced by anodizing via the openings in the Si_3N_4 mask (HF/ethanol etching solution). The porous silicon is thermally oxidized for an electrical insulation and as etch stop for the subsequent deep reactive ion etching (DRIE). The resulting hollows used for supplying H_2 and O_2 are covered by glass wafers that are anodically bonded to the silicon wafer. The catalytic Pt/Ti electrodes are structured using lift-off processes. Two such wafers then form a complete cell. Here, the polymer electrolyte membrane connects the two wafers. Hot molding is used to generate them.

[36] A silicon-based air-breathing micro direct methanol fuel cell using MEMS technology. Agilent Technologies. 2006.

[37] Min, K.-B., Tanaka, S., Esashi, M. (2006) Fabrication of novel MEMS-based polymer electrolyte fuel cell architectures with catalytic electrodes supported on porous SiO_2. *Journal of Micromachining and Microengineering* 16, pp. 505–11.

Currently, there are two approaches to the further development of fuel cells:

1. use of Si process technology in order to create integrated structures, and
2. micro-assembly of individual miniaturized components.

The latter approach has the advantage that for the reactor's specific areas other materials than silicon can be used.

Further microfluidic components (channel structures, pumps, valves) are required for completing micro fuel cells. The complex structure and the multifaceted requirements for the individual function components continue to make the packaging of micro fuel cells costly, which means that new and better structural concepts are needed.

REFERENCES

[AKIN05] Akin, T. (2005) CMOS-based thermal sensors. In: *Advanced Micro- and Nanosystems*. Vol. 2: *CMOS-MEMS*. Weinheim, Wiley-VCH Verlag, pp. 479–512.

[ALBRECHT90] Albrecht, T. R., Akamine, S., Carver, T. E., Quate, C. F. (1990) Microfabrication of cantilever styli for the atomic force microscope. *Journal of Vacuum Science and Technology* A 8(4), pp. 3386–96.

[BINNIG99] Binnig, G., Rohrer, H. (1999) In touch with atoms. *Review of Modern Physics* 71(2), pp. S324–S330.

[BCC03] Business Communications Company, Inc.: Microfluidics Technology–Updated Edition. Publication No.RGB-226R, 2003, Norwalk, CT.

[CANIOU99] Caniou, J. (1999) *Passive Infrared Detection – Theory and Application*. Boston: Kluwer.

[CETINER04] Cetiner, B. A., Jafarkhani, H., Qian, J., Yoo, H. J. *et al.* (2004) Multifunctional reconfigurable MEMS integrated antennas for adaptive MIMO systems. *IEEE Communications Magazine*, 40(2), pp. 62–72.

[DAY]05 Day, D. R., Butler, M. A., Smith, M. C., McAllister, A., Deutsch, E. R. *et al.* (2005) Diffractive-MEMS implementation of a Hadamard near-infrared spectrometer. *Transducers '06, The 13th International Conference on Solid-State Sensors, Actuators and Microsystems*, Korea 2005, pp. 1246–9.

[ELOY06] Eloy, J. C. (2006) Status of the MEMS industry in 2006. *Sensors & Transducers Magazine* (S&T e-Digest) 66 (4), pp. 521–5.

[GABBAY00] Gabbay, L., Mehner, J., Senturia, S. D. (2000) Computer-aided generation of nonlinear reduced – order dynamic macromodels. *IEEE Journal of Microelectromechanical Systems*, 9, pp. 262–8.

[GRAF06] Graf, A., Arndt, M., Sauer, M., Gerlach, G. (2006) Review of micromachined thermopiles for infrared detection. *Measurement Science and Technology*, 18, pp. R59–R75.

[KIMATA98] Kimata, M. (1998) Infrared focal plane arrays. In: *Sensors Update*, 4(1), Weinheim: Wiley-VCH Verlag, pp. 53–79.

[LASER04] Laser, D. J., Santiago, J. G. (2004) A review of micropumps. *Journal of Micromachining and Microengineering* 14(6), R35–R64.

[MANZ90] Manz, A., Graber, N. Widmer, H. M. (1990) Miniaturized total chemical analysis systems: a novel concept for chemical sensing. *Sensors and Actuators* B1, pp. 244–8.

[MOTAMEDI05] Motamedi, M. E. (2005) *MOEMS Micro-Opto-Electro-Mechanical Systems*. SPIE Press.

[NGUYEN02] Nguyen, N. T., Huang, X. Y., Toh, K. C. (2002) MEMS-micropumps: a review. *ASME Journal of Fluids Engineering*, 124(2), pp. 384–92.

[OH06] Oh, K. W, Ahn, C. H. (2006) A review of microvalves. *Journal of Micromachining and Microengineering* 16 (5), R13–R39.

[POLY] Polychromix spectrometer technology: http://www.polychromix.com

[RANGELOW05] Rangelow, I. W. (2005) Piezoresistive scanning proximity probes for nanoscience. *Technisches Messen* 72, pp. 103–10.

[SOMMER06] Sommer-Larsen, P., Kornbluh, R. (2006) Overview and recent advances in polymer actuators. *Proceedings of the 10th International Conference on New Actuators*. Bremen, pp. 86–96.

[TANG05] Tang, M., Liu, A. Q., Agarwal, A., Liu, Z. S., Lu, C. (2005) A single-pole double-throw (SPDT) circuit using lateral metal-contact micromachined switches. *Sensors and Actuators* A121(1), pp. 187–96.

[TEXINS] Texas Instruments digital light processing technology: http://dlp.com

[TRISNADI04] Trisnadi, J. I., Carlisle, C. B., Monteverde, R. (2004) Overview and application of grating light valve based optical engines for high-speed digital imaging. Photonics West, paper 5348-05.

[VARADAN03] Varadan, V. K. (2003) *RF MEMS and Their Applications*. John Wiley & Sons, Inc. New York.

[YURISH05] Yurish, S. Y., Kirianaki, N. V., Myshkin, I. L. (2005) World sensors and MEMS markets: analysis and trends. *Sensors & Transducers Magazine* (S&T e-Digest) 62.

Appendix A

Physical Constants

Electric unit charge	$e = 1.6022 \cdot 10^{-19}$ As
Electric field constant	$\varepsilon_0 = 8.8542 \cdot 10^{-12}$ As/Vm
Magnetic field constant	$\mu_0 = 1.2566 \cdot 10^{-6}$ Vs/Am
PLANCK constant	$h = 6.6261 \cdot 10^{-34}$ Js
BOLTZMANN constant	$k = 1.3807 \cdot 10^{-23}$ J/K
STEFAN-BOLTZMANN constant	$\sigma = 5.6705 \cdot 10^{-8}$ W/m^{2}K^4

Introduction to Microsystem Technology: A Guide for Students Gerald Gerlach and Wolfram Dötzel

Appendix B

Coordinate Transformation

In microsystem technology, silicon is used for a large number of actuators and sensors. Regarding the mechanical and electric properties of these microsystem components and their manufacturing, the function elements are often not oriented in the basic direction of the silicon lattice (e.g. in [1 0 0]–direction), but in any direction $[h\,k\,l]$ (see Sections 4.6 and 7.2.5). The tensor properties for this direction can be calculated based on their basic orientation, using coordinate transformation relations.

Starting point is the rotation of the original coordinate system xyz by angles ψ, φ and ν into coordinate system $x'y'z'$ (Figure B.1).

The transformation equations thus become

$$\begin{bmatrix} e_x \\ e_y \\ e_z \end{bmatrix} = \begin{bmatrix} \cos(e_x e_{x'}) & \cos(e_x e_{y'}) & \cos(e_x e_{z'}) \\ \cos(e_y e_{x'}) & \cos(e_y e_{y'}) & \cos(e_y e_{z'}) \\ \cos(e_z e_{x'}) & \cos(e_z e_{y'}) & \cos(e_z e_{z'}) \end{bmatrix} \cdot \begin{bmatrix} e_{x'} \\ e_{y'} \\ e_{z'} \end{bmatrix} = \begin{bmatrix} l_1 & l_2 & l_3 \\ m_1 & m_2 & m_3 \\ n_1 & n_2 & n_3 \end{bmatrix} \begin{bmatrix} e_{x'} \\ e_{y'} \\ e_{z'} \end{bmatrix} \quad \text{(B.1)}$$

or respectively

$$\begin{bmatrix} e_{x'} \\ e_{y'} \\ e_{z'} \end{bmatrix} = \begin{bmatrix} l_1 & m_1 & n_1 \\ l_2 & m_2 & n_2 \\ l_3 & m_3 & n_3 \end{bmatrix} \cdot \begin{bmatrix} e_x \\ e_y \\ e_z \end{bmatrix}. \quad \text{(B.2)}$$

The resulting direction cosines between the two coordinate systems are

$$\begin{aligned} l_1 &= \cos\psi\cos\varphi - \cos\nu\sin\psi\sin\varphi \\ l_2 &= -\cos\psi\sin\varphi - \cos\nu\sin\psi\cos\varphi \\ l_3 &= \sin\nu\sin\psi \\ m_1 &= \sin\psi\cos\varphi + \cos\nu\cos\psi\sin\varphi \\ m_2 &= -\sin\psi\sin\varphi + \cos\nu\cos\psi\cos\varphi. \\ m_3 &= -\sin\cos\psi \\ n_1 &= \sin\nu\sin\varphi \\ n_2 &= \sin\nu\cos\varphi \\ n_3 &= \cos\nu \end{aligned} \quad \text{(B.3)}$$

Introduction to Microsystem Technology: A Guide for Students Gerald Gerlach and Wolfram Dötzel

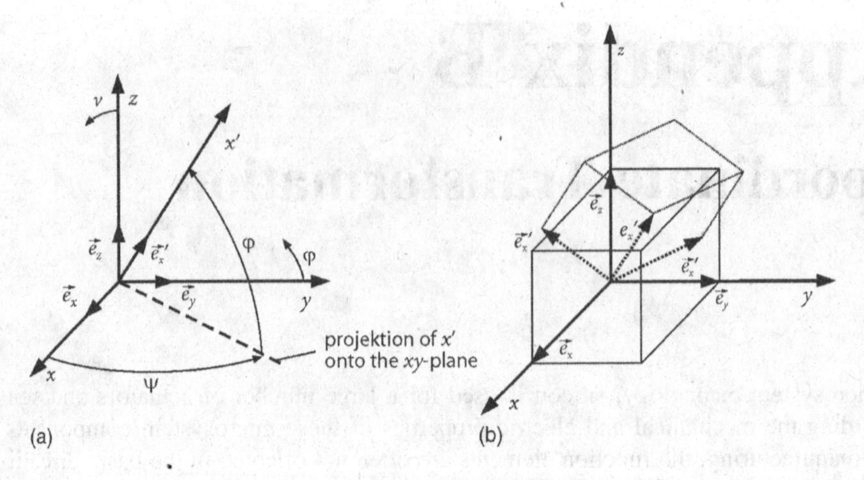

Figure B.1 Transformation of the coordinate system by rotating with EULER angles ψ, φ and ν (presented for the rotation of the x-axis; y- and z-axis are rotated by the corresponding angles ψ, φ and ν)

B.1 ELASTIC COEFFICIENTS

The relationship between strain tensor ε_i ($\varepsilon_1 \ldots \varepsilon_3$ normal strains; $\varepsilon_4 \ldots \varepsilon_6$ shear strains) and stress tensor σ_i ($\sigma_1 \ldots \sigma_3$ normal stresses; $\sigma_4 \ldots \sigma_6$ shear stresses) is determined in VOIGT notation via the following relations:

$$\varepsilon_i = s_{\mathrm{ij}} \cdot \sigma_j \tag{B.4a}$$

or, correspondingly,

$$\sigma_i = c_{\mathrm{ij}} \cdot \varepsilon_j \tag{B.4b}$$

with

$$s_{\mathrm{ij}} = c_{\mathrm{ij}}^{-1}. \tag{B.4c}$$

Here, s_{ij} are the elastic coefficients and c_{ij} the elastic moduli. In the rotated coordinate system, it results that

$$\varepsilon_i' = s_{\mathrm{ij}}' \cdot \sigma_j' \tag{B.5}$$

with

$$s_{\mathrm{ij}}' = \begin{bmatrix} s_{11} - sF_1 & s_{12} + (1/2)sF_{12} & s_{12} + (1/2)sF_{13} & sG_{231} & sG_{31} & sG_{21} \\ s_{12} + (1/2)sF_{12} & s_{11} - sF_2 & s_{12} + (1/2)sF_{23} & sG_{32} & sG_{132} & sG_{12} \\ s_{12} + (1/2)sF_{13} & s_{12} + (1/2)sF_{23} & s_{11} - sF_3 & sG_{23} & sG_{13} & sG_{123} \\ sG_{231} & sG_{32} & sG_{23} & s_{44} + 2sF_{23} & 2sG_{123} & 2sG_{312} \\ sG_{31} & sG_{132} & sG_{13} & 2sG_{123} & s_{44} + 2sF_{31} & 2sG_{231} \\ sG_{21} & sG_{12} & sG_{123} & 2sG_{312} & 2sG_{231} & s_{44} + 2sF_{12} \end{bmatrix} \tag{B.6}$$

and

$$\begin{aligned} s &= 2s_{11} - 2s_{12} - s_{44} \\ F_i &= l_i^2 m_i^2 + m_i^2 n_i^2 + n_i^2 l_i^2 \\ F_{ij} &= l_i^2 l_j^2 + m_i^2 m_j^2 + n_i^2 n_j^2 \\ G_{ij} &= l_i l_j^3 + m_i m_j^3 + n_i n_j^3 \\ G_{ijk} &= l_i l_j l_k^2 + m_i m_j m_k^2 + n_i n_j n_k^2. \end{aligned} \tag{B.7}$$

For each coordinate system, silicon can be assumed to be quasi-isotropic. This way, we can introduce the corresponding YOUNG's and shear moduli E' and G' as well as POISSON's ratios ν':

$$\frac{1}{E'} = s'_{11} = s_{11} - s(l_1^2 m_1^2 + m_1^2 n_1^2 + l_1^2 n_1^2) \tag{B.8}$$

$$\frac{1}{G'} = s'_{44} = s_{44} + 2s(l_1^2 l_2^2 + m_1^2 m_2^2 + n_1^2 n_2^2) \tag{B.9}$$

$$\nu' = -\frac{s'_{12}}{s'_{11}} = -\frac{2s_{12} + s(l_1^2 l_2^2 + m_1^2 m_2^2 + n_1^2 n_2^2)}{2s_{11} - 2s(l_1^2 m_1^2 + m_1^2 n_1^2 + l_1^2 n_1^2)}. \tag{B.10}$$

For <1 1 0>-direction in (1 0 0)-silicon, it results that $\nu = 0°, \psi = 45°, \varphi = 0°$, and applying the relations in Equation (B.3),

$$\begin{aligned} l_1 &= -l_2 = m_1 = m_2 = 1/\sqrt{2} \\ l_3 &= m_3 = n_1 = n_2 = n_3 = 0. \end{aligned}$$

From Equation (B.8), it thus results that

$$E_{(1\ 0\ 0)<1\ 1\ 0>} = \frac{1}{s_{11} - (2s_{11} - 2s_{12} - s_{44})/4} = 1.69 \cdot 10^9 \mathrm{N/m^2} \tag{B.11}$$

and from Equation (B.10)

$$\nu_{(1\ 0\ 0)<1\ 1\ 0>} = -\frac{2s_{12} + s/4}{2s_{11} - s/2} = -\frac{s_{11} + s_{12} - s_{44}/2}{s_{11} + s_{12} + s_{44}/2} = 0.063 \tag{B.12}$$

(see Table 3.8).

In the different direction of the (1 0 0)-area ($\nu = 0°, \varphi = 0°$) of silicon, YOUNG's modulus becomes

$$E_{(1\ 0\ 0)} = \frac{1}{s_{11} - (2s_{11} - 2s_{12} - s_{44}) l_1^2 m_1^2} \tag{B.13}$$

and

$$\nu_{(1\ 0\ 0)} = -E_{(1\ 0\ 0)}[s_{12} - (2s_{11} - 2s_{12} - s_{44}) \cdot (l_1^2 l_2^2 + m_1^2 m_2^2)] \tag{B.14}$$

Biaxial YOUNG's modulus $E_{(1\ 0\ 0)}/(1 - \nu_{(1\ 0\ 0)})$ is thus independent of the orientation:

$$\left(\frac{E}{1-\nu}\right)_{(1\ 0\ 0)} = \frac{1}{s_{11} + s_{12}}. \tag{B.15}$$

The elastic properties of polysilicon results from the average value of all existing orientations. A completely random distribution of the crystallites results in a mean YOUNG's modulus [SENTURIA01]

$$\overline{E} \approx 1/(0.6s_{11} + 0.4s_{12} + 0.25s_{44}) = 1.45 \text{ GPa}. \tag{B.16}$$

For specific textures, e. g. if the distribution of the crystallites is not even any longer, but has a preferential orientation, Equation (B.16) changes.

The calculation of the mean value of POISSON's ratio is extremely complicated, as the mean value has to be calculated across all orientations with additional consideration of the rectangular position of longitudinal and transversal axis. [SENTURIA01] suggests the value $\overline{\nu} = 0.25$ as a good approximation of most amorphous and polycrystalline silicon films.

Table B.1 comprises the calculated values of the elastomechanical parameters und for the piezoresistive coefficients of different textures in polysilicon.[1]

Table B.1 Electromechanical and piezoresistive coefficients of poly-Si without and with texture

Texture	YOUNG's modulus $\overline{E}$ in GPa	POISSON's ratio $\overline{\nu}$	Piezoresistive longitudinal coefficient in 10^{-11} m^{2}N^{-1}	
			p-Si	n-Si
without	160	0.27	+53.1	−55.8
[1 0 0]	150	0.25	+39.4	−68.0
[1 1 1]	171	0.21	+72.8	−31.6
[3 1 1]	156	0.23	+49.5	−56.9
[1 1 0]	167	0.22	+64.5	−40.9
[3 3 1]	168	0.22	+66.7	−38.2

B.2 PIEZORESISTIVE COEFFICIENTS

In a resistor without mechanical stress load ($\sigma_j = 0$), current (current density J_i) only flows in the direction in which electrical field strength E_i acts:

$$\begin{bmatrix} E_1 \\ E_2 \\ E_3 \end{bmatrix} = \begin{bmatrix} \rho_1 & 0 & 0 \\ 0 & \rho_2 & 0 \\ 0 & 0 & \rho_3 \end{bmatrix} \cdot \begin{bmatrix} J_1 \\ J_2 \\ J_3 \end{bmatrix}. \tag{B.17}$$

Mechanical stress σ_j changes the symmetry and thus, current can also flow in the other directions.

[1] French, P. J. (2002) Polysilicon: a versatile material for microsystems. *Sensors and Actuators* A 99, pp. 3–12.

The generalization of Equation (7.112) results in

$$\frac{1}{\rho_{11}}E_1 + \frac{1}{\rho_{12}}E_2 + \frac{1}{\rho_{13}}E_3 = (1+\pi_{1m}\sigma_m)J_1 + \pi_{6m}\sigma_m J_2 + \pi_{5m}\sigma_m J_3$$

$$\frac{1}{\rho_{12}}E_1 + \frac{1}{\rho_{22}}E_2 + \frac{1}{\rho_{23}}E_3 = \pi_{6m}\sigma_m J_1 + (1+\pi_{2m}\sigma_m)J_2 + \pi_{4m}\sigma_m J_3$$

$$\frac{1}{\rho_{13}}E_1 + \frac{1}{\rho_{23}}E_2 + \frac{1}{\rho_{33}}E_3 = \pi_{5m}\sigma_m J_1 + \pi_{4m}\sigma_m J_2 + (1+\pi_{3m}\sigma_m)J_3. \quad \text{(B.18)}$$

The six mechanical stress components are represented by σ_m; using VOIGT's sum notation for these equations, the summation can be carried out via index $m \in \{1\ldots6\}$. Equation (B.18) contains $6 \times 6 = 36$ piezoresistive coefficients π_{im}. The crystals that are of technical interest have a comparatively large symmetry, though. That means that many of the coefficients are identical or zero. For cubic crystals, it results that

$$\pi_{11} = \pi_{22} = \pi_{33}$$

$$\pi_{12} = \pi_{21} = \pi_{13} = \pi_{31} = \pi_{23} = \pi_{32}$$

$$\pi_{44} = \pi_{55} = \pi_{66}. \quad \text{(B.19)}$$

All remaining piezoresistive coefficients are zero. Matrix $[\pi_{im}]$ adopts thus the following form:

$$[\pi_{im}] = \begin{bmatrix} \pi_{11} & \pi_{12} & \pi_{12} & 0 & 0 & 0 \\ \pi_{12} & \pi_{11} & \pi_{12} & 0 & 0 & 0 \\ \pi_{12} & \pi_{12} & \pi_{11} & 0 & 0 & 0 \\ 0 & 0 & 0 & \pi_{44} & 0 & 0 \\ 0 & 0 & 0 & 0 & \pi_{44} & 0 \\ 0 & 0 & 0 & 0 & 0 & \pi_{44} \end{bmatrix}. \quad \text{(B.20)}$$

In the rotated coordinate systems, all values in Equation (B.18) have to be replaced with the transformed values ($E \rightarrow E'$; $\rho \rightarrow \rho$; $\pi \rightarrow \pi'$; $\sigma \rightarrow \sigma'$; $J \rightarrow J'$).

The matrix in Equation (B.20) changes for the new coordinate system to

$$[\pi_{im}] = \begin{bmatrix} \pi_{11} - 2\pi F_1 & \pi_{12} + 2\pi F_{12} & \pi_{12} + \pi F_{13} & 2\pi G_{231} & 2\pi G_{31} & 2\pi G_{21} \\ \pi_{12} + 2\pi F_{12} & \pi_{11} - 2\pi F_2 & \pi_{12} + 2\pi F_{23} & 2\pi G_{32} & 2\pi G_{132} & 2\pi G_{12} \\ \pi_{12} + 2\pi F_{13} & \pi_{12} + 2\pi F_{23} & \pi_{11} - 2\pi F_3 & 2\pi G_{23} & 2\pi G_{13} & 2\pi G_{123} \\ \pi G_{231} & \pi G_{32} & \pi G_{23} & \pi_{44} + 2\pi F_{23} & 2\pi G_{123} & 2\pi G_{312} \\ \pi G_{31} & \pi G_{132} & \pi G_{13} & 2\pi G_{123} & \pi_{44} + 2\pi F_{31} & 2\pi G_{231} \\ \pi G_{23} & \pi G_{12} & \pi G_{123} & 2\pi G_{312} & 2\pi G_{231} & \pi_{44} + 2\pi F_{12} \end{bmatrix} \quad \text{(B.21)}$$

with

$$\pi = \pi_{11} - \pi_{12} - \pi_{44}. \quad \text{(B.22)}$$

For longitudinal and transversal coefficients π_L and π_T, it applies – due to the correspondence of π_{im} to π'_{im} – to the transformed Equation (B.18) in the rotated coordinate system

$$\pi_L \equiv \pi'_{11} \tag{B.23}$$

$$\pi_T \equiv \pi'_{12}. \tag{B.24}$$

Table B.2 shows the piezoresistive coefficients π_L and π_T for the selected orientations in silicon.

Table B.2 Piezoresistive longitudinal and transversal coefficients of selected crystallographic orientations in silicon

Wafer orientation	Current direction	$\pi_L = \pi'_{11}$	Transversal direction	$\pi_T = \pi'_{12}$
(1 0 0)	[1 1 0]	$\frac{\pi_{11}+\pi_{12}+\pi_{44}}{2}$	[1 $\bar{1}$ 0]	$\frac{\pi_{11}+\pi_{12}-\pi_{44}}{2}$
(1 1 0)	[0 0 1]	π_{11}	[1 1 0]	π_{12}
	[1 1 1]	$\frac{\pi_{11}+2\pi_{12}+2\pi_{44}}{3}$	[1 1 0]	$\frac{\pi_{11}+2\pi_{12}-\pi_{44}}{3}$
(1 1 1)	[1 1 0]	$\frac{\pi_{11}+\pi_{12}+\pi_{44}}{2}$	[1 1 2]	$\frac{\pi_{11}+5\pi_{12}-\pi_{44}}{6}$

REFERENCES

[BAO05] Bao, M. (2005) *Analysis and Design Principles of MEMS Devices*. New York: Elsevier.

[LENK75] Lenk, A. (1975) *Elektromechanische Systeme* (Electromechanical Systems). Vol. 3: *Systeme mit Hilfsenergie* (Systems with Energy Supply; in German). Berlin. Verlag Technik.

[MESCHEDER04] Mescheder, U. (2004) *Mikrosystemtechnik* (Microsystem Technology; in German). 2. Ed. Stuttgart, Leipzig: B.G. Teubner.

[SENTURIA01] Senturia, S.D. (2001) *Microsystem Design*. Boston: Kluwer Academic Publishers.

Appendix C

Properties of Silicon Dioxide and Silicon Nitride Layers

The properties of SiO_2 and Si_3N_4 thin layers are extremely influenced by the layer deposition technique (Tables C.1, C.2).

Table C.1 Properties of SiO_2 layers

Deposition technique	PECVD	$SiH_4 + O_2$	TEOS	$SiCl_2H_2 + N_2O$	Thermal
Deposition temperature in °C	200	450	700	900	1100
Composition[1]	$SiO_{1,9}(H)$	$SiO_2(H)$	SiO_2	$SiO_2(Cl)$	SiO_2
Edge coverage	not conformal	not conformal	conformal	conformal	conformal
Thermal stability	loses H	compression	stable	loses Cl	excellent
Density in g/cm³	2.3	2.1	2.2	2.2	2.2
Refractive index	1.47	1.44	1.46	1.46	1.46
Intrinsic stress in MPa[2]	−300...+300	+300	−100	−100	−300
Dielectric strength in V/μm	300...600	800	1000	1000	1000
Etch rate in nm/min (H_2O : HF = 100 : 1)	40	6	3	3	3

[1] in brackets: incorporated atoms;
[2] + tensile stress, – compressive stress.

Table C.2 Properties of Si_3N_4 layers

Deposition technique	LPCVD	PECVD
Deposition temperature in °C	700...800	250...350
Composition[1]	$Si_3N_4(H)$	SiN_xH_y
Si/N ratio	0.75	0.6...1.2
Portion of H atoms	4...8	20...25
Refractive index	2.01	1.8...2.5
Density in g/cm³	2.9...3.1	2.4...2.8
Resistivity in Ω · cm	10^{16}	$10^{6}...10^{15}$
Dielectric strength in V/μm	1.100	500
Intrinsic stress in MPa[2]	+1000	−200...+500

[1] in brackets: incorporated atoms;
[2] + tensile stress, – compressive stress.

REFERENCES

[SZE83] Sze, S.M. (ed.) (1983) *VLSI-Technology*. New York: McGraw-Hill.
[KOVACZ98] Kovacz, G.T.A. (1998) *Transducers Sourcebook*. Boston: WCB/McGraw-Hill.

Appendix D

Nomenclature of Thin-film Processes

There are a large number of procedures for depositing and structuring thin layers. They have different goals, comprise different physical processes and show a large variability regarding the contributing particles. Due to the wide range of techniques and their simplified distinction, often only abbreviations (acronyms) are used. In general, these denominations are structured as follows:

$X_1X_2Y(B)Z$.

They denominate the following

Z	main goal of the procedure
Y	type of contributing particles
B	if applicable, whether the particle flow is collimated (B from beam)
X_1, X_2	specificities of the technology

Table D.1 presents the most important variants of Z, Y and X.

Table D.1 Abbreviations in the denominations of thin-film processes

Z Main goal	Y Particles	X_1X_2
E Etching[b]	N Neutrals	R Reactive
D Deposition	R Radicals	F Focused
M Modification	I Ions	A Assisted[a]
	E Electrons	TA Thermally Assisted
	P Photons	CA Chemically Assisted
	T Phonons	RA Reactively Assisted
		D Deep
		CV Chemical Vapor
		PV Physical Vapor
		PL Pulsed Laser

[a]'assisted' is used if, in addition to the energy form of the concerned particles, other energy sources are included;
[b]seldom also Evaporation.

Examples

IBE	Ion Beam Etching
FIBE	Focused Ion Beam Etching
DRIE	Deep Reactive Ion Etching
CARIBE	Chemically Assisted Reactive Ion Beam Etching (flow of reactive ions, assisted by a chemical gas flow)
CVD	Chemical Vapor Deposition

In practice, there are many exceptions or fuzzy denominations, though (Table D.2):

Table D.2 Exceptions or fuzzy denominations of acronyms in thin-film technology

In use:	Preferably:
RIE Reactive Ion Etching	RAIBE Reactively Assisted Ion Beam Etching
PE Plasma Etching (at high temperatures)	TARE Thermally Assisted Radical Etching

Applying the nomenclature in Table D.1, it is possible to derive acronyms for a number of techniques, that are however not used (Table D.3):

Table D.3 Denominations according to nomenclature in D.1 for known techniques

Denomination in use	Abbreviation according to nomenclature in Table D.1 (not used)
Photolithography	PBM Photon Beam Modification
Sandblasting	NBE Neutral Beam Etching
Electron beam evaporation	EBE Electron Beam Etching
Thermal decomposition	TE Thermal Etching
Laser ablation	PBE Photon Beam Etching

REFERENCE

[BERENSCHOT96] Berenschot, E., Jansen, H., Burger, G.-J., Gardeniers, H., Elwenspoek, M. (1996) Thermally assisted ion beam etching of polytetrafluoroethylene – A new technique for high aspect ratio etching of MEMS. In: *Proceedings of the Ninth Annual International Workshop on Micro Electro Mechanical Systems, February 11-15, 1996, San Diego, California, USA*. IEEE, pp. 277–84.

Appendix E

Adhesion of Surface Micromechanical Structures

Adhesion of flexible structures in surface micromachining is caused by capillary forces in thin gaps that result from aqueous solutions used for etching thin sacrificial layers (Section 4.10, Figure 4.45).

E.1 CAPILLARY FORCES

Figure E.1 shows a liquid drop in a capillary gap between two fixed plates. If contact angle φ between liquid and gap is smaller than 90°, the internal pressure of the drop becomes smaller than the surrounding air. This results in attraction force F between the plates, which – for the static state of equilibrium – would have to be contributed as an external counterforce. Pressure difference Δp_{FA} at the boundary between liquid and air is proportional to surface tension γ_{FA}:

$$\Delta p_{FA} = p_F - p_A = \frac{\gamma_{FA}}{r}. \tag{E.1}$$

Radius r of the liquid meniscus is negative, if the curvature is concave (Figure E.1a). It results from the geometry to

$$r = -\frac{d}{2} \cdot \frac{1}{\cos\varphi}. \tag{E.2}$$

External counterforce F, which is required for the equilibrium, thus becomes

$$F = -\Delta p_{FA} \cdot A = \frac{2 \cdot A \cdot \gamma_{FA} \cdot \cos\varphi}{d}. \tag{E.3}$$

At equilibrium, contact angle φ is determined by the surface tensions liquid-air (index FA) and solid-liquid (index SF) which contribute to the surface tension between fixed

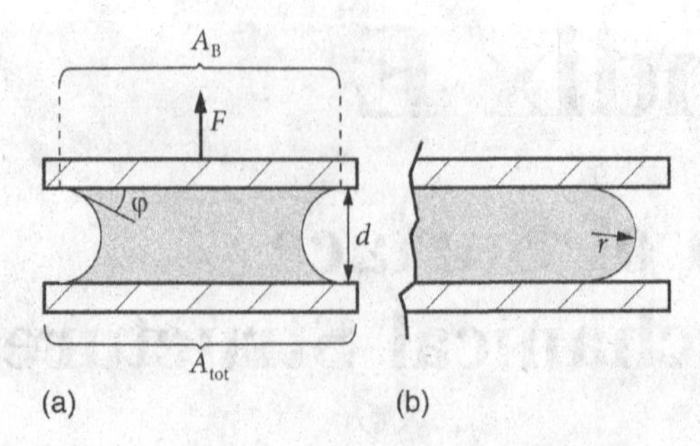

Figure E.1 Capillary forces at a thin gap filled with liquid: (a) concave; (b) convex curvature of the liquid meniscus. φ contact angle, A_B wetted area, A_{tot} total area, d gap width, F force, r radius of the meniscus

plates and air (index SA):

$$\gamma_{SA} = \gamma_{SF} + \gamma_{FA}\cos(\varphi). \tag{E.4}$$

For the arrangement in Figure E.1, the surface energy E_O becomes

$$E_O = A_{tot}\gamma_{SA} - A_B(\gamma_{SA} - \gamma_{SF}) = 2(A_G\gamma_{SA} - A_B\gamma_{FA}\cos\varphi). \tag{E.5}$$

Total area A_{tot} and surface tension γ_{SA}, γ_{SF} and γ_{FA} are constant, which means that E_O can be described as a function of the wetted area A_B:

$$E_O = C - 2A_B\gamma_{FA}\cos\varphi. \tag{E.6}$$

C is a constant and amounts to $C = 2(A_{tot}\gamma_{SA})$. The expression $\gamma_{FA}\cos\varphi$ describes the adhesive tension.

E.2 CRITICAL LENGTH OF CANTILEVER SPRINGS

Mechanically movable, surface-micromachined structures adhere via a liquid film to the substrate surface if the total energy there reaches a minimum. Here, total energy E_{tot} consists of elastic deformation energy E_D and surface energy E_O:

$$E_{tot} = E_D + E_O. \tag{E.7}$$

The deformation energy of the cantilever spring in Figure E.2 is:

$$E_D = \frac{Eh^3d^2b}{2x^3}. \tag{E.8}$$

Here, b is the cantilever spring width and E the YOUNG's modulus.
Surface energy E_O of adhering area $A_B = (l - x)b$ results according to Equation (E.6) in

$$E_O = C - \gamma_S(l - x)w \tag{E.9}$$

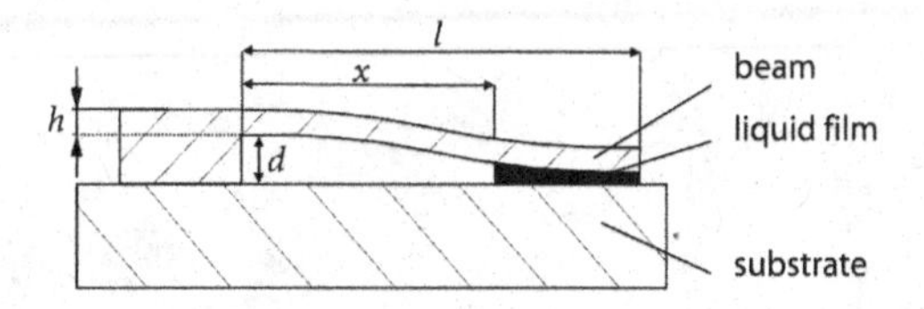

Figure E.2 Deformation of a cantilever spring

with $\gamma_S = 2\gamma_{FA}\cos\varphi$ as the area-related adhesion energy. The equilibrium of forces occurs for the minimum of E_{tot}:

$$\frac{\partial E_{tot}}{\partial x} = \frac{\partial E_D}{\partial x} + \frac{\partial E_O}{\partial x} = 0. \tag{E.10}$$

Inserting Equations (E.8) and (E.9) in Equation (E.10) results in

$$-\frac{3}{2} \cdot \frac{Eh^3 d^2 b}{x_{EQUI}^4} + \gamma_S b = 0, \tag{E.11}$$

where x_{EQUI} is the equilibrium length of the cantilever that does not adhere to the substrate.

Critical cantilever length l_c, at which adhesion occurs, results from the condition $l_c > x_{EQUI}$:

$$l_c = \sqrt[4]{\frac{3}{2} \cdot \frac{Eh^3 d^2}{\gamma_S}}. \tag{E.12}$$

For other mechanical structures, the corresponding deformation energy has to be used in Equation (E.8), and the respective adhesion area has to be inserted into Equation (E.6) or (E.9).

REFERENCE

[TAS96] Tas, N., Sonnenberg, T., Jansen, H.. Legtenberg, R., Elwenspoek, M. (1996) Stiction in surface micromachining. *Journal of Micromechanics and Microengineering* 6, pp. 385–97.

Index